7

SERIES IN THEORETICAL AND APPLIED MECHANICS

Edited by R.K.T. Hsieh

SERIES IN THEORETICAL AND APPLIED MECHANICS

Editor: R. K. T. Hsieh

Published

Volume 1: Nonlinear Electromechanical Effects and Applications
by G. A. Maugin

Volume 2: Lattice Dynamical Foundations of Continuum Theories
by A. Askar

Volume 3: Heat and Mass Transfer in MHD Flows
by E. Blūms, Yu. Mikhailov, and R. Ozols

Volume 5: Inelastic Mesomechanics
by V. Kafka

Volume 6: Design Technology of Fusion Reactors
edited by M. Akiyama

Volume 8: Mechanics of Porous and Fractured Media
by V. N. Nikolaevskij

Volume 9: Aspects of Non-Equilibrium Thermodynamics
by W. Muschik

Volume 10: Fragments of the Theory of Anisotropic Shells
by S. A. Ambartsumian

Volume 11: Diffusion Processes during Drying of Solids
by K. N. Shukla

Forthcoming

Volume 4: Mechanics of Continuous Media
by L. Sedov

Inhomogeneous Waves in Solids and Fluids

Giacomo Caviglia
Angelo Morro

World Scientific
Singapore • New Jersey • London • Hong Kong

Authors

Giacomo Caviglia
Department of Mathematics
University of Genoa
via L. B. Alberti 4
16132, Genoa
Italy

Angelo Morro
DIBE
University of Genoa
via Opera Pia 11a
16145 Genoa
Italy

Series Editor-in-Chief

R. K. T. Hsieh
Department of Mechanics, Royal Institute of Technology
S-10044 Stockholm, Sweden

Published by

World Scientific Publishing Co. Pte. Ltd.
P O Box 128, Farrer Road, Singapore 9128
USA office: Suite 1B, 1060 Main Street, River Edge, NJ 07661
UK office: 73 Lynton Mead, Totteridge, London N20 8DH

Library of Congress Cataloging-in-Publication Data is available.

INHOMOGENEOUS WAVES IN SOLIDS AND FLUIDS

ISSN 0218-0235
ISBN 981-02-0804-9

Printed in Singapore

PREFACE

The subject of linear time-harmonic acoustic waves has been investigated for more than a century, so that a wide number of outstanding results are now easily available. Nevertheless it seems that some topics still deserve further attention and a deeper understanding is desirable. In this regard we mention that, in writing this book, we had in mind two purposes: the elaboration of more realistic mathematical models for wave propagation phenomena and the development of mathematical techniques for the determination of the solution to specific wave propagation problems that arise in applied sciences.

Underlying ideas. To our mind the following features characterize this book within the literature on wave propagation. First, we observe that the propagation of mechanical disturbances is often modelled within the scalar theory of acoustics. There is good reason for this, because of the inherent, intrinsic simplicity of the scalar theory and because wave propagation in inviscid fluids is adequately described in terms of a scalar field, to be identified with the perturbing pressure or the velocity potential. Meanwhile, though mechanical waves in (linear, elastic) solids are at the outset a vector phenomenon, if a disturbance propagates over a sufficiently large range then the compressional and shear modes decouple as a consequence of their different propagation speeds. Hence there are circumstances where the scalar acoustic wave equation provides a sound model for elastic waves. However, recourse to the vector theory cannot be avoided when specific boundary conditions are assigned or the underlying medium is not homogeneous, in particular when a discontinuity in the material parameters occurs at an interface. That is why here we are concerned with vector wave propagation.

Second, dissipative bodies, solids or fluids, are rarely under consideration for wave propagation problems. The scant attention to dissipation may be explained by the superficial idea that, apart from the damping of the amplitude, the wave behaviour in dissipative materials is not qualitatively different to that in non-dissipative materials. Throughout the book we show that such is not always the case.

Third, even when dissipation is considered, the modelling looks very poor in comparison with the variety of models for dissipative materials. In this sense it should come as no surprise that no connection is established between thermodynamic restrictions and wave propagation properties. Here, instead, we take advantage of certain results of thermodynamic character to provide qualitative and quantitative properties about the amplitude decay.

Fourth, waves in a prestressed medium are of interest in a number of situations. For example, this is the case when a a time-harmonic wave perturbs a solid in equilibrium under the action of suitable boundary tractions and body forces. The occurrence of a prestress produces significant qualitative changes on wave propagation. Here we show that the prestress induces anisotropic effects that allow (inhomogeneous) plane wave propagation only at certain priviledged directions and make rays in solids be no longer orthogonal to surfaces of constant phase.

Scope. The aim of this book is to make some progress into these subjects by setting up the general framework and deriving basic results about wave propagation in dissipative materials. In our approach we had the idea that recent achievements on the thermodynamics of dissipative materials should be essential for the understanding of many aspects of wave propagation. It is really so, at least to our mind, and throughout the book the appropriate connections are emphasized.

In the choice of the material contained in the book, emphasis has been given to topics related to applications in various fields of research. After a preliminary investigation of bulk waves and surface waves, we examine propagation in discretely and continuosly stratified media, scattering from an obstacle, perturbation methods and ray methods in heterogeneous media. All these arguments are regarded as fundamental when wave models are introduced, for example, in seismology, non-destructive testing of materials, or ocean acoustics. In our analysis, techniques and procedures that are often used in the study of elastic wave propagation are extended so as to incorporate the effects of dissipation and prestress.

Inhomogeneous waves. The waves are taken to be time-harmonic. The materials are described through linear or linearized constitutive models (viscoelastic solids, viscoelastic or viscous fluids); since we aim at bringing into evidence the effects of dissipation, these materials are chosen as isotropic and homogeneous, although heterogeneous media are also considered. In this scheme, inhomogeneous waves are the bed-rock upon which our analysis is performed. Essentially, inhomogeneous waves may be represented as the more familiar (homogeneous) plane waves, where however the amplitude and the wave vector are both complex-valued. As a consequence, planes of constant phase do not coincide with planes of constant amplitude, and the amplitude decays in the direction of propagation. Though such waves are not new in the literature, certainly they deserve more attention than they are customarily given especially when wave propagation is multi-dimensional in character. Then, to make the book self-contained, we develop a preliminary analysis of inhomogeneous waves as such and, particularly, as bulk waves in dissipative bodies.

Contents. In essence, the contents of the book may be described as follows. The first three chapters develop general aspects of wave propagation in linear dissipative media. The wave behaviour at a plane interface is examined in Chapters 4 and 5. The remaining chapters may be regarded as an introduction to applications of wave phenomena to various kinds of dissipative heterogeneous media. Such heterogeneities may result from the juxtaposition of homogeneous layers or the presence of a finite obstacle; otherwise it is assumed that deviations from homogeneity may be regarded as small or that they cannot be appreciated in the space of a wavelength. A concise account of the topics investigated is given as follows.

Chapter 1 is meant as an introduction to general properties of inhomogeneous waves, where emphasis is given to those features that have no analogue in the more familiar framework of (homogeneous) plane waves.

Chapter 2 contains a short review of basic principles of continuum mechanics that allows investigating the connections between inhomogeneous waves and dissipation. In particular viscoelastic solids and fluids are introduced as the prototype of dissipative media, where dissipation is modelled via a dependence of the Cauchy stress on the whole history of the related deformations. The restrictions placed by the second law of thermodynamics on the lossy medium are then examined. Perturbation equations that describe wave propagation in prestressed bodies are also derived.

Chapter 3 is devoted to bulk wave propagation in dissipative media but the analysis is confined to time-harmonic waves. The representations of inhomogeneous longitudinal and transverse waves in terms of complex scalar and vector potentials are derived. Then the energy flux vector and the energy flux intensity for lossy media are investigated, and the pertinent results are applied to the determination of the energy content of inhomogeneous waves. Finally, effects of constraints and body forces on wave propagation are examined.

The behaviour of inhomogeneous waves at a plane interface is studied in Chapter 4. New features that have no analogue in the equivalent elastic problem are outlined. Degenerate cases are extensively analyzed, also with a view to applications to wave propagation in stratified media. The framework of inhomogeneous waves can be adopted to describe reflection and refraction between elastic half-spaces, leading to a general scheme that works also when incidence occurs beyond the critical angle. These results are strengthened by an analysis of the energy flux intensity of incident, reflected, and transmitted waves at the interface between an inviscid fluid and a viscoelastic solid.

Chapter 5 introduces surface waves. A procedure for the determination of Rayleigh surface waves at the free boundary of a viscoelastic body is exhibited. The result is that dissipation allows for the existence of two Rayleigh waves. The algorithm is then modified to analyse surface waves at a plane interface between an inviscid fluid and a viscoelastic body. A specific example is discussed in detail and the admissible waves are classified. In the last section we investigate the effect of body force on surface waves on a viscoelastic half-space.

In Chapter 6 we consider discretely and continuosly stratified media embedded between two parallel homogeneous half-spaces. Such models are often encountered in the study of seismic waves. It is assumed that an inhomogeneous wave is incident and the waves reflected and transmitted by the stratification are determined. Particular attention is devoted to the analysis of the effects of dissipation on wave propagation within thick layers. A procedure for the elimination of singularities in the transfer matrices is also outlined. In the case of continuously stratified layers, definite results are exhibited in terms of fundamental systems of solutions.

Chapter 7 contains a self-consistent treatment of scattering by a viscoelastic obstacle immersed in a lossy, solid, homogeneous matrix. This chapter may also be regarded as an introduction to the scattering by inclusions in solid bodies. The incoming wave is taken to be inhomogeneous. An integral representation for the scattered field is found and the behaviour at infinity of the scattered field is determined. In particular, effects of dissipation on the high-frequency limit of the far field are evaluated, and difficulties which arise in connection with the scattering cross-section are discussed.

The Born approximation and the WKB method are presented in Chapter 8. After an outline of the general scheme, the Born approximation is applied to the determination of the displacement field generated when an inhomogeneous wave travels within a region that contains small heterogeneities. The WKB method is applied to the study of the behaviour of waves in an Epstein layer; particular emphasis is given to the effects of turning points.

In Chapter 9 ray methods are applied to the study of wave propagation in prestressed viscoelastic solids and fluids. Qualitative changes in the eikonal equation arising from the prestress are pointed out. As expected, dissipation results in a decay of the amplitude along rays. The behaviour of rays at an interface is also examined in detail.

Limitations. We are aware of some aspects that might have been considered. It is of interest to analyse other models of dissipative materials, especially in connection with electromagnetic phenomena, porosity, and effects due to anisotropy. We have not considered the numerical implementation of the pertinent algorithms although numerical approaches should be useful in dealing with scattering phenomena and propagation in layered media. Only a limited number of examples have been proposed, but relevant applications can be found in the cited references. Also, a description of wave phenomena related to acoustic beams might be in order. We have chosen to present the various topics with all necessary details, with the purpose of bringing into evidence new aspects related to dissipation, rather than providing an exhaustive account of wave propagation. Then some subjects have been dropped out. In this sense the scope of the book is rather limited.

Prerequisites. The reader is supposed to be acquainted with the basic notions of Continuum Mechanics, though Chapter 2 provides a concise but self-consistent review of concepts and results for the understanding of subsequent developments.

Notation. The meaning of any symbol is given in the appropriate place when the symbol is introduced. However, we list here some symbols. $\mathbb{R}$ is the set of reals, $\mathbb{R}^+$ the set of positive reals, and $\mathbb{R}^{++}$ the set of strictly positive reals. Sym denotes the set of symmetric tensors, sym the symmetric part, and tr the trace. As usual in standard books of continuum mechanics, we denote vectors and tensors by boldface letters; recourse to the indicial notation is made only when ambiguities may arise. For any tensor, the superscript † denotes the transpose. The volume and surface elements indicate the pertinent variable(s) of integration. For instance, if $\mathbf{x}$ is the current position vector then dx is the volume element and da_x the surface element. The symbol ∇ stands for the gradient operator; $\nabla\cdot$ and $\nabla\times$ are the divergence and curl operators. For any vector (or tensor) $\mathbf{w}$, $\nabla\mathbf{w}$ or $\nabla\otimes\mathbf{w}$ denote the gradient of $\mathbf{w}$. When this is the case, we specify the vector variable with respect to which the gradient is evaluated, such as $\nabla_{\mathbf{x}}$, $\nabla_{\mathbf{y}}$. The symbols $\mathbf{1}$ and $\mathbb{1}$ denote the identity tensor and the identity matrix in the appropriate space. Usually $\simeq$ means approximately equal to; sometimes it means proportionality. Numbers in square brackets denote the corresponding References listed at the end of the book. The abbreviation Ch. and the symbol § stand for Chapter and Section.

Acknowledgments. The research leading to this book has been partially supported by the Italian CNR, through the Research Project MADESS and under contract number 91.01320.CT01. Also the Italian Ministry of Universities and Research in Science and Technology (MURST) provided support through the Research Project "Evolution problems in fluids and solids". All supports are gratefully acknowledged. Finally, we wish to thank Mr. R. Santeramo for preparing the figures.

CONTENTS

Inhomogeneous Waves in Solids and Fluids

1 INHOMOGENEOUS WAVES

The wave propagation problems developed in this book involve material behaviours expressed by linear, or linearized, constitutive equations. The linearity of the pertinent equations in the unknown fields gives importance, through superposition and Fourier analysis, to the sinusoidally varying time dependence. That is why, throughout this book, only single-frequency waves are investigated.

Wave propagation is taken to occur in dissipative bodies, solids or fluids. Even in homogeneous bodies, the dissipativity results in the inhomogeneity of the waves. This chapter exhibits the general properties of inhomogeneous waves which prove useful in the subsequent investigations of wave propagation problems. Emphasis is given to the major novelties of inhomogeneous waves versus the customary (homogeneous) plane waves.

1.1 Introduction to inhomogeneous waves

Consider a sinusoidal time dependence with (angular) frequency ω. For technical convenience we express the dependence in the form $\exp(-i\omega t)$, where t is the time and i is the imaginary unit. It is usual to call such a dependence time-harmonic.

Let $\mathbf{x}$ be any displacement vector in the Euclidean space $\mathcal{E}^3$. Plane waves are described in the form

$$\boldsymbol{a} = \mathbf{A}\exp[i(\mathbf{k}\cdot\mathbf{x} - \omega t)] \tag{1.1}$$

where the vectors $\mathbf{A}$ and $\mathbf{k}$ are real; $\mathbf{A}$ is called the amplitude (or polarization) and $\mathbf{k}$ the wave vector. The frequency ω is taken to be real; indeed, to fix ideas, if no explicit mention is made we regard ω as positive. The meaning of $\boldsymbol{a}$ depends on the phenomenon under investigation. Anyway, it is the real part of (1.1) that has a physical relevance or, as shown later, the combination of terms relative to any ω with the corresponding terms relative to $-\omega$. By elementary notions of wave propagation we say that if $\mathbf{n} = \mathbf{k}/|\mathbf{k}|$ then the function (1.1) describes propagation in the direction $\mathbf{n}$ (if $\omega > 0$) with phase speed $c = \omega/|\mathbf{k}|$. Of course, if a one-dimensional scheme is considered then the exponential in (1.1) specializes to $\exp[i(kx - \omega t)]$ where x is the pertinent Cartesian coordinate in the direction of $\mathbf{k}$ and k the corresponding component of $\mathbf{k}$. Usually k is referred to as the wavenumber.

If, as it will generally be the case, $\mathbf{A}$ is a complex-valued vector then we write

$$\mathbf{A} = \mathbf{c} + i\mathbf{d}$$

where $\mathbf{c}$ and $\mathbf{d}$ are real-valued vectors. Likewise real-valued vectors, complex-valued vectors may be combined according to the usual rules of vector algebra [83]. If the material sustaining wave propagation is (linear and) dissipative, the simplest wave solutions have the form (1.1) with complex-valued wave vector $\mathbf{k}$ and amplitude $\mathbf{A}$. Such waves are called inhomogeneous as we make it precise in a moment. Waves of the type (1.1) with real $\mathbf{A}$ and $\mathbf{k}$ are referred to as homogeneous. Unless otherwise specified, it is understood that the waves under consideration are inhomogeneous.

The need for inhomogeneous waves is recognized by observing that (1.1) satisfies

$$\Delta \boldsymbol{a} - \frac{\mathbf{k}\cdot\mathbf{k}}{\omega^2}\frac{\partial^2 \boldsymbol{a}}{\partial t^2} = 0,$$

where Δ denotes the Laplacian operator. Thus $\boldsymbol{a}$ is a solution to the wave equation with complex coefficient $\mathbf{k}\cdot\mathbf{k}/\omega^2$. This observation indicates that $\mathbf{k}$ is generally a complex-valued vector. Indeed, the coefficient of $\partial^2\boldsymbol{a}/\partial t^2$ in the pertinent wave equation is determined by the constitutive properties of the medium and is *complex* if the medium is *dissipative.* This implies that solutions of the form (1.1) hold for the wave equation if $\mathbf{k}\cdot\mathbf{k}$ is complex-valued. Even though the scalar $\mathbf{k}\cdot\mathbf{k}$ happens to be real-valued, we have to allow for complex-valued wave vectors $\mathbf{k}$, which is the case if $\mathrm{Re}\,\mathbf{k}$ and $\mathrm{Im}\,\mathbf{k}$ are orthogonal. This is why inhomogeneous waves are regarded as basic solutions in many fields of theoretical and applied research concerned with wave propagation; in this regard we mention optics [19], electromagnetism [161, 113], elasticity, viscoelasticity [89], porous media [146, 114] viscous [21, 33], viscoelastic [44], and thermoviscous fluids [143], seismology [29, 30] analysis of living tissues [49] and of interface waves [144], plane wave decomposition of acoustic beams [22] and scalar wave fields [58]. These papers or books, in turn, provide extensive lists of references on the subject.

To begin with we consider the essential features of inhomogeneous waves. Denote by $\mathbf{k}_1$ and $\mathbf{k}_2$ the real and imaginary parts of $\mathbf{k}$, namely

$$\mathbf{k} = \mathbf{k}_1 + i\mathbf{k}_2.$$

Substitution into (1.1) gives the representation

$$\boldsymbol{a} = \mathbf{A}\exp(-\mathbf{k}_2\cdot\mathbf{x})\exp[i(\mathbf{k}_1\cdot\mathbf{x} - \omega t)].$$

This shows that $\mathbf{k}_1$ enters the argument of the imaginary exponent and is related to phase propagation; more precisely, $\mathbf{k}_1$ determines the propagation speed (or phase speed) as $c = \omega/k_1$, $k_1 = |\mathbf{k}_1|$ while the condition $\mathbf{k}_1\cdot\mathbf{x} = \text{const.}$ characterizes planes of constant phase. Meanwhile $\mathbf{k}_2$ accounts for a position-dependent amplitude of the wave in the form $\exp(-\mathbf{k}_2\cdot\mathbf{x})$. Hence, $\mathbf{k}_2\cdot\mathbf{x} = \text{const.}$ yields planes of constant amplitude. Later on we will show how $\mathbf{k}_2$ describes the amplitude decay for waves in dissipative media.

Depending on the constitutive properties of the material, $\mathbf{k}$ may be affected by ω in a nonlinear way so that c itself is a function of ω (dispersive waves). Also in dissipative

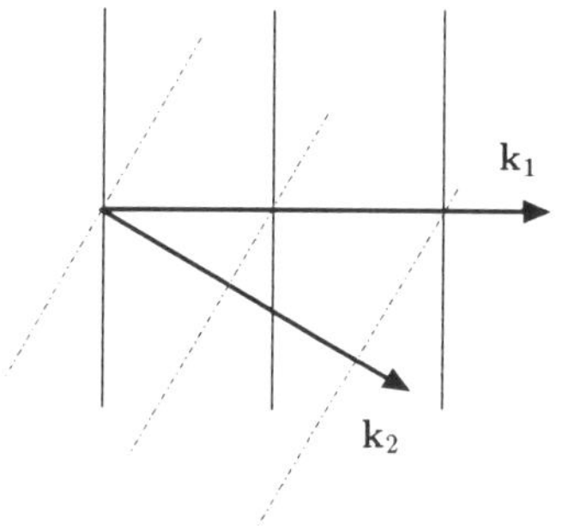

Fig. 1.1 At inhomogeneous waves the planes of constant phase (continuous lines) are not parallel to the planes of constant amplitude (dotted lines).

materials described by linear memory functionals, the corresponding waves turn out to be dispersive.

It is of interest to sketch chronologically some views about inhomogeneous waves in dissipative materials. Up to the sixties, it was customary to account for the wave behaviour by letting angles be complex-valued. Regardless of how natural this view may appear, the motivation was due to the will of keeping Snell's law valid at interfaces between two, or at least one, dissipative media. To understand this point, look at electrically conducting media and let k, k_0 be the two wavenumbers and θ, θ_0 the corresponding angles of the direction of propagation with respect to the normal to the interface. To fix ideas let k_0, θ_0 characterize the incident wave. Because of electrical conduction the wavenumbers turn out to be complex (cf. [161], §9.8). Snell's law in the form

$$\sin\theta = \frac{k_0}{k}\sin\theta_0$$

shows that if θ_0 is real but at least one of k, k_0 is complex then θ must be complex too. The evaluation of $\sin\theta$ and $\cos\theta$ thus leads to a phase shift of the transmitted wave along with an attenuation factor induced by the imaginary parts of $\sin\theta$ and $\cos\theta$.

To clarify this view let z be the axis orthogonal to the interface and x the axis along the (plane) interface in the plane of the direction of propagation of the incident wave. Let k_0 be real and $k = \alpha + i\beta$, $\alpha, \beta \in \mathbb{R}$. Then

$$\sin\theta = (a - ib)\sin\theta_0$$

where $a = \alpha k_0/|k|$, $b = \beta k_0/|k|$. Then

$$\cos\theta = \sqrt{1 - (a^2 - b^2 - 2iab)\sin^2\theta_0}.$$

Let $\rho\exp(i\gamma) = \cos\theta$ with γ real and ρ positive. It then follows that the transmitted wave is proportional to

$$\exp\{i[-(q+ip)x + k_0 \sin\theta_0\, z - \omega t]\}$$

where $p = \rho(\alpha\cos\gamma + \alpha\sin\gamma)$, $q = \rho(\alpha\cos\gamma - \beta\sin\gamma)$. Accordingly, the surfaces of constant amplitude are the planes x = constant and the surfaces of constant phase are planes $-qx + k_0 z \sin\theta_0$ = constant. More formally, we can merely say that a representation of the form (1.1) is still allowed provided we let $\mathbf{k}$ be complex-valued.

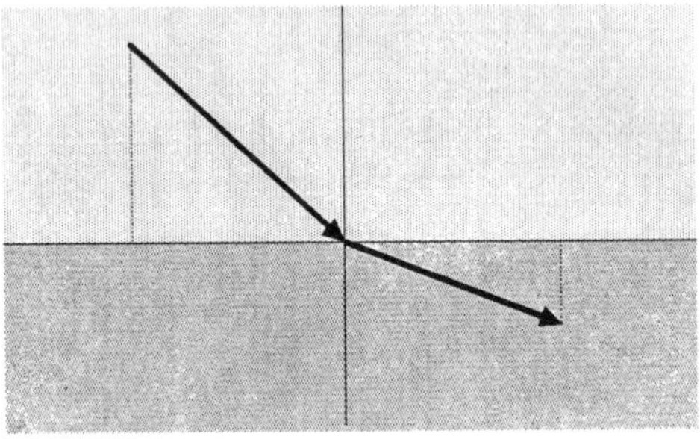

Fig. 1.2 If the wave vector is real then Snell's law is the classical statement that the projection of the wave vector on the pertinent interface is continuous across the interface itself.

Of course, even for dissipative materials, wave propagation has long been studied. The review by Hunter [94], which appeared in 1960, may be a preliminary reference on the subject especially for the one-dimensional approximation and the connection with experimental work. Yet, some salient features of the pertinent waves have been emphasized in recent years also because they are latent in a one-dimensional description. A paper by Lockett [119], published in 1962, gives a clear account of wave modes. Seemingly, it is such a paper that firstly clarified how, in general, the direction of attenuation of the wave is not in the direction of propagation (cf. also [53]). A second feature, which is best seen in the multi-dimensional description, is the dispersive character of the waves whereby propagation speed, attenuating constant, and pertinent angles are in general dependent on the frequency.

We are accustomed to the feature that time-harmonic waves in dissipative materials have a complex-valued wavenumber. Less obvious is the fact that there can be waves with a complex-valued wave vector, with real and imaginary parts $\mathbf{k}_1, \mathbf{k}_2$ not necessarily parallel. Nevertheless, looking at $\mathbf{k}$ as a complex vector, with $\mathbf{k}_1$ and $\mathbf{k}_2$ not necessarily parallel, is the most suitable way to the investigation of waves in dissipative media. That is why we follow such a view throughout.

Non-parallel vectors $\mathbf{k}_1$, $\mathbf{k}_2$ were considered by Brekhovskikh (cf. [22], first edition) in the limit case of non-dissipative materials, where $\mathbf{k}_1 \cdot \mathbf{k}_2 = 0$, and in connection with analysis of acoustic beams. In the electromagnetic context the view that the wave vector is in fact complex was made clear by Landau and Lifchitz ([113] , §63). Later Mott [133], while examining reflection and refraction at a fluid-solid interface, emphasized that $\mathbf{k}_1$ and $\mathbf{k}_2$ may be neither parallel nor orthogonal. To our knowledge, though, it was only in the seventies that researchers started describing systematically plane waves in dissipative materials as waves with a complex-valued wave vector. In this regard we mention the works of Buchen [26] and Hayes and Rivlin [89].

Generally the adjective "inhomogeneous" is involved to denote variations of properties in space. Quite consistently, it is customary to call inhomogeneous waves those with a complex-valued wave vector $\mathbf{k}$ in that the amplitude decreases with distance (cf., e.g., [150, 22, 167]). More precisely, we adhere to a now standard terminology [143] and call *inhomogeneous waves* those with a complex-valued $\mathbf{k}$ such that $\mathbf{k}_1$ and $\mathbf{k}_2$ are not parallel. So, the inhomogeneity of the wave consists in the breaking of the plane symmetry in the planes of constant phase. Meanwhile a comment is in order about the use of the adjective "plane". Its original meaning served to define a wave whose amplitude and phase were constant on a plane, the normal to which was the propagation direction. Strictly speaking, as soon as the wave is inhomogeneous the use of "plane" is not correct. Anyway, quite often plane and inhomogeneous are used at the same time to denote waves with a complex-valued wave vector.

Complex vectors, in the form $\mathbf{a} + i\mathbf{b}$, were called *bivectors* by Hamilton [80]. In this sense a thorough investigation of the geometric properties of bivectors was performed by Gibbs ([74], cf. [20]); in the next section some geometric properties are given which are strictly connected with inhomogeneous waves. It is worth remarking, though, that the term bivector is used with different meanings in other contexts. In particular, often (cf. [141]) bivectors are meant as (double) skew-symmetric tensors. To avoid any ambiguity, the term bivector is ignored throughout this book.

1.2 Geometric aspects of inhomogeneous waves

It has already been observed that a physical interpretation is ascribed to the real part of $\boldsymbol{a}$, say $\mathrm{Re}\,\boldsymbol{a}$. In this section polarization of travelling waves is studied by examining the locus described by the vector $\mathrm{Re}\,\boldsymbol{a}$ regarded as a function of t at a fixed point $\mathbf{x}$. Further comments on longitudinal and transverse waves, and the related polarizations will also follow.

As a basic property we show that any complex vector $\mathbf{c} + i\mathbf{d}$, with $\mathbf{c} \times \mathbf{d} \neq 0$, may be written in the form

$$\mathbf{c} + i\mathbf{d} = \exp(i\beta)(\mathbf{a} + i\mathbf{b}), \qquad \mathbf{a} \cdot \mathbf{b} = 0. \tag{2.1}$$

For, given the vector $\mathbf{c} + i\mathbf{d}$, with $\mathbf{c}$ and $\mathbf{d}$ not parallel, consider the vector

$$\mathbf{a} + i\mathbf{b} = \exp(-i\beta)(\mathbf{c} + i\mathbf{d})$$

and determine β so that the inner product $\mathbf{a} \cdot \mathbf{b}$ vanish. Now, the inner product $(\mathbf{a} + i\mathbf{b}) \cdot (\mathbf{a} + i\mathbf{b})$ gives

$$\mathbf{a} \cdot \mathbf{a} - \mathbf{b} \cdot \mathbf{b} + 2i\mathbf{a} \cdot \mathbf{b} = \exp(-2i\beta)(\mathbf{c} \cdot \mathbf{c} - \mathbf{d} \cdot \mathbf{d} + 2i\mathbf{c} \cdot \mathbf{d}).$$

The requirement $\mathbf{a} \cdot \mathbf{b} = 0$ and the assumption $\mathbf{c} \cdot \mathbf{c} \neq \mathbf{d} \cdot \mathbf{d}$ yield

$$\tan 2\beta = \frac{2\mathbf{c} \cdot \mathbf{d}}{\mathbf{c} \cdot \mathbf{c} - \mathbf{d} \cdot \mathbf{d}}.$$

If, instead, $\mathbf{c} \cdot \mathbf{c} = \mathbf{d} \cdot \mathbf{d}$ then $(\mathbf{c} + \mathbf{d}) \cdot (\mathbf{c} - \mathbf{d}) = 0$. This property in turn provides the identity

$$\mathbf{c} + i\mathbf{d} = \frac{1}{\sqrt{2}} \exp(-i\pi/4)[\mathbf{c} - \mathbf{d} + i(\mathbf{c} + \mathbf{d})],$$

which corresponds to $\mathbf{a} = (\mathbf{c} - \mathbf{d})/\sqrt{2}$ and $\mathbf{b} = (\mathbf{c} + \mathbf{d})/\sqrt{2}$.

This observation allows the vector $\mathbf{k}_1 + i\mathbf{k}_2$ to be written as

$$\mathbf{k}_1 + i\mathbf{k}_2 = \exp(i\gamma)(\mathbf{h}_1 + i\mathbf{h}_2), \qquad \mathbf{h}_1 \cdot \mathbf{h}_2 = 0.$$

Hence

$$\mathbf{k}_1 = \cos\gamma\, \mathbf{h}_1 - \sin\gamma\, \mathbf{h}_2,$$

$$\mathbf{k}_2 = \sin\gamma\, \mathbf{h}_1 + \cos\gamma\, \mathbf{h}_2.$$

By means of (2.1) we can give a suggestive representation of the wave (1.1). Let $\mathbf{A} = \mathbf{c} + i\mathbf{d}$, $\mathbf{c} \times \mathbf{d} \neq 0$, and consider the (physical) real part

$$\mathbf{u} = \mathrm{Re}\,\boldsymbol{a} = \mathrm{Re}\{(\mathbf{c} + i\mathbf{d}) \exp[i(\mathbf{k} \cdot \mathbf{x} - \omega t)]\}.$$

Since ω is real, the argument $i(\mathbf{k} \cdot \mathbf{x} - \omega t)$ of the exponential allows us to write

$$\exp[i(\mathbf{k} \cdot \mathbf{x} - \omega t)] = \exp(-\mathbf{k}_2 \cdot \mathbf{x}) \exp[i(\mathbf{k}_1 \cdot \mathbf{x} - \omega t)]. \tag{2.2}$$

Then, by (2.1) it follows that

$$\mathbf{u} = \mathrm{Re}\{\exp(-\mathbf{k}_2 \cdot \mathbf{x})\,(\mathbf{a} + i\mathbf{b}) \exp[i(\mathbf{k}_1 \cdot \mathbf{x} - \omega t + \beta)]\}, \qquad \mathbf{a} \cdot \mathbf{b} = 0.$$

Letting $\zeta = \mathbf{k}_1 \cdot \mathbf{x} - \omega t + \beta$ we have

$$\mathbf{u} = \exp(-\mathbf{k}_2 \cdot \mathbf{x})(\mathbf{a} \cos\zeta - \mathbf{b} \sin\zeta).$$

The orthogonality of **a** and **b** allows us to obtain

$$\frac{u_a^2}{[a\exp(-\mathbf{k}_2\cdot\mathbf{x})]^2}+\frac{u_b^2}{[b\exp(-\mathbf{k}_2\cdot\mathbf{x})]^2}=1$$

where u_a, u_b are the components of **u** along $\mathbf{a},\mathbf{b}$ and a, b are the moduli of $\mathbf{a},\mathbf{b}$. This means that at any point **x** the vector **u** describes an ellipse which has $\mathbf{a}\exp(-\mathbf{k}_2\cdot\mathbf{x}), \mathbf{b}\exp(-\mathbf{k}_2\cdot\mathbf{x})$ as a pair of semidiameters. Accordingly, if **c** and **d** are not parallel then the wave is elliptically polarized. The semidiameters depend on **x** as $a\exp(-\mathbf{k}_2\cdot\mathbf{x})$ and $b\exp(-\mathbf{k}_2\cdot\mathbf{x})$; the plane of polarization is determined by the vectors **c** and **d**. If in addition $|\mathbf{a}|=|\mathbf{b}|$ then the wave is circularly polarized. The requirements $\mathbf{a}\cdot\mathbf{b}=0$ and $|\mathbf{a}|^2=|\mathbf{b}|^2$ are satisfied if

$$(|\mathbf{c}|^2-|\mathbf{d}|^2)^2=-4(\mathbf{c}\cdot\mathbf{d})^2,$$

that is, if $\mathbf{c}\cdot\mathbf{d}=0$ and $\mathbf{c}\cdot\mathbf{c}=\mathbf{d}\cdot\mathbf{d}$. Some details on this subject can be found in [83].

If, instead, $\mathbf{c}\times\mathbf{d}=0$ then we can use the polar form

$$\mathbf{c}+i\mathbf{d}=\rho\exp(i\theta)\mathbf{a}$$

where ρ,θ are real-valued scalars and **a** is a real unit vector. Hence we have

$$\begin{aligned}\mathbf{u}=&\mathrm{Re}\{(\mathbf{c}+i\mathbf{d})\exp[i(\mathbf{k}\cdot\mathbf{x}-\omega t)]\}\\=&\mathrm{Re}\{\rho\mathbf{a}\exp(-\mathbf{k}_2\cdot\mathbf{x})\exp[i(\mathbf{k}_1\cdot\mathbf{x}-\omega t+\theta)]\}\\=&\rho\mathbf{a}\cos\zeta\,\exp(-\mathbf{k}_2\cdot\mathbf{x})\end{aligned}$$

where, again, $\zeta=\mathbf{k}_1\cdot\mathbf{x}-\omega t+\theta$. Accordingly, the parallelism of **c** and **d** means that the wave is linearly polarized.

The complex amplitude $\mathbf{A}=\mathbf{c}+i\mathbf{d}$ determines the planes (or the direction) of polarization of $\mathbf{u}=\mathrm{Re}\,\boldsymbol{\mathfrak{a}}$. We now introduce longitudinal and transverse inhomogeneous waves and then we examine the positions of the corresponding polarization planes (directions) relative to the complex wave vector **k**.

By analogy with the behaviour of homogeneous waves we characterise an inhomogeneous wave as longitudinal or transverse according as

$$\mathbf{A}=\Phi\mathbf{k}\quad\text{or}\quad\mathbf{A}=\mathbf{k}\times\boldsymbol{\Psi},$$

with Φ and $\boldsymbol{\Psi}$ complex scalar and complex vector. It follows at once that

$$\mathbf{A}=\Phi\mathbf{k}\quad\Longleftrightarrow\quad\mathbf{A}\times\mathbf{k}=0,\tag{2.3}$$

thus providing an alternative, equivalent definition of longitudinal waves. As regards transverse waves, we make use of the vector identity

$$\mathbf{k}\times(\mathbf{A}\times\mathbf{k})=(\mathbf{k}\cdot\mathbf{k})\mathbf{A}-(\mathbf{k}\cdot\mathbf{A})\mathbf{k},$$

valid also for complex-valued vectors, to conclude that

$$\mathbf{A} = \mathbf{k} \times \mathbf{\Psi} \quad \Longleftrightarrow \quad \mathbf{A} \cdot \mathbf{k} = 0, \tag{2.4}$$

where, if $\mathbf{A} \cdot \mathbf{k} = 0$, the corresponding $\mathbf{\Psi}$ is given by

$$\mathbf{\Psi} = \frac{\mathbf{A} \times \mathbf{k}}{\mathbf{k} \cdot \mathbf{k}}.$$

By (2.4) we can also define transverse waves as inhomogeneous waves satisfying $\mathbf{A} \cdot \mathbf{k} = 0$.

We consider now the description of polarization and outline two remarkable differences from the behaviour of homogeneous waves. First we observe that, quite paradoxically, a longitudinal wave is allowed to be elliptically polarized; indeed, the polarization plane coincides with the plane spanned by $\mathbf{k}_1$ and $\mathbf{k}_2$, provided $\mathbf{k}_1 \times \mathbf{k}_2 \neq 0$, in view of (2.3). Second, the polarization plane of a transverse wave is neither orthogonal nor parallel, in general, to the complex vector $\mathbf{k}$. The results are a consequence of (2.4), to which we apply the following remarks concerning the vanishing of the scalar product between two complex-valued vectors.

Given two complex vectors $\mathbf{k} = \mathbf{k}_1 + i\mathbf{k}_2$ and $\mathbf{h} = \mathbf{h}_1 + i\mathbf{h}_2$ the inner product $\mathbf{k} \cdot \mathbf{h}$ is given by

$$\mathbf{k} \cdot \mathbf{h} = \mathbf{k}_1 \cdot \mathbf{h}_1 - \mathbf{k}_2 \cdot \mathbf{h}_2 + i(\mathbf{k}_1 \cdot \mathbf{h}_2 + \mathbf{k}_2 \cdot \mathbf{h}_1).$$

By analogy with the real case, we might think that if $\mathbf{k} \cdot \mathbf{h} = 0$ then $\mathbf{k}$ and $\mathbf{h}$ are orthogonal. In the complex case it is not simply so. Setting apart the non-trivial case when either $\mathbf{k} = 0$ or $\mathbf{h} = 0$, we examine the content of $\mathbf{k} \cdot \mathbf{h} = 0$.

The vanishing of $\mathbf{k} \cdot \mathbf{h}$ implies that

$$\mathbf{k}_1 \cdot \mathbf{h}_1 = \mathbf{k}_2 \cdot \mathbf{h}_2, \qquad \mathbf{k}_1 \cdot \mathbf{h}_2 = -\mathbf{k}_2 \cdot \mathbf{h}_1. \tag{2.5}$$

Moreover, by (2.1), there exist complex numbers $\tilde{k}, \tilde{h}$ and real vectors $\mathbf{m}, \mathbf{n}, \mathbf{p}, \mathbf{q}$ such that

$$\mathbf{k} = \tilde{k}(\mathbf{m} + i\mathbf{n}), \qquad \mathbf{m} \cdot \mathbf{n} = 0,$$

$$\mathbf{h} = \tilde{h}(\mathbf{p} + i\mathbf{q}), \qquad \mathbf{p} \cdot \mathbf{q} = 0.$$

If $\mathbf{k}$ and $\mathbf{h}$ are coplanar, namely $\mathbf{m}, \mathbf{n}$ are in the plane of $\mathbf{p}, \mathbf{q}$, we can write

$$\mathbf{h} = \tilde{h}[\mu\mathbf{m} + \lambda\mathbf{n} + i(\nu\mathbf{m} + \eta\mathbf{n})],$$

where $\mu, \lambda, \nu, \eta \in \mathbb{R}$, and hence

$$\mathbf{k} \cdot \mathbf{h} = \tilde{k}\tilde{h}[(\mu + i\nu)\mathbf{m} \cdot \mathbf{m} + i(\lambda + i\eta)\mathbf{n} \cdot \mathbf{n})].$$

Then $\mathbf{k} \cdot \mathbf{h} = 0$ implies that

$$\mu\mathbf{m} \cdot \mathbf{m} = \eta\mathbf{n} \cdot \mathbf{n}, \qquad \nu\mathbf{m} \cdot \mathbf{m} + \lambda\mathbf{n} \cdot \mathbf{n} = 0.$$

Substitution yields

$$\mathbf{h} = \tilde{h}(\eta - i\lambda)\Big(\frac{\mathbf{m}\cdot\mathbf{m}}{\mathbf{n}\cdot\mathbf{n}}\mathbf{m} + i\mathbf{n}\Big).$$

Now consider the case when $\mathbf{k}$ and $\mathbf{h}$ are not coplanar. Observe that $\mathbf{k}_1 \times \mathbf{k}_2$ and $\mathbf{h}_1 \times \mathbf{h}_2$ are orthogonal to the planes of $\mathbf{k}$ and $\mathbf{h}$, respectively. Incidentally, by (2.5) we have

$$\begin{aligned}(\mathbf{k}_1 \times \mathbf{k}_2)\cdot(\mathbf{h}_1 \times \mathbf{h}_2) =& (\mathbf{k}_1\cdot\mathbf{h}_1)(\mathbf{k}_2\cdot\mathbf{h}_2) - (\mathbf{k}_2\cdot\mathbf{h}_1)(\mathbf{k}_1\cdot\mathbf{h}_2)\\ =& (\mathbf{k}_1\cdot\mathbf{h}_1)^2 + (\mathbf{k}_2\cdot\mathbf{h}_1)^2 > 0\end{aligned}$$

and then the planes of $\mathbf{k}$ and $\mathbf{h}$ are not orthogonal. Letting $\mathbf{r}$ be orthogonal to $\mathbf{m}$ and $\mathbf{n}$ we can write $\mathbf{h} = \rho\mathbf{m} + \xi\mathbf{n} + \zeta\mathbf{r}$. Then $\mathbf{k}\cdot\mathbf{h} = 0$ yields $\rho\mathbf{m}\cdot\mathbf{m} + i\xi\mathbf{n}\cdot\mathbf{n} = 0$ and hence

$$\mathbf{h} = -i\xi\Big(\frac{\mathbf{n}\cdot\mathbf{n}}{\mathbf{m}\cdot\mathbf{m}}\mathbf{m} + i\mathbf{n}\Big) + \zeta\mathbf{r}.$$

As a last remark on geometric properties, we consider an inhomogeneous wave of the general form (1.1) and observe that the condition of irrotationality $\nabla \times \boldsymbol{a} = 0$ is equivalent to the geometric requirement $\mathbf{k} \times \mathbf{A} = 0$. This in turn means that the wave is longitudinal. Similarly, we say that the field (1.1) is solenoidal if and only if $\nabla \cdot \boldsymbol{a} = 0$, which is equivalent to the geometric constraint $\mathbf{k}\cdot\mathbf{A} = 0$. This means that a solenoidal inhomogeneous wave is always a transverse wave.

1.3 Damping of inhomogeneous waves

Consider an inhomogeneous wave in the typical form (1.1). As shown on many occasions, the condition for the existence of wave solutions (propagation condition) results in a relation between the wave vector $\mathbf{k}$ and the amplitude (or polarization) $\mathbf{A}$. Meanwhile $\mathbf{k}$ is characterized by a relation of the form

$$\mathbf{k}\cdot\mathbf{k} = \varrho + iv \tag{3.1}$$

where $\varrho, v \in \mathbb{R}$ depend on material properties and are parameterized by the frequency ω. Moreover it will be shown that, as a consequence of thermodynamics, $v \geq 0$; we can say that $v > 0$ characterizes dissipative materials. Now, (3.1) amounts to

$$k_1^2 - k_2^2 = \varrho, \tag{3.2}$$

$$\mathbf{k}_1\cdot\mathbf{k}_2 = \tfrac{1}{2}v. \tag{3.3}$$

By (3.3) we conclude that the angle between $\mathbf{k}_1$ and $\mathbf{k}_2$ is smaller than $\pi/2$. Of course, given ϱ and v the system (3.2)-(3.3) is underdetermined; we cannot find $\mathbf{k}_1$ and $\mathbf{k}_2$ or even

k_1, k_2 and the angle between them unless further information is provided as is the case in reflection-refraction problems.

Now, a wave $\mathbf{A}\exp[i(\mathbf{k}\cdot\mathbf{x} - \omega t)]$ may in fact be viewed as a wave propagating in the direction $\mathbf{k}_1$ with speed ω/k_1 and (effective) amplitude $\mathbf{A}\exp(-\mathbf{k}_2\cdot\mathbf{x})$. Accordingly the amplitude decreases exponentially and the maximum rate is in the direction of $\mathbf{k}_2$. Since $\mathbf{k}_1\cdot\mathbf{k}_2 > 0$, this means that (cf. [130]), in the direction of (phase) propagation $\mathbf{x} = x\,\mathbf{k}_1/k_1$, $x > 0$, the amplitude decreases as

$$\mathbf{A}\exp\Big(-\frac{\mathbf{k}_2\cdot\mathbf{k}_1}{k_1}x\Big).$$

Represent the position $\mathbf{x}'$ of any point at the plane of constant phase through $\mathbf{x} = x\mathbf{k}_1/k_1$ as $\mathbf{x}' = x\mathbf{k}_1/k_1 + \mathbf{k}_1\times\mathbf{y}$ for any vector $\mathbf{y}$. Then we have

$$\mathbf{A}\exp(-\mathbf{k}_2\cdot\mathbf{x}') = \mathbf{A}\exp\Big(-\frac{\mathbf{k}_2\cdot\mathbf{k}_1}{k_1}x\Big)\exp[-(\mathbf{k}_2\times\mathbf{k}_1)\cdot\mathbf{y}].$$

In words, at planes of constant phase the amplitude decreases exponentially in the direction of $\mathbf{k}_2\times\mathbf{k}_1$.

Non-dissipative bodies may be viewed as those for which $v = 0$. In such a case (3.3) becomes

$$\mathbf{k}_1\cdot\mathbf{k}_2 = 0,$$

namely $\mathbf{k}_1$ and $\mathbf{k}_2$ are orthogonal. Then the amplitude is constant along the direction of (phase) propagation but exponentially decreasing, at planes of constant phase, along the direction of $\mathbf{k}_2\times\mathbf{k}_1$. For instance this is the case in elastic solids and inviscid fluids. Of course non-dissipative bodies allow also for the particular type of solutions $k_2 = 0$, $k_1^2 = \varrho$. In such a case only the plane of constant phase is meaningful and the amplitude is constant at those planes.

Especially in connection with surface waves at fluid-solid interfaces, it is often claimed that attenuation of the surface wave is related to leaking of energy into the fluid and this determines the exponential *increase* of amplitude (of the wave in the fluid) with distance from the interface. For this reason often such waves are termed leaky [22] while sometimes the term leaky is referred to the surface wave (which leaks energy into the fluid) [128, 144]. In this regard we have to avoid the type of ambiguity sometimes appearing in the literature. The ambiguity is at the basis of an objection, mainly put forward among experimentalists and concerning the wave in the fluid due to radiation from the leaky surface wave. The objection is: a wave whose amplitude increases exponentially to infinity with distance from the interface is not physically admissible. First we recall that, as a consequence of thermodynamics, the amplitude decays in the direction of propagation and then we cannot think of leaky waves as waves whose amplitude increases in the direction of propagation. Rather, in (inviscid) fluids $\mathbf{k}_1$ and $\mathbf{k}_2$ are mutually orthogonal and then the amplitude varies exponentially at the planes of constant phase in the direction of $\mathbf{k}_2$. If $\mathbf{k}_1$ is *not* orthogonal to the interface and the angle between the z-axis (orthogonal to the interface) and $\mathbf{k}_2$ is greater than $\pi/2$ then at $\mathbf{x} = z\mathbf{e}_z$ the pertinent field is given by

$$\mathbf{A}\exp[-\mathbf{k}_2 \cdot \mathbf{e}_z\, z]\exp[i(\mathbf{k}_1 \cdot \mathbf{e}_z\, z - \omega t)].$$

where $\mathbf{k}_2 \cdot \mathbf{e}_z < 0$. Then, obviously the amplitude increases with z. Geometrically (cf. Fig. 1.3), this is immediately seen if we observe that the value at any point is propagated, without attenuation, along lines parallel to $\mathbf{k}_1$.

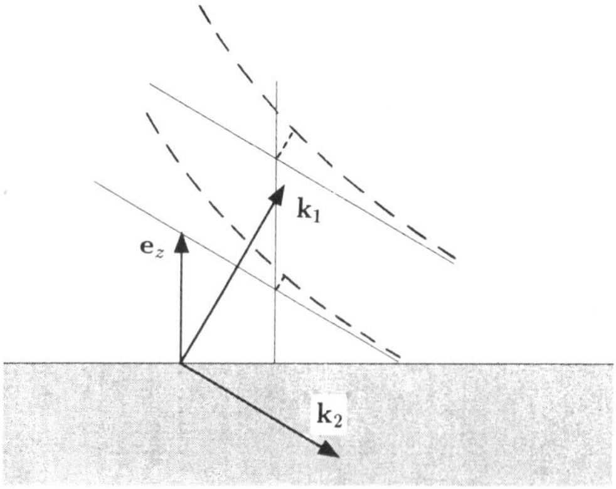

Fig. 1.3 Dashed lines represent the amplitude behaviour at different planes of constant phase. If $\mathbf{k}_2 \cdot \mathbf{e}_z < 0$ the amplitude increases along $\mathbf{e}_z$.

It is worth appending a comment on terminology. Sometimes (cf. [133, 144]) waves with complex-valued wave vectors are called *evanescent* if $\mathbf{k}_1 \cdot \mathbf{k}_2 = 0$ in that they do not involve energy dissipation. Quite naturally these authors denote by attenuating the waves whose amplitude decreases as they propagate. If $\mathbf{k}_1$ is parallel to $\mathbf{k}_2$ the corresponding waves are called homogeneous, although damping effects are present; really, amplitude is constant on planes of constant phase, but varies in the course of propagation. The combination of *attenuating* and *evanescent* would represent the general wave solution in dissipative bodies [133]. It has been proposed [143] to call heterogeneous those waves with $\mathbf{k}_1$ and $\mathbf{k}_2$ neither parallel nor perpendicular, while inhomogeneous waves should encompass both evanescent and heterogeneous waves. In other contexts (cf. [76]) the adjective evanescent is ascribed to the waves that have a component of $\mathbf{k}$ purely imaginary. For example, k_x, k_y real with $k_x^2 + k_y^2 > k^2$ and then k_z purely imaginary. It seems then prudent not to use the adjective evanescent for inhomogeneous waves. As already remarked, we prefer to denote by the adjective inhomogeneous all wave solutions with a complex-valued wave vector $\mathbf{k}$ such that $\mathbf{k}_1 \times \mathbf{k}_2 \neq 0$. However, it is well understood that evanescent waves and homogeneous waves are recovered as limit cases of inhomogeneous waves.

1.4 Properties from Fourier analysis

Consider a time-dependent field $\mathbf{u}(\mathbf{x},t)$ on $\Omega \times \mathbb{R}$, Ω being any region in $\mathcal{E}^3$. Assume that $\mathbf{u}(\mathbf{x},\cdot) \in L_1(\mathbb{R})$ and piecewise smooth in $\mathbb{R}$ for any position $\mathbf{x} \in \Omega$. Then its Fourier transform

$$\mathbf{u}_F(\mathbf{x},\omega) = \int_{-\infty}^{\infty} \exp(-i\omega t)\, \mathbf{u}(\mathbf{x},t)\, dt$$

is well defined for any $\omega \in \mathbb{R}$. Of course $\mathbf{u}$ is taken to be a real-valued vector and then it follows that

$$\mathbf{u}_F^*(\mathbf{x},\omega) = \mathbf{u}_F(\mathbf{x},-\omega) \tag{4.1}$$

where $*$ denotes the complex conjugate. The condition (4.1) in turn ensures that

$$\mathbf{u}(\mathbf{x},t) = \frac{1}{2\pi} \int_{-\infty}^{\infty} \exp(-i\omega t)\, \mathbf{u}_F(\mathbf{x},\omega)\, d\omega \tag{4.2}$$

provides a real-valued field for any time $t \in \mathbb{R}$.

In fact, a physical wave is hardly a time-harmonic (or single-frequency) oscillation. Yet, because of linearity we can evaluate the effects of time-harmonic oscillations and then superposition in the form (4.2) yields the physical field. This allows us to consider time-harmonic oscillations

$$\mathbf{u}(\mathbf{x},t;\omega) = \mathbf{u}_F(\mathbf{x},\omega) \exp(-i\omega t)$$

whose values are complex vectors.

For later convenience we consider the inhomogeneous wave

$$\mathbf{u}(\mathbf{x},t;\omega) = \mathbf{A} \exp[i(\mathbf{k}\cdot\mathbf{x} - \omega t)]$$

and examine the consequences of (4.1) on the amplitude (polarization) $\mathbf{A}(\omega)$ and the wave vector $\mathbf{k}(\omega)$. Let $\mathbf{A} = \mathbf{A}_1 + i\mathbf{A}_2$, with $\mathbf{A}_1, \mathbf{A}_2$ real. Then it follows that

$$\mathbf{u}^*(\mathbf{x},t;\omega) = \mathbf{u}(\mathbf{x},t;-\omega)$$

for any $\mathbf{x} \in \Omega$ if and only if

$$\mathbf{A}_1(\omega) = \mathbf{A}_1(-\omega), \qquad \mathbf{A}_2(\omega) = -\mathbf{A}_2(-\omega)$$

and

$$\mathbf{k}_1(\omega) = -\mathbf{k}_1(-\omega), \qquad \mathbf{k}_2(\omega) = \mathbf{k}_2(-\omega). \tag{4.3}$$

This shows that $\mathbf{A}_1$ and $\mathbf{k}_2$ are even functions in ω while $\mathbf{A}_2$ and $\mathbf{k}_1$ are odd. The result (4.3) is especially useful in deriving the solution for $\mathbf{k}$ as roots of suitable polynomials.

We recall that $\mathbf{k}_1 \cdot \mathbf{k}_2 > 0$, $\omega > 0$, means the amplitude decay in the direction of propagation. Examine what happens as $\omega \to -\omega$. By (4.3), to the inequality $(\mathbf{k}_1 \cdot \mathbf{k}_2)(\omega) > 0$ there corresponds the inequality $(\mathbf{k}_1 \cdot \mathbf{k}_2)(-\omega) < 0$. However, under the change $\omega \to -\omega$ we have $\mathbf{k}_1(\omega)\cdot\mathbf{x} - \omega t \to \mathbf{k}_1(-\omega)\cdot\mathbf{x} + \omega t = -\mathbf{k}_1(\omega)\cdot\mathbf{x} + \omega t$ and then the component relative to $-\omega$ propagates in the direction of $-\mathbf{k}_1(-\omega) = \mathbf{k}_1(\omega)$. Accordingly, also the $-\omega$ component decays in the direction of propagation.

2 MODELLING OF DISSIPATIVE MEDIA

For useful reference in later developments, this chapter outlines the essentials of continuum mechanics and thermodynamics which are applied extensively in the analysis of wave propagation in dissipative, continuous media. Also to introduce the pertinent notation, a quick review is given of the description of deformation and motion for continuous bodies. Then the balance equations are examined. Since thermal effects are disregarded, attention is mainly addressed to balance mass and linear momentum but the essential aspects of (the second law of) thermodynamics are also outlined.

Constitutive properties of some continuous media are modelled. Of course, any connection between mathematical models and experiments is always suggestive. In this sense it is of interest to see how data on creep tests in metals single out a simple model of viscoelastic body and, moreover, determine the pertinent parameters characterizing the model. Viscoelasticity is the natural framework for modelling dissipation in continuous media. Roughly speaking, it describes dissipation through memory effects and reduces to elasticity in equilibrium conditions. Both schemes of linear viscoelastic solid and viscoelastic fluid are exhibited along with the relevant restrictions placed by thermodynamics. Such restrictions, which are the subject of recent research, prove of fundamental importance in the study of wave propagation.

Interesting topics arise in connection with the superposition of a motion, or a wave, on a deformation of a body at equilibrium under the action of given body forces and stresses. The initial deformation is allowed to be finite while the superimposed motion is taken to be small, in a proper sense. To take advantage of the smallness of the superimposed motion and make the problem analytically tractable, an appropriate form of the equation of motion is required. This topic is re-visited here and, for both solids and fluids, detailed expressions of the equation of motion are determined.

2.1 Preliminaries on deformation and motion

Following the standard scheme of continuum mechanics we regard a body as a continuous distribution of material points or particles. A body may occupy different regions of the Euclidean space $\mathcal{E}^3$ at different times. It is convenient to choose one of these regions as reference. We denote by $\mathcal{B} \subset \mathcal{E}^3$ such a reference region, usually called *reference placement,* and label the material points by the position vector $\mathbf{X}$ they occupy in $\mathcal{B}$, relative to some arbitrarily chosen origin. The region $\mathcal{B}$ is taken to be regular and possibly unbounded ([78], §6). Incidentally, in the literature the word configuration is often used instead of placement.

The motion of the body is given by a function

$$\mathbf{x} = \mathbf{x}(\mathbf{X}, t) \tag{1.1}$$

which assigns to each material point $\mathbf{X} \in \mathcal{B}$, its position vector $\mathbf{x}$ in the Euclidean space $\mathcal{E}^3$ at time t; when the component form is convenient we refer $\mathcal{B}$ and $\mathcal{E}^3$ to Cartesian coordinates. We assume that the functions under consideration are continuously differentiable with respect to their arguments as many times as required; this holds in particular for the representation (1.1). At fixed $\mathbf{X}$, equation (1.1) yields the trajectory of the corresponding particle while at fixed time t it describes a *deformation* of the given continuum. Letting $\mathcal{B}_t$ be the image of $\mathcal{B}$, at the time t, we regard the map $\mathbf{x}(\cdot, t) : \mathcal{B} \to \mathcal{B}_t$, $t \in \mathbb{R}$, as a diffeomorphism; $\mathcal{B}_t$ is called the *current* (or present) placement. It follows that the transformation (1.1) from $\mathbf{X}$ to $\mathbf{x}$ can be inverted to give

$$\mathbf{X} = \mathbf{X}(\mathbf{x}, t). \tag{1.2}$$

All pertinent fields on the body can be thought of as functions of the material points and the time, that is of $\mathbf{X}$ and t. In view of (1.2) they can also be described by functions of $\mathbf{x}$ and t. As t varies, in the first case we follow the evolution of the quantity under consideration at a fixed material point (material or Lagrangian description), in the second one we follow the evolution at a fixed point in the Euclidean space $\mathcal{E}^3$ (spatial or Eulerian description). For example, any scalar field which is given by $\phi(\mathbf{X}, t)$ in the material description and by $\varphi(\mathbf{x}, t)$ in the spatial one. Because of (1.2) the two functions are related by

$$\varphi(\mathbf{x}, t) = \phi(\mathbf{X}(\mathbf{x}, t), t).$$

For formal simplicity, however, we will not use different symbols for the two functions.

In correspondence with any given motion $\mathbf{x}(\mathbf{X}, t)$ we can determine the spatial position of all material points of the body at each time t through (1.1). Then comparison between the current placement $\mathcal{B}_t$ and the reference placement $\mathcal{B}$ shows the deformation undergone by the continuum. Here we are confined to the essential geometric and kinematic aspects that are needed for later developments.

The fundamental geometric object underlying the local analysis of deformation is the *deformation gradient* defined as

$$\mathbf{F} = \nabla_{\mathbf{X}} \mathbf{x}^{\dagger}(\mathbf{X}, t), \tag{1.3}$$

which in components reads $F_{iA} = x_{i,A}$, where the operator $\nabla_{\mathbf{X}}$ denotes the gradient with respect to the material coordinates, a superscript † denotes the transpose, small (capital) latin indices refer to spatial (material) coordinates and a comma stands for partial differentiation. Here we call the attention of the reader to two notational peculiarities relative to the literature where, usually, $\mathbf{F} = \nabla \mathbf{x}$ (cf. [78], §6). First, so as to give the reader an immediate, unambiguous, interpretation of the symbols we distinguish between $\partial/\partial\mathbf{X}$ and $\partial/\partial\mathbf{x}$ and write $\nabla_{\mathbf{X}}$ for $\partial/\partial\mathbf{X}$ and ∇ for $\partial/\partial\mathbf{x}$. Second, again to simplify the interpretation, we place the vector quantities in the right position due to the tensor properties of the pertinent expression; so, $F_{iA} = (\nabla_{\mathbf{X}}\, \mathbf{x}^{\dagger})_{iA} = x_{i,A}$.

In terms of $\mathbf{F}$ the jacobian J of the transformation (1.1) is given by

$$J = \det \mathbf{F}. \tag{1.4}$$

Under the assumption of invertibility of (1.1) it turns out that $J \neq 0$ and hence $\mathbf{F}$ may be inverted too; without loss of generality we may also assume $J > 0$.

By the chain rule of differentiation we find

$$x_{i,A} X_{A,j} = \delta_{ij} \tag{1.5}$$

whence $\mathbf{F}^{-1}$ is such that

$$F^{-1}_{Aj} = X_{A,j}.$$

The matrix $X_{A,j}$, the inverse of $x_{i,A}$, may be represented as

$$X_{A,j} = \frac{1}{2J} \eta_{jpq} \eta_{ABC} x_{p,B} x_{q,C} \tag{1.6}$$

where η_{jpq} is the Levi-Civita permutation symbol. This result is proved by showing that replacement of $X_{A,j}$ into (1.5) with the expression given by (1.6) yields an identity. For, by the definition of determinant we have

$$J\eta_{ijq} = \eta_{ABC} x_{i,A} x_{j,B} x_{q,C}$$

while the permutation symbol obeys the conditions

$$\eta_{ijp}\eta_{irs} = \delta_{jr}\delta_{ps} - \delta_{js}\delta_{pr}, \qquad \eta_{ijp}\eta_{ijs} = 2\delta_{ps}, \qquad \eta_{ijp}\eta_{ijp} = 3! \tag{1.7}$$

and similarly for η_{ABC}. Thus we find that

$$x_{i,A} X_{A,j} = \frac{1}{2J} \eta_{jpq} \eta_{ABC} x_{i,A} x_{p,B} x_{q,C} = \frac{1}{2} \eta_{jpq} \eta_{ipq} = \delta_{ij}.$$

From the skewness of η_{ABC} and (1.6) we also get the useful identity

$$(J X_{A,i})_{,A} = 0. \tag{1.8}$$

By interchanging the roles of reference and current placement we obtain the dual relation

$$(x_{i,A}/J)_{,i} = 0. \tag{1.8'}$$

Consider a vector $\mathbf{W}$ defined at $\mathbf{X}$; its image in space at the time t is given by the vector

$$\mathbf{w}_t = \mathbf{F}\mathbf{W} \tag{1.9}$$

applied at $\mathbf{x}(\mathbf{X}, t)$. This is easily seen by considering any curve $\boldsymbol{\gamma}(s)$ through $\mathbf{X}$ with tangent vector $\mathbf{W} = d\boldsymbol{\gamma}/ds$, and taking as $\mathbf{w}_t$ the tangent to the image of the curve at the point $\mathbf{x}(\mathbf{X}, t)$, that is $\mathbf{w}_t = d\mathbf{x}(\mathbf{X}(s), t)/ds$. Repetition of this procedure at varying t yields a vector field, still denoted by $\mathbf{w}_t$, defined over the trajectory of the particle $\mathbf{X}$. Denote by $\mathbf{Z}$ another vector at $\mathbf{X}$ and by $\mathbf{z}_t$ its image. In view of (1.9) we have

$$\mathbf{z}_t \cdot \mathbf{w}_t = \mathbf{Z}^\dagger \mathbf{F}^\dagger \mathbf{F}\mathbf{W}. \tag{1.10}$$

The symmetric tensor

$$\mathbf{C} = \mathbf{F}^\dagger \mathbf{F}, \tag{1.11}$$

whose components are $C_{AB} = x_{i,A} x_{i,B}$, is the *right Cauchy-Green strain* tensor; as follows from (1.10) $\mathbf{C}$ is used to measure deformed lengths and angles. In geometric terms, $\mathbf{C}$ is the pull-back of the spatial metric [122]. Notice also that the representation (1.11) has to be changed, in a straightforward way, if the coordinate patch is not Cartesian. Similarly, the (symmetric) *left Cauchy-Green strain* tensor $\mathbf{B}$ is defined by $\mathbf{B} = \mathbf{F}\mathbf{F}^\dagger$.

With a view to applications to (linear) viscoelasticity it is convenient to introduce the *Lagrangian strain tensor*

$$\boldsymbol{\mathcal{E}} = \tfrac{1}{2}(\mathbf{C} - \mathbf{1}) \tag{1.12}$$

or, in components, $\mathcal{E}_{AB} = (C_{AB} - \delta_{AB})/2$. Comparison with (1.9)-(1.11) gives

$$\mathbf{W}^T \boldsymbol{\mathcal{E}} \mathbf{W} = \tfrac{1}{2}(\mathbf{w}_t \cdot \mathbf{w}_t - \mathbf{W} \cdot \mathbf{W}), \tag{1.13}$$

thus showing that $\boldsymbol{\mathcal{E}}$ can be used just to measure variations in length.

It is also of interest to examine the effect of a deformation on a material surface. Describe the surface by the parameters u, v and let $\mathbf{X} = \mathbf{X}(u, v)$ be the position, of any point of the surface, in $\mathcal{B}$. Denote by $\mathbf{W}$ and $\mathbf{Z}$, respectively, the tangent vectors $\partial \mathbf{X}/\partial u$ and $\partial \mathbf{X}/\partial v$ at an arbitrary point. The normal unit vector $\mathbf{N}$ is parallel to $\mathbf{W} \times \mathbf{Z}$, while the area dA of the surface element generated by a change du, dv of the parameters is $|\mathbf{W} \times \mathbf{Z}|$ times $du dv$. Then we have

$$\mathbf{N} dA = \mathbf{W} \times \mathbf{Z}\, du dv.$$

Similarly, by use of (1.9), the corresponding quantity $\mathbf{n}\, da$ in the deformed surface (image) is given by

$$\mathbf{n}\, da = \mathbf{w}_t \times \mathbf{z}_t\, du dv = \mathbf{F}\mathbf{W} \times \mathbf{F}\mathbf{Z}\, du dv.$$

To find an explicit connection between $\mathbf{N}$ and $\mathbf{n}$ it is convenient to consider the component form, namely

$$n_j da = \eta_{jpq} x_{p,A} x_{q,B} W_A Z_B\, du dv.$$

Then we observe that (1.6), (1.7) and the skewness of η yield

$$JX_{A,j}\eta_{AHK} = \frac{1}{2}\eta_{jpq}\eta_{ABC}\eta_{AHK}x_{p,B}x_{q,C} = \eta_{jpq}x_{p,H}x_{q,K},$$

whence it follows that

$$n_j da = JX_{A,j}\eta_{AHK}W_H Z_K\, dudv.$$

In compact notation,

$$\mathbf{n}da = J\mathbf{F}^{-\dagger}\mathbf{N}dA, \tag{1.14}$$

where $\mathbf{F}^{-\dagger}$ stands for $(\mathbf{F}^{-1})^{\dagger}$, with inverse

$$\mathbf{N}\,dA = \frac{1}{J}\mathbf{F}^{\dagger}\mathbf{n}da. \tag{1.14'}$$

Up to now we have been concerned with geometric aspects of deformation when really deformations depend on time, possibly in a smooth way, and then also some aspects of kinematics are in order. Denote by a superposed dot, or d/dt, the *material* or *Lagrangian* time derivative

$$\dot{\phi} = \frac{d}{dt}\phi := \frac{\partial\phi(\mathbf{X},t)}{\partial t}$$

for every field ϕ on $\mathcal{B}\times\mathbb{R}$. Accordingly we define the *velocity* as the material time derivative of the spatial position vector,

$$\mathbf{v} = \frac{\partial\mathbf{x}(\mathbf{X},t)}{\partial t} \tag{1.15}$$

and the *acceleration* as

$$\mathbf{a} = \frac{\partial^2\mathbf{x}(\mathbf{X},t)}{\partial t^2}. \tag{1.16}$$

We can say that (1.15) and (1.16) give the velocity and acceleration of the particle $\mathbf{X}$ occupying the space position $\mathbf{x}$ at time t. To find the material time derivative of a spatial field $\phi(\mathbf{x},t)$ we consider the corresponding material representation $\phi(\mathbf{x}(\mathbf{X},t),t)$; the use of the chain rule yields

$$\dot{\phi} = \frac{\partial\phi}{\partial t} + \mathbf{v}\cdot\nabla\phi. \tag{1.17}$$

for every field ϕ on $\mathcal{B}_t\times\mathbb{R}$.

The tensor field

$$\mathbf{L} = \nabla\mathbf{v}^{\dagger}, \tag{1.18}$$

of Cartesian components $L_{ij} = v_{i,j}$, is called the *velocity gradient*. By applying the material gradient $\nabla_{\mathbf{X}}$ to $\mathbf{v}$ and using the chain rule for partial derivatives we obtain the relation

$$\dot{\mathbf{F}} = \mathbf{L}\,\mathbf{F} \tag{1.19}$$

whence $\mathbf{L} = \dot{\mathbf{F}}\mathbf{F}^{-1}$. As an immediate application of (1.19) we derive an expression for the time derivative of J. Observe that, by (1.3), (1.4) and the coordinate version of (1.19),

$$\dot{J} = \frac{d}{dt}(\det x_{i,A}) = \frac{\partial J}{\partial x_{i,A}}x_{i,At} = \frac{\partial J}{\partial x_{i,A}}v_{i,j}x_{j,A}.$$

In view of the identity $x_{i,A}X_{A,j} = \delta_{ij}$, it follows that $JX_{A,i}$ is the cofactor of $x_{i,A}$, which means that $\partial J/\partial x_{i,A} = JX_{A,i}$. Upon substitution we have the desired result

$$\dot{J} = J\,\nabla \cdot \mathbf{v}. \tag{1.20}$$

We end the set of kinematic properties by considering the *stretching tensor* $\mathbf{D}$, namely the symmetric part of the velocity gradient,

$$\mathbf{D} = \tfrac{1}{2}(\mathbf{L} + \mathbf{L}^{\dagger}). \tag{1.21}$$

Then, on recalling that $\mathbf{C} = \mathbf{F}^{\dagger}\mathbf{F}$ and evaluating the time derivative of (1.12) we have

$$\dot{\boldsymbol{\mathcal{E}}} = \tfrac{1}{2}(\dot{\mathbf{F}}^{\dagger}\,\mathbf{F} + \mathbf{F}^{\dagger}\,\dot{\mathbf{F}}).$$

Substitution of (1.19) and use of (1.21) yields

$$\dot{\boldsymbol{\mathcal{E}}} = \mathbf{F}^{\dagger}\mathbf{D}\mathbf{F}, \tag{1.22}$$

If $\mathbf{F} \simeq \mathbf{1}$ then $\mathbf{D} \simeq \dot{\boldsymbol{\mathcal{E}}}$, which justifies why $\mathbf{D}$ is also called the *rate-of-strain* tensor. Finally, apply the material time derivative to (1.13) and observe that $\mathbf{W}$ is independent of t. Comparison with (1.22) shows that

$$\frac{1}{2}\frac{d}{dt}(\mathbf{w}_t \cdot \mathbf{w}_t) = \mathbf{w}_t \cdot \mathbf{D}\mathbf{w}_t$$

thus providing the meaning of deformation rate of lengths for $\mathbf{D}$.

2.2 Balance laws

Any model of a body must comply with general principles of continuum mechanics. Among them are the principles that mass, momentum, and energy are conserved, which means that appropriate balance laws must hold. With a view to the applications we have in mind, here we disregard energetic (thermal) aspects and review merely the essential topics concerning the balance of mass and momentum.

Consider an arbitrary motion $\mathbf{x}(\mathbf{X}, t)$. A *mass density* field $\rho(\mathbf{x}, t)$ is defined such that, at every time t, the mass of any portion of continuum determined by a subregion Ω of $\mathcal{B}$ is given by

$$m_t(\Omega) = \int_{\Omega_t} \rho\, dx$$

where Ω_t denotes the image of Ω at the time t. The balance (conservation) of mass is expressed by saying that $m_t(\Omega)$ is constant in time. Then the observation that

$$\int_{\Omega_t} \phi\, dx = \int_{\Omega} \phi\, J\, dX$$

for any function ϕ on $\mathcal{B}_t \times \mathbb{R}$ or $\mathcal{B} \times \mathbb{R}$ and the condition that $dm_t(\Omega)/dt = 0$ show that ρJ is a function of $\mathbf{X}$ only. We write

$$\rho J = \rho_0 \tag{2.1}$$

where $\rho_0 = \rho_0(\mathbf{X})$ takes the meaning of mass density in $\mathcal{B}$. Time differentiation of (2.1) and use of (1.20) yield

$$\dot{\rho} + \rho \nabla \cdot \mathbf{v} = 0. \tag{2.2}$$

Alternatively, (2.2) can be written as

$$\frac{\partial \rho}{\partial t} + \nabla \cdot (\rho \mathbf{v}) = 0. \tag{2.3}$$

So we have three local forms of balance of mass; (2.1) provides directly the mass density in the current configuration, ρ, in terms of the reference mass density ρ_0 and the deformation (through J) while (2.2) and (2.3) are the differential forms in the Lagrangian and Eulerian description, respectively.

The rate of change of momentum of each material region $\Omega \subset \mathcal{B}$ is equal to the total force acting on Ω. This total force is thought to consist of two contributions: *body forces*, exerted on the interior points by the environment of the body through long range effects such as gravity; *contact forces*, exerted on the boundary $\partial\Omega$ by the remaining region $\mathcal{B} \setminus \Omega$ of the body and, possibly, the environment of the body. The body force per unit volume is taken as $\rho \mathbf{b}(\mathbf{x}, t)$. Following Cauchy's hypothesis we assume that the contact forces are expressed by a surface force density $\mathbf{f}(\mathbf{n}, \mathbf{x}, t)$, $\mathbf{n}$ being the unit outward normal to $\partial\Omega$ and $\mathbf{x}$ the position vector on $\partial\Omega$. Then for any region $\Omega \subset \mathcal{B}$ we write the balance of linear momentum as

$$\frac{d}{dt}\int_{\Omega_t} \rho\, \mathbf{v}\, dx = \int_{\partial\Omega_t} \mathbf{f}(\mathbf{n}) da + \int_{\Omega_t} \rho\, \mathbf{b}\, dx.$$

To obtain a local form for the balance of momentum we recall that for any C^1 field Φ and region $\Omega \subset \mathcal{B}$, use of (1.20) and (2.2) yields

$$\frac{d}{dt}\int_{\Omega_t} \rho\, \Phi\, dx = \frac{d}{dt}\int_{\Omega} \rho\, \Phi\, J\, dX = \int_{\Omega} (\dot{\rho}\Phi + \rho\dot{\Phi} + \rho\Phi\nabla \cdot \mathbf{v}) J\, dX = \int_{\Omega} \rho\dot{\Phi} J\, dX = \int_{\Omega_t} \rho\, \dot{\Phi}\, dx.$$

whence it follows Reynold's transport theorem

$$\frac{d}{dt}\int_{\Omega_t} \rho\, \Phi\, dx = \int_{\Omega_t} \rho\, \dot{\Phi}\, dx. \tag{2.4}$$

Also, Cauchy's theorem ([78], §14) proves the existence of the *Cauchy stress* tensor $\mathbf{T}(\mathbf{x}, t)$ such that

$$\mathbf{f}(\mathbf{n}) = \mathbf{T}\, \mathbf{n}\ .$$

As a consequence, use of (2.4), application of the divergence theorem to the integral on $\partial\Omega_t$ and the arbitrariness of Ω yield the local form as

$$\rho\mathbf{a} = \nabla \cdot \mathbf{T} + \rho\,\mathbf{b} \tag{2.5}$$

namely the *equation of motion* in the spatial description. Hereafter the divergence of a second-order tensor is meant to be relative to the second index. In particular, the component form of (2.5) is $\rho a_i = T_{ij,j} + \rho b_i$.

An analogous integral balance law for angular momentum and use of (2.5) leads to the conclusion that $\mathbf{T}$ is a symmetric tensor, namely $\mathbf{T} = \mathbf{T}^\dagger$.

For later convenience we examine also the corresponding material formulation of the balance of momentum. Integrate each term of (2.5) over Ω_t. The change of coordinates $\mathbf{x} \to \mathbf{X}$, whereby $\Omega_t \to \Omega$, yields

$$\int_\Omega \rho\, J\, \mathbf{a}\, dX = \int_\Omega J\nabla \cdot \mathbf{T}\, dX + \int_\Omega \rho\, J\mathbf{b}\, dX, \tag{2.6}$$

where the integrands are regarded as functions of $\mathbf{X}$. By (2.1) and the arbitrariness of Ω we have the local form

$$\rho_0\mathbf{a} = J\nabla \cdot \mathbf{T} + \rho_0\mathbf{b}. \tag{2.7}$$

The term $J\nabla \cdot \mathbf{T}$ can be given a more genuine material form as follows. Observe that $(J\nabla \cdot \mathbf{T})_i = JT_{ij,A}X_{A,j}$ and that, by (1.8),

$$(J\nabla \cdot \mathbf{T})_i = (JT_{ij}X_{A,j})_{,A}.$$

Then in terms of the *first Piola-Kirchhoff stress* tensor

$$\mathbf{S} = J\mathbf{T}\mathbf{F}^{-\dagger} \tag{2.8}$$

we have

$$J\nabla \cdot \mathbf{T} = \nabla_{\mathbf{X}} \cdot \mathbf{S}.$$

Substitution into (2.7) provides the equation of motion in the material form

$$\rho_0\ddot{\mathbf{x}} = \nabla_{\mathbf{X}} \cdot \mathbf{S} + \rho_0\mathbf{b} \tag{2.9}$$

or, in components, $\rho_0\ddot{x}_i = S_{iA,A} + \rho_0 b_i$.

The meaning of $\mathbf{S}$ is easily seen by considering the surface traction $\mathbf{t} = \mathbf{T}\mathbf{n}$ at a surface element of area da and normal $\mathbf{n}$, so that $\mathbf{T}\mathbf{n}da$ yields the corresponding force. Substitution of $\mathbf{n}$ from (1.14) and use of (2.8) gives

$$\mathbf{t}da = \mathbf{T}\mathbf{n}da = J\mathbf{T}\mathbf{F}^{-\dagger}\mathbf{N}dA = \mathbf{S}\mathbf{N}dA,$$

thus showing that $\mathbf{S}$ generates the surface traction referred to by the material surface element $\mathbf{N}\,dA$. This is consistent with the fact that, according to (2.6), that is

$$\int_\Omega \rho_0\, \mathbf{a}\, dX = \int_{\partial\Omega} \mathbf{S}\,\mathbf{N}\, dA + \int_\Omega \rho_0\, \mathbf{b}\, dX,$$

the flux of $\mathbf{S}$ through $\partial\Omega$ yields the total contact force exerted on Ω. Sometimes it is convenient to describe the stress in terms of the *second Piola-Kirchhoff stress* tensor $\mathbf{Y}$ which is related to $\mathbf{T}$ and $\mathbf{S}$ by

$$\mathbf{Y} = \mathbf{F}^{-1}\mathbf{S} = J\mathbf{F}^{-1}\mathbf{T}\mathbf{F}^{-\dagger}. \tag{2.10}$$

Also because we disregard thermal effects, the balance equations (2.1) and (2.5), or (2.9), are enough for the analysis of wave propagation in many circumstances. There are cases, though, where the expressions (2.5) or (2.9) are not appropriate. This is typically the case when (small amplitude) wave propagation is considered to occur in a body which is predeformed or prestressed. For these circumstances a more sophisticated version is in order.

We consider a body which is at equilibrium under the action of suitable external forces. Then we let the configuration of the body be slightly changed in the sense that a small motion is superposed. It is our purpose to find the equations to be obeyed by the superposed motion. To this end it is convenient to make use of three placements, namely

- the *reference* placement $\mathcal{B}$;
- the *intermediate*, equilibrium placement $\mathcal{B}_i$;
- the *current* placement $\mathcal{B}_t$.

We may view these placements as the result of subsequent deformations or motions as

$$\mathbf{X} \in \mathcal{B} \quad \longrightarrow \quad \mathbf{x} \in \mathcal{B}_i \quad \longrightarrow \quad \tilde{\mathbf{x}} \in \mathcal{B}_t.$$

Denote by

$$\mathbf{u}(\mathbf{X},t) = \tilde{\mathbf{x}}(\mathbf{X},t) - \mathbf{x}(\mathbf{X})$$

the displacement of the particle $\mathbf{X}$, at time t, due to the motion. By the assumed invertibility of the deformation $\mathbf{x}(\mathbf{X})$ we can also express $\mathbf{u}$ as $\mathbf{u}(\mathbf{x},t)$. Then we can regard both the placement $\mathcal{B}$ and the placement $\mathcal{B}_i$ as reference.

Consider the displacement gradient $\mathbf{H} = \nabla\mathbf{u}^\dagger$ or, in Cartesian components, $H_{ij} = u_{i,j}$. We follow the approximation that $\mathbf{H}$ remains small, i.e. $|\mathbf{H}| \ll 1$ at every $\mathbf{x} \in \mathcal{B}_i$ and $t \in \mathbb{R}$. Accordingly we neglect quadratic terms in $\mathbf{H}$ and higher.

First we examine the equilibrium condition at $\mathcal{B}_i$. Denote by a superscript 0 the value of a quantity at equilibrium. Then we write the equilibrium equation as

$$\nabla_{\mathbf{X}} \cdot \mathbf{S}^0 + \rho_0 \mathbf{b} = 0, \tag{2.11}$$

or, in Cartesian components, $S^0_{iK,K} + \rho_0 b_i = 0$. Here ρ_0 is the mass density in $\mathcal{B}$ and is related to the mass density in $\mathcal{B}_i$, ρ, by $\rho_0 = \rho J$. Let $\mathbf{S}$ be the first Piola-Kirchhoff stress corresponding to the motion $\tilde{\mathbf{x}}(\mathbf{X}, t)$. Since $\ddot{\tilde{\mathbf{x}}} = \ddot{\mathbf{u}}$ we can write

$$\rho_0 \ddot{\mathbf{u}} = \nabla_{\mathbf{X}} \cdot \mathbf{S} + \rho_0 \mathbf{b}. \tag{2.12}$$

Now, if $\mathbf{b}$ is known as a function of the position in space then

$$\mathbf{b}(\tilde{\mathbf{x}}) - \mathbf{b}(\mathbf{x}) = (\mathbf{u} \cdot \nabla)\mathbf{b} + o(|\mathbf{u}|).$$

Subtraction of (2.11) from (2.12) and neglect of $o(|\mathbf{u}|)$ yields

$$\rho_0 \ddot{\mathbf{u}} = \nabla_{\mathbf{X}} \cdot (\mathbf{S} - \mathbf{S}^0) + \rho_0 (\mathbf{u} \cdot \nabla)\mathbf{b}. \tag{2.13}$$

In terms of the second Piola-Kirchhoff stresses $\mathbf{Y}^0, \mathbf{Y}^0 + \hat{\mathbf{Y}}$ corresponding to $\mathbf{x}(\mathbf{X})$, $\tilde{\mathbf{x}}(\mathbf{X}, t)$ we have

$$\mathbf{S}^0 = \mathbf{F}\mathbf{Y}^0, \qquad \mathbf{S} = (\mathbf{F} + \nabla_{\mathbf{X}}\, \mathbf{u}^\dagger)(\mathbf{Y}^0 + \hat{\mathbf{Y}})$$

where $\hat{\mathbf{Y}}$ is the effect of the perturbation $\tilde{\mathbf{x}} - \mathbf{x}$. Concerning $\hat{\mathbf{Y}}$ we observe that it is small inasmuch as $\mathbf{H}$ is small, $|\mathbf{H}| \ll 1$, and hence we neglect quadratic terms in $\hat{\mathbf{Y}}$ and higher. Then (2.13) becomes

$$\rho_0 \ddot{\mathbf{u}} = \nabla_{\mathbf{X}} \cdot (\mathbf{H}\mathbf{F}\mathbf{Y}^0 + \mathbf{F}\hat{\mathbf{Y}}) + \rho_0 (\mathbf{u} \cdot \nabla)\mathbf{b}. \tag{2.14}$$

The next step consists in establishing the analogue of (2.14) when the intermediate (equilibrium) placement $\mathcal{B}_i$ is taken as reference. Divide by J, observe that $\nabla_{\mathbf{X}}\, \mathbf{u}^\dagger = \mathbf{H}\mathbf{F}$ and make use of the identity $(F_{iK}/J)_{,i} = 0$. Then (2.14) can be written as

$$\rho \ddot{\mathbf{u}} = \nabla \cdot \left(\frac{1}{J}\mathbf{H}\mathbf{F}\mathbf{Y}^0\mathbf{F}^\dagger + \frac{1}{J}\mathbf{F}\hat{\mathbf{Y}}\mathbf{F}^\dagger\right) + \rho(\mathbf{u} \cdot \nabla)\mathbf{b}$$

where ρ stands for the mass density ρ_i in the intermediate placement $\mathcal{B}_i$, and then is independent of the displacement field $\mathbf{u}$. Now, as follows from (2.10), $\mathbf{F}\mathbf{Y}^0\mathbf{F}^\dagger/J$ is just the Cauchy stress $\mathbf{T}^0$ in the intermediate placement $\mathcal{B}_i$ and hence

$$\rho \ddot{\mathbf{u}} = \nabla \cdot \left(\mathbf{H}\mathbf{T}^0 + \frac{1}{J}\mathbf{F}\hat{\mathbf{Y}}\mathbf{F}^\dagger\right) + \rho(\mathbf{u} \cdot \nabla)\mathbf{b} \tag{2.15}$$

or, in components,

$$\rho \ddot{u}_i = \left(H_{ip}T^0_{pq} + \frac{1}{J}F_{iK}\hat{Y}_{KH}F_{qH}\right)_{,q} + \rho u_p b_{i,p}.$$

Accordingly, in the intermediate placement $\mathcal{B}_i$, the equilibrium stress $\mathbf{T}^0$ enters the equation of motion through the equivalent stress $\mathbf{H}\mathbf{T}^0$. Further, an additional term, due to the perturbation $\mathbf{x} \to \tilde{\mathbf{x}} = \mathbf{x} + \mathbf{u}$, occurs in the form $\mathbf{F}\hat{\mathbf{Y}}\mathbf{F}^\dagger/J$ which may be viewed as the effect of $\hat{\mathbf{Y}}$ on the Cauchy stress, in $\mathcal{B}_i$. Of course, if $\mathbf{b}$ is the (constant) gravity acceleration then the last term vanishes.

Two remarks are in order about (2.15). First, the tensor $\mathbf{HT}^0$ is generally non-symmetric. This should come as no surprise in that (2.15) has a structure of material balance equation, relative to $\mathcal{B}_i$. Second, a detailed expression for $\mathbf{F}\hat{\mathbf{Y}}\mathbf{F}^T$ can be given only when the constitutive equation for $\mathbf{T}$, and then for $\mathbf{Y}$, is specified.

Thermodynamics always plays a central role in the elaboration of constitutive models. Although here we disregard thermal effects, thermodynamic conditions will prove crucial versus wave propagation. It is then worth giving an outline of the theory which underlies the derivation of thermodynamic restrictions. Detailed accounts of thermodynamics of viscoelastic bodies can be found in [70, 129, 130].

Denote by ϵ the internal energy density so that $\frac{1}{2}\mathbf{v}^2 + \epsilon$ is the total energy per unit mass. The balance of energy is stated by saying that the time rate of the total energy of each material region $\Omega \subset \mathcal{B}$ equals the rate of work made by all forces plus the rate of energy supplied from the remaining part $\mathcal{B} \setminus \Omega$ and the environment. This is specified formally as

$$\frac{d}{dt}\int_{\Omega_t} (\tfrac{1}{2}\rho\mathbf{v}^2 + \rho\epsilon)\, dx = \int_{\partial\Omega_t} [\mathbf{v}\cdot\mathbf{T}\mathbf{n} - q(\mathbf{n})] da + \int_{\Omega_t} (\rho\mathbf{b}\cdot\mathbf{v} + \rho r)\, dx$$

where $-q$ is the rate of energy per unit area and r is the heat supply (from the environment). By analogy with Cauchy's theorem it follows the existence of a vector field $\mathbf{q}$, called the heat flux, such that $q = \mathbf{q}\cdot\mathbf{n}$. By use of the equation of motion (2.5) and the arbitrariness of Ω we obtain the local balance of energy in the form

$$\rho\dot{\epsilon} = \mathbf{T}\cdot\mathbf{D} - \nabla\cdot\mathbf{q} + \rho\, r \tag{2.16}$$

where $\mathbf{T}\cdot\mathbf{D} = T_{ij}D_{ij}$.

Let η be the entropy density and θ the absolute temperature. The second law of thermodynamics is often expressed by saying that the Clausius-Duhem inequality

$$\rho\dot{\eta} + \nabla\cdot\big(\frac{1}{\theta}\mathbf{q}\big) - \frac{\rho\, r}{\theta} \geq 0$$

must hold for every evolution of the body compatible with the balance equations. This results into (thermodynamic) restrictions for the admissible constitutive equations. It is a drawback of this formulation of the second law that the existence of the entropy function is assumed at the outset. An integral (in time) formulation of the second law, free from this assumption, involves cyclic processes: for any cyclic process between 0 and d the Clausius inequality

$$\int_0^d \big[-\frac{1}{\rho}\nabla\cdot\big(\frac{1}{\theta}\mathbf{q}\big) + \frac{r}{\theta}\big]dt \leq 0 \tag{2.17}$$

holds. In view of the balance of energy (2.16), for isothermal processes (θ = constant) the term $\dot{\epsilon}$ is ineffective and then the Clausius inequality (2.17) simplifies to

$$\int_0^d \frac{1}{\rho} \mathbf{T} \cdot \mathbf{D}\, dt \geq 0. \tag{2.18}$$

Further, by the linearity of the theory we let ρ in (2.18) be constant. It is the inequality (2.18) that yields the restrictions for viscoelastic solids and fluids. Based on a characterization of reversible processes, we say that equality in (2.18) holds only for (trivial) processes such that $\mathbf{D}$ vanishes identically on $[0, d)$.

2.3 Elementary models of dissipative bodies

Models which date back to the XVIII century describe the (elastic and) dissipative properties of solids and fluids in terms of various configurations of springs and dashpots. Among them, those named after Maxwell and Kelvin (or Kelvin-Voigt) (cf. [158], §8.5) have the advantage of providing the essential features of the material behaviour albeit they consist of the minimal number of components, namely one spring and one dashpot. Here we recall the essentials of such models and then we analyse how they, or their appropriate combinations, yield an efficient description of material behaviour.

Maxwell's model accounts for the behaviour of a material element through a spring and a dashpot joined in series. Let E be Young's modulus (elastic constant) of the spring and η the viscosity coefficient of the dashpot. Then the relation between the Cauchy stress T and the strain e in the standard one-dimensional form can be written as

$$\dot{T} + \frac{E}{\eta} T = E\, \dot{e} \tag{3.1}$$

in the unknown function $T(t)$, $t \in \mathbb{R}$. The trivial integration of (3.1) on $(-\infty, t)$, for any finite value of the time t and $T(-\infty)$, yields

$$T(t) = E \int_0^\infty \exp(-s/\tau)\, \dot{e}(t-s)\, ds$$

where $\tau = \eta/E$. An integration by parts gives

$$T(t) = E\, e(t) - \frac{E}{\tau} \int_0^\infty \exp(-s/\tau)\, e(t-s)\, ds. \tag{3.2}$$

Kelvin's model describes the behaviour of a material element through a spring and a dashpot joined in parallel. This amounts to letting T and e be related by

$$\dot{e} + \frac{1}{\tau} e = \frac{1}{\tau E} T.$$

Incidentally, tables of E for common solids, and particularly for metals, are easily available. In contrast, it is quite difficult to find tables of values of τ. Then experiments, besides substantiating the plausibility of the model under consideration, should provide estimates for the time constant τ.

One of the standard experiments for investigating material properties is the creep test: a specimen is subjected to a constant load for a relatively long period of time and the extension is measured. Examine briefly how a creep test can determine the material constants of the model.

Consider a creep test with

$$T = \begin{cases} 0, & t < 0, \\ T_0, & t \geq 0. \end{cases}$$

With reference to Maxwell's model, by (3.2) we have $e(0) = T_0/E$ and then, by (3.1),

$$e(t) = \frac{T_0}{E}(1 + \frac{t}{\tau}), \qquad t \in \mathbb{R}^+. \tag{3.3}$$

In view of (3.3), the initial value of the $e - t$ curve provides Young's modulus E for the material. Then the slope of the curve yields the relaxation time τ. Experimental results usually provide the extension as a function of time. Via obvious scale transformations, the desired values of E and τ can be derived.

Such a procedure leads quite often to reasonable results. Yet it is known that joining a Maxwell element and a Kelvin element in series "gives rise to creep curves that are more representative of the experimentally obtained curves" [96]. In fact, experiments show a relatively long transient which is not described by (3.3) and, what is more, affects the values of E and τ.

Consider the two elements joined in series and denote by the subscripts M and K the quantities pertaining to the Maxwell and Kelvin elements. By the two differential equations

$$\dot{e}_M = \frac{1}{E_M}\dot{T} + \frac{1}{\tau_M E_M}T,$$

$$\dot{e}_K + \frac{1}{\tau_K}e_K = \frac{1}{\tau_K E_K}T$$

and the condition $e = e_K + e_M$ we find the single equation

$$\ddot{e} + \frac{1}{\tau_K}\dot{e} = \frac{1}{E_M}\ddot{T} + (\frac{1}{\tau_M E_M} + \frac{1}{\tau_K E_K} + \frac{1}{\tau_K E_M})\dot{T} + \frac{1}{\tau_K \tau_M E_M}T. \tag{3.4}$$

In connection with the creep test we have

$$\ddot{e} + \frac{1}{\tau_K}\dot{e} = \frac{1}{\tau_K \tau_M E_M}T_0, \qquad t \in \mathbb{R}^+,$$

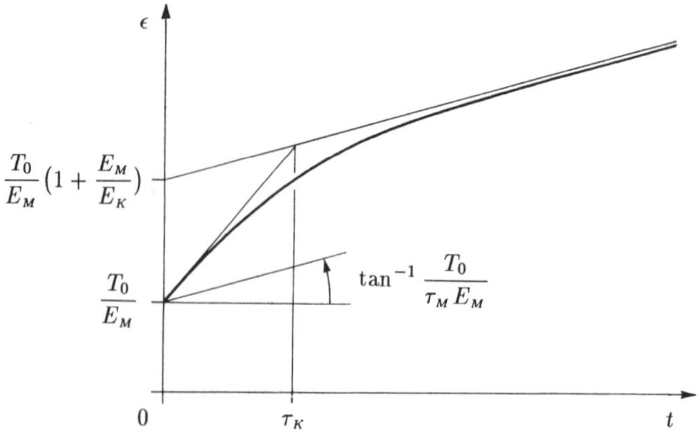

Fig. 2.1 Response of the material to a creep test.

$$e(0) = \frac{T_0}{E_M}, \qquad \dot{e}(0) = (\frac{1}{\tau_K E_K} + \frac{1}{\tau_M E_M})T_0$$

whence

$$e(t) = T_0\{\frac{1}{E_K}[1 - \exp(-\frac{t}{\tau_K})] + \frac{1}{E_M}(1 + \frac{t}{\tau_M})\}.$$

This solution fits very well experimental data for a great many materials. Figure 2.1 shows how the solution, and the data, enable us to determine the material parameters τ_K, E_K, τ_M, E_M. In many cases it turns out that $E_K \gg E_M$; in those cases it is reasonable to model the behaviour of the solid through a Maxwell element. Moreover, if as usual Young's modulus is available from tables then the evaluation of the slope of the asymptote to the curve yields the value of τ_M. This is important also in regard to the fact that often the data for small t do not provide the limit of $e(t)$, as $t \to 0$, in a precise way.

As an example, we can consider some creep curves on annealed copper [120] for which also the limit value $e(0)$ is neatly represented. Upon the procedure indicated above we can find that

$$\tau_M = 6\ 10^5 \text{ s}, \qquad \tau_K = 2 \text{ s}.$$

It is worth remarking that it is a hazard to take the value of E as given by available tables. This is so because E depends markedly on the applied stress. As Fig. 12.16 of [120] shows for annealed copper, $e(0)$ increases by 23 times when T_0 increases by 4 times.

Besides being of interest on its own, the passage from a spring and dashpot model to a memory functional is convenient in many circumstances. Let

$$\sum_{k=1}^{n} p_k \frac{d^k}{dt^k} T(t) = \sum_{h=1}^{m} q_h \frac{d^h}{dt^h} e(t)$$

be the differential equation describing the behaviour of the body, p_k, q_h being real parameters. A theorem by Gurtin and Sternberg ([79], Th. 4.4) allows us to say that if $n \geq m$ then there exists a unique relaxation function $G \in C^n(\mathbb{R}^{++})$ such that the differential equation is equivalent to a linear function of the history of e. On the basis of this theorem now we look for the specific functional form which corresponds to the model (3.4).

Letting τ_1, $\tau_2 > 0$ consider the stress functional

$$T(t) = Ee(t) + a\int_0^\infty \left[\exp\left(-\frac{s}{\tau_1}\right) + \alpha\exp\left(-\frac{s}{\tau_2}\right)\right]e(t-s)\,ds$$

where E, a, and α are real parameters. In fact we expect that $E > 0$ and $a < 0$. Time differentiation and integration by parts yields

$$\dot{T}(t) = E\dot{e}(t) + a(1+\alpha)e(t) - a\int_0^\infty \left[\frac{1}{\tau_1}\exp\left(-\frac{s}{\tau_1}\right) + \frac{\alpha}{\tau_2}\exp\left(-\frac{s}{\tau_2}\right)\right]e(t-s)\,ds$$

whence

$$\dot{T}(t) + \frac{1}{\tau_2}T(t) = E\dot{e}(t) + \left[a(1+\alpha) + \frac{E}{\tau_2}\right]e(t) + a\left(\frac{1}{\tau_2} - \frac{1}{\tau_1}\right)\int_0^\infty \exp\left(-\frac{s}{\tau_1}\right)e(t-s)\,ds.$$

A further time differentiation and the analogous procedure give

$$\ddot{T} + \left(\frac{1}{\tau_1} + \frac{1}{\tau_2}\right)\dot{T} + \frac{1}{\tau_1\tau_2}T = E\ddot{e} + \left[a(1+\alpha) + E\left(\frac{1}{\tau_1} + \frac{1}{\tau_2}\right)\right]\dot{e} + \left[a\left(\frac{1}{\tau_2} + \frac{\alpha}{\tau_1}\right) + \frac{E}{\tau_1\tau_2}\right]e.$$

Comparison with (3.4) yields

$$\frac{1}{\tau_1} + \frac{1}{\tau_2} = \frac{1}{\tau_M} + \frac{1}{\tau_K} + \frac{E_M}{E_K\tau_K},$$

$$\tau_1\tau_2 = \tau_K\tau_M,$$

$$E = E_M,$$

$$a(1+\alpha) + E\left(\frac{1}{\tau_1} + \frac{1}{\tau_2}\right) = \frac{E_M}{\tau_K},$$

$$a\left(\frac{1}{\tau_2} + \frac{\alpha}{\tau_1}\right) + \frac{E}{\tau_1\tau_2} = 0.$$

Let τ_1 be the minimal solution and τ_2 the maximal solution of

$$\tau_1 + \tau_2 = \tau_K + \tau_M + \frac{E_M}{E_K}\tau_M,$$

$$\tau_1\tau_2 = \tau_K\tau_M;$$

of course $\tau_1 < \min(\tau_K, \tau_M)$, $\tau_2 > \max(\tau_K, \tau_M)$. Then we have

$$a = -\frac{E_M}{\tau_1 + \alpha\tau_2},$$

$$\alpha = \frac{\tau_K\tau_M - \tau_1(\tau_K + E_M\tau_M/E_K)}{\tau_2(\tau_K + E_M\tau_M/E_K) - \tau_K\tau_M}.$$

It is worth remarking that, because

$$\tau_K\tau_M - \tau_1(\tau_K + \frac{E_M}{E_K}\tau_M) = \tau_1(\tau_M - \tau_1) > 0,$$

we have $\alpha > 0$ whence $a < 0$. In conclusion we can write the functional version of (3.4) in the form

$$T(t) = E_M\, e(t) - \frac{E_M(\tau_2 - \tau_M)}{\tau_1(\tau_2 - \tau_1)} \int_0^\infty \left[\exp\left(-\frac{s}{\tau_1}\right) + \frac{\tau_1(\tau_M - \tau_1)}{\tau_2(\tau_2 - \tau_M)} \exp\left(-\frac{s}{\tau_2}\right)\right] e(t-s)\, ds. \quad (3.5)$$

The previous picture is one-dimensional in character. The passage to the three-dimensional case can be performed in much the same way as we do in linear elasticity. In this regard we recall that for a (linear, isotropic) elastic solid we have

$$\mathbf{T} = 2\mu\mathbf{E} + \lambda(\operatorname{tr}\mathbf{E})\mathbf{1} \quad (3.6)$$

where

$$\mathbf{E} = \tfrac{1}{2}(\nabla\mathbf{u} + \nabla\mathbf{u}^\dagger)$$

is the *infinitesimal strain* tensor, which is related to $\boldsymbol{\mathcal{E}}$ by

$$\boldsymbol{\mathcal{E}} = \mathbf{E} + \tfrac{1}{2}\nabla\mathbf{u}^\dagger\nabla\mathbf{u},$$

and μ, λ are called the Lamé constants. The inverse of (3.6) is

$$\mathbf{E} = \frac{1}{2\mu}\mathbf{T} - \frac{\lambda}{2\mu(2\mu + 3\lambda)}(\operatorname{tr}\mathbf{T})\mathbf{1}. \quad (3.7)$$

Assume that a uniform tension is applied to a rod. Letting x_3 be the direction of the tension we can write

$$T_{33} \neq 0, \qquad T_{ik} = 0 \quad \text{otherwise.}$$

By (3.7) we have

$$E_{33} = \frac{\mu + \lambda}{\mu(2\mu + 3\lambda)}T_{33}, \qquad E_{11} = E_{22} = -\frac{\lambda}{2\mu(2\mu + 3\lambda)}T_{33},$$

and $E_{ik} = 0$ otherwise. Then we have

$$T_{33} = E\,E_{33}, \qquad E_{11} = E_{22} = -\nu E_{33}$$

by identifying *Young's modulus* E and *Poisson's ratio* ν as

$$E = \frac{\mu(2\mu + 3\lambda)}{\mu + \lambda}, \qquad \nu = \frac{\lambda}{2(\mu + \lambda)}.$$

Conversely

$$\mu = \frac{E}{2(1+\nu)}, \qquad \lambda = \frac{E\nu}{(1+\nu)(1-2\nu)}. \tag{3.8}$$

The positive definiteness of the elastic energy holds if and only if $\mu > 0$, $\lambda + \frac{2}{3}\mu > 0$, which means $-1 < \nu < \frac{1}{2}$. In fact experiments indicate that $0 < \nu < \frac{1}{2}$; e. g., ν equals 0.29 for carbon steel, 0.33 for copper, 0.25 for glass.

By (3.8), once we have determined ν and E_M, a, α, we can write the three-dimensional version of (3.5) as

$$\begin{aligned}\mathbf{T}(t) = {} & \frac{E_M}{1+\nu}\,\mathbf{E}(t) + \frac{\nu E_M}{(1+\nu)(1-2\nu)}(\operatorname{tr}\mathbf{E})(t)\,\mathbf{1}\\ & - \frac{1}{\tau_1 + \alpha\tau_2}\int_0^\infty \Big[\exp\Big(-\frac{s}{\tau_1}\Big) + \alpha\exp\Big(-\frac{s}{\tau_2}\Big)\Big]\frac{E_M}{1+\nu}\,\mathbf{E}(t-s)\,ds\\ & - \frac{1}{\tau_1 + \alpha\tau_2}\int_0^\infty \Big[\exp\Big(-\frac{s}{\tau_1}\Big) + \alpha\exp\Big(-\frac{s}{\tau_2}\Big)\Big]\frac{\nu E_M}{(1+\nu)(1-2\nu)}(\operatorname{tr}\mathbf{E})(t-s)\,ds\,\mathbf{1}.\end{aligned} \tag{3.9}$$

The result (3.9) provides an explicit, operative model for describing dissipative bodies through constitutive relations in the form of a functional. Indeed, the linear theory of viscoelasticity is based on a generalization of the constitutive relation (3.9).

2.4 Viscoelastic solids

Now we set aside arguments based on the behaviour of spring-and-dashpot systems and go back to general considerations, which in essence trace back to Boltzmann, for modelling dissipative (solid) materials. Of course, comparison with (3.9) proves instructive.

Consider a body relative to a reference (unstressed) placement $\mathcal{B}$. At any point $\mathbf{X} \in \mathcal{B}$ of the body, the stress at any time t depends upon the strain at all preceding times. If the strain at all preceding times is in the same direction then the effect is to reduce the corresponding stress. The influence of a previous strain on the stress depends on the time elapsed since the strain occurred and is weaker for those strains occurred long ago. Such properties make the model envisaged by Boltzmann a material with (fading) memory. In addition Boltzmann made the assumption that a superposition of the influence of previous strains holds, which means that the stress-strain relation is linear. Also, since we disregard thermal effects, constitutive properties are all embodied in the constitutive equation for the stress.

Owing to the superposition principle and the underlying linearity assumption, we consider displacement fields $\mathbf{u}(\mathbf{X},t) = \mathbf{x}(\mathbf{X},t) - \mathbf{X}$ subject to

$$\sup_{\mathbf{X},t} \left| \nabla_{\mathbf{X}} \mathbf{u}(\mathbf{X},t) \right| \ll 1.$$

Accordingly we approximate the Lagrangian strain tensor $\boldsymbol{\mathcal{E}}$ with the linearized strain tensor $\mathrm{sym}\nabla_{\mathbf{X}}\mathbf{u}$. Then the constitutive equation is in fact expressed by a functional of $\mathrm{sym}\nabla_{\mathbf{X}}\mathbf{u}$ which provides the value of the stress at the pertinent point $\mathbf{X}$. Really, because of the linear approximation we identify the dependence on the reference position $\mathbf{X}$ with that on the present position $\mathbf{x}$. Hence we let $\mathbf{x} \in \mathcal{B}$ and identify $\boldsymbol{\mathcal{E}}$ with the infinitesimal strain tensor $\mathbf{E} = \mathrm{sym}\nabla\mathbf{u}$. In the same approximation we identify the first Piola-Kirchhoff stress tensor $\mathbf{S}$ with the Cauchy stress tensor $\mathbf{T}$ and the power of the stress $\mathbf{T}\cdot\mathbf{D}$ (or $\mathbf{S}\cdot\dot{\mathbf{F}}$) with $\mathbf{T}\cdot\dot{\mathbf{E}}$.

Boltzmann's model is naturally expressed by saying that the stress $\mathbf{T}$, at a point $\mathbf{x}$ and time t, depends linearly on the history of $\mathbf{E}$ up to time t, at the same point $\mathbf{x}$. To save writing, we usually omit specifying the dependence on $\mathbf{x}$. Then we represent Boltzmann's model of *viscoelastic solid* through the constitutive equation

$$\mathbf{T}(t) = \mathbf{G}_0\mathbf{E}(t) + \int_0^\infty \mathbf{G}'(s)\,\mathbf{E}(t-s)ds \tag{4.1}$$

where $\mathbf{G}_0 \in \mathrm{Lin(Sym)}$ and $\mathbf{G}' : \mathbb{R}^+ \to \mathrm{Lin(Sym)}$. As usual,

$$\mathbf{G}(s) = \mathbf{G}(0) + \int_0^s \mathbf{G}'(v)\,dv$$

is called the *relaxation function*. Indeed, if the strain $\mathbf{E}$ vanishes in the past infinity, $\mathbf{E}(-\infty) = 0$, then an obvious integration by parts yields

$$\mathbf{T}(t) = \int_0^\infty \mathbf{G}(s)\,\dot{\mathbf{E}}(t-s)ds,$$

which makes evident why $\mathbf{G}$, rather than $\mathbf{G}'$, is called the relaxation function. The right-hand side of (4.1) is the stress functional, $\boldsymbol{\mathcal{T}}(\mathbf{E}^t)$ say, which maps every history $\mathbf{E}^t$ into the corresponding stress $\mathbf{T}(t)$.

The fourth-order tensor $\mathbf{G}_0 := \mathbf{G}(0)$ is called the *instantaneous elastic modulus* and governs the response to instantaneous changes in strain. The limit of $\mathbf{G}(s)$, as $s \to \infty$, is supposed to exist and

$$\mathbf{G}_\infty = \lim_{s\to\infty} \mathbf{G}(s)$$

is called the *equilibrium elastic modulus* in that the constancy of the history $\mathbf{E}^t$, viz. $\mathbf{E}^t(s) = \mathbf{E}$ as $s \in \mathbb{R}^+$, gives

$$\mathbf{T} = \mathbf{G}_\infty \mathbf{E}.$$

The assumption

$$\mathbf{G}_\infty > 0$$

is characteristic of solids; it means that a non-zero, constant strain is supposed to induce a non-zero stress such that $\mathbf{T} \cdot \mathbf{E}$ is strictly positive. Without any loss of generality we let $\mathbf{G}$ satisfy the symmetry conditions (in indicial notation)

$$G_{ijkl} = G_{jikl} = G_{jilk}.$$

Of course the solid may be heterogeneous and $\mathbf{G}$ depends on $\mathbf{x}$. This possible dependence has remarkable consequences on wave propagation. Whenever appropriate, the dependence on $\mathbf{x}$ is indicated explicitly.

If the viscoelastic solid is isotropic then the relaxation tensor function $\mathbf{G}$ is represented by two scalar functions only. For convenience we let

$$\mu(s) = \mu_0 + \int_0^s \mu'(\tau)\,d\tau, \qquad \lambda(s) = \lambda_0 + \int_0^s \lambda'(\tau)\,d\tau, \quad s \in \mathbb{R}^+,$$

and represent $\mathbf{G}_0$ and $\mathbf{G}'$ as

$$(G_0)_{ijkl} = \lambda_0 \delta_{ij}\delta_{kl} + \mu_0(\delta_{ik}\delta_{jl} + \delta_{il}\delta_{jk})$$

and

$$G'_{ijkl}(s) = \lambda'(s)\delta_{ij}\delta_{kl} + \mu'(s)(\delta_{ik}\delta_{jl} + \delta_{il}\delta_{jk}), \qquad s \in \mathbb{R}^+.$$

The function $\mu(s)$ is called the *relaxation modulus in shear* while $\kappa(s) = \lambda(s) + 2\mu(s)/3$ is called the *relaxation modulus in dilatation.* Upon substitution we see that, for isotropic solids, the equation of motion (2.5) takes the form

$$\rho\ddot{\mathbf{u}} = \nabla \cdot \left\{\lambda_0 \operatorname{tr}\mathbf{E}\,\mathbf{1} + 2\mu_0\mathbf{E} + \int_0^\infty [\lambda'(s)\operatorname{tr}\mathbf{E}(t-s)\,\mathbf{1} + 2\mu'(s)\mathbf{E}(t-s)]ds\right\} + \rho\mathbf{b}.$$

If μ and λ depend on the position $\mathbf{x}$ then the equation of motion becomes

$$\begin{aligned}\rho\ddot{\mathbf{u}} &= (\lambda_0 + \mu_0)\nabla(\nabla\cdot\mathbf{u}) + \mu_0\Delta\mathbf{u} + (\nabla\cdot\mathbf{u})\nabla\lambda_0 + \nabla\mathbf{u}\,\nabla\mu_0 + (\mu_0\cdot\nabla)\mathbf{u}\\ &+ \int_0^\infty \left[(\lambda' + \mu')\nabla(\nabla\cdot\mathbf{u}^t) + \mu'\Delta\mathbf{u}^t + (\nabla\cdot\mathbf{u}^t)\nabla\lambda' + \nabla\mathbf{u}^t\,\nabla\mu' + (\nabla\mu'\cdot\nabla)\mathbf{u}^t\right](s)ds + \rho\mathbf{b}\end{aligned}$$

and reduces to

$$\rho\ddot{\mathbf{u}} = (\lambda_0 + \mu_0)\nabla(\nabla\cdot\mathbf{u}) + \mu_0\Delta\mathbf{u} + \int_0^\infty \left[(\lambda' + \mu')\nabla(\nabla\cdot\mathbf{u}^t) + \mu'\Delta\mathbf{u}^t\right]ds + \rho\mathbf{b}$$

if the solid is homogeneous (λ and μ independent of the position).

Although thermal effects are disregarded, thermodynamics plays an important role in the characterization of the model for viscoelastic solids. The functional $\boldsymbol{T}(\mathbf{E}^t)$, expressed by (4.1), is required to obey the second law of thermodynamics and this places remarkable restrictions on the relaxation function $\mathbf{G}$. Upon the statement of the second law through the Clausius inequality (2.18) we say that the constitutive functional $\boldsymbol{T}(\mathbf{E}^t)$ is compatible with the second law of thermodynamics if

$$\int_0^d \boldsymbol{T}(\mathbf{E}^t) \cdot \dot{\mathbf{E}}(t)\, dt > 0 \tag{4.2}$$

for any non-trivial cycle starting at $t = 0$ and ending at $t = d > 0$. Here non-trivial cycle means a one-parameter family of histories $\mathbf{E}^\tau$, the parameter τ running on $[0, d)$, with $\dot{\mathbf{E}}(\tau)$ piecewise continuous, such that $\mathbf{E}^d = \mathbf{E}^0$ but $\mathbf{E}(\tau) \neq \mathbf{E}^0$ in a subinterval of $[0, d)$. By considering a function $\mathbf{E}(\cdot)$ in the form

$$\mathbf{E}(t) = \mathbf{E}_1 \cos\omega t + \mathbf{E}_2 \sin\omega t, \qquad \omega > 0, \quad \mathbf{E}_1, \mathbf{E}_2 \in \mathrm{Sym},$$

with $d = 2\pi/\omega$, the analysis of (4.2) shows that $\mathbf{G}$ must satisfy the symmetry conditions

$$\mathbf{G}_0 = \mathbf{G}_0^\dagger, \tag{4.3}$$

$$\mathbf{G}_\infty = \mathbf{G}_\infty^\dagger, \tag{4.4}$$

and

$$\int_0^\infty [\mathbf{E}_1 \cdot \mathbf{G}'(s)\mathbf{E}_1 + \mathbf{E}_2 \cdot \mathbf{G}'(s)\mathbf{E}_2] \sin\omega s\, ds + \int_0^\infty \mathbf{E}_1 \cdot [\mathbf{G}'(s) - \mathbf{G}'^\dagger(s)]\mathbf{E}_2 \cos\omega s\, ds < 0 \tag{4.5}$$

for every $\omega > 0$ and every pair of symmetric tensors $\mathbf{E}_1, \mathbf{E}_2$. As shown in [70], the relations (4.3) to (4.5) are also sufficient for the validity of (4.2) and, what is more, they are necessary and sufficient for the validity of the second law in the more general case of approximate cycles.

Assume that $\mathbf{G}'$ is absolutely integrable on $\mathbb{R}^+$ and let $\mathbf{G}'_s$ on $\mathbb{R}^+$ be the half-range Fourier sine transform of $\mathbf{G}'$, namely

$$\mathbf{G}'_s(\omega) = \int_0^\infty \mathbf{G}'(v) \sin\omega v\, dv.$$

The inequality (4.5) implies that, in Sym,

$$\mathbf{G}'_s(\omega) < 0, \qquad \omega \in \mathbb{R}^{++}, \tag{4.6}$$

while $\mathbf{G}'_s(0) = 0$ is identically true. The physical meaning of (4.6) is made apparent by observing that the energy dissipated in one period $[0, d]$ for a strain function $\mathbf{E}(t) = \mathbf{E}\sin\omega t$ is

$$\int_0^d \boldsymbol{T}(\mathbf{E}^t) \cdot \dot{\mathbf{E}}(t)\, dt = -\pi\, \mathbf{E} \cdot \mathbf{G}'_s(\omega)\mathbf{E}.$$

So $\pi\, \mathbf{E} \cdot \mathbf{G}'_s(\omega)\mathbf{E}$ is the energy dissipated per period when the amplitude is $\mathbf{E}$ and, accordingly, $-\mathbf{G}'_s(\omega)$ is often called the *loss modulus.*

For isotropic bodies (4.3) and (4.4) are trivially true. Moreover $\mathbf{G}' = \mathbf{G}'^{\dagger}$ and then (4.5) is equivalent to (4.6). Upon substitution in (4.6) we have

$$\mathbf{E} \cdot \mathbf{G}'_s(\omega)\mathbf{E} = 2\,\mu'_s(\omega)\mathbf{E} \cdot \mathbf{E} + \lambda'_s(\omega)(\operatorname{tr}\mathbf{E})^2$$

or

$$\mathbf{E} \cdot \mathbf{G}'_s(\omega)\mathbf{E} = 2\,\mu'_s(\omega)\, \mathring{\mathbf{E}} \cdot \mathring{\mathbf{E}} + \kappa'_s(\omega)(\operatorname{tr}\mathbf{E})^2$$

where $\kappa' = \lambda' + 2\,\mu'/3$ and a superposed ring denotes the trace-free part. Then (4.6) holds if and only if

$$\mu'_s(\omega) < 0, \qquad \kappa'_s(\omega) < 0, \qquad \omega \in \mathbb{R}^{++}. \tag{4.7}$$

Henceforth we confine our attention to isotropic solids and then the restrictions placed by thermodynamics are in fact the two inequalities (4.7).

There seem to be motivations for models of viscoelastic solids with $\mathbf{G}'$ non-integrable and $\mathbf{G}_0$ infinite (cf. [135]). Here we disregard such possibilities also because they rule out wave propagation.

2.5 Dissipative fluids

The most natural models of dissipative fluids are the viscous and viscoelastic fluids. We regard the viscous fluid model as satisfactory whenever memory properties are inessential. Both models are usually applied in the linear approximation.

A fluid at rest cannot sustain shear stresses and differences in normal stresses. Then, at rest,

$$\mathbf{T} = -p\,\mathbf{1},$$

p being regarded as the pressure. Of course, by Galilean invariance the same occurs if the fluid undergoes a rigid motion. For the general case of non-rigid fluid motions we can write

$$\mathbf{T} = -p\,\mathbf{1} + \boldsymbol{\tau} \tag{5.1}$$

where $\boldsymbol{\tau}$ is the extra-stress tensor. The symmetry of $\mathbf{T}$ implies that $\boldsymbol{\tau}$ too is symmetric. The splitting (5.1) of $\mathbf{T}$ is made operative by saying that p depends on the motion through the density ρ, at most, while $\boldsymbol{\tau}$ depends on kinematical variables and vanishes if the motion is rigid.

Examine first viscous fluids. Linearly *viscous fluids* are characterized by letting $\boldsymbol{\tau}$ be a linear (isotropic) function of the rate-of-strain tensor $\mathbf{D}$, namely

$$\boldsymbol{\tau} = 2\,\mu\,\mathbf{D} + \lambda(\mathrm{tr}\,\mathbf{D})\mathbf{1}. \tag{5.2}$$

Alternatively we can write

$$\boldsymbol{\tau} = 2\,\mu\,\overset{\circ}{\mathbf{D}} + \kappa(\mathrm{tr}\,\mathbf{D})\mathbf{1} \tag{5.3}$$

where $\overset{\circ}{\mathbf{D}}$ is the trace-free part of $\mathbf{D}$ and $\kappa = \lambda + 2\mu/3$; μ and κ are called the *shear* and *bulk viscosity coefficients.* Of course it may well happen that μ and λ depend on the position (or on the particle) in which case we say that the fluid is heterogeneous.

If the fluid is compressible then p is a (known) function of the mass density ρ and the balance equations take the form

$$\frac{\partial\rho}{\partial t} = -\nabla\cdot(\rho\mathbf{v}),$$

$$\rho\Big(\frac{\partial}{\partial t} + \mathbf{v}\cdot\nabla\Big)\mathbf{v} = -p_\rho\nabla\rho + \nabla\cdot[\mu(\nabla\mathbf{v} + \nabla\mathbf{v}^\dagger)] + \nabla[\lambda\nabla\cdot\mathbf{v}] + \rho\mathbf{b},$$

the unknowns being the density field $\rho(\mathbf{x},t)$ and the velocity field $\mathbf{v}(\mathbf{x},t)$. If, instead, the fluid is incompressible then ρ is time-independent and the unknowns $p(\mathbf{x},t), \mathbf{v}(\mathbf{x},t)$ satisfy the differential equations

$$\nabla\cdot\mathbf{v} = 0,$$

$$\rho\Big(\frac{\partial}{\partial t} + \mathbf{v}\cdot\nabla\Big)\mathbf{v} = -\nabla p + \nabla\cdot[\mu(\nabla\mathbf{v} + \nabla\mathbf{v}^\dagger)] + \nabla[\lambda\nabla\cdot\mathbf{v}] + \rho\mathbf{b}.$$

In the particular case of μ, λ constant, the last equation is usually referred to as Navier-Stokes equation. Often the linearly viscous, incompressible fluid is called Newtonian.

The viscoelastic fluid may be viewed as a generalization of the viscous fluid, the generalization consisting in the account for memory effects. Here we give an outline of the model and its derivation by following the lines of [61]. We start from the constitutive functional

$$\mathbf{T}(t) = -p(\rho(t))\mathbf{1} + \int_0^\infty \mu'(s;\rho)[\mathbf{C}_t(t-s) - \mathbf{1}]ds + \tfrac{1}{2}\Big\{\int_0^\infty \lambda'(s;\rho)\mathrm{tr}[\mathbf{C}_t(t-s) - \mathbf{1}]\,ds\Big\}\mathbf{1}$$

where μ' and λ' are scalar functions on $\mathbb{R}^+$ parameterized by the present value ρ of the mass density. The subscript t means that the current placement is taken as reference. Then $\mathbf{F}_t(t) = \mathbf{1}$ and

$$\boldsymbol{\mathcal{E}}_t(t) = \tfrac{1}{2}[\mathbf{C}_t(t) - \mathbf{1}] = 0.$$

Hence we can write

$$\boldsymbol{\tau}(t) = \boldsymbol{\tau}(\boldsymbol{\mathcal{E}}_t^t, \rho),$$

which means that the extra-stress $\boldsymbol{\tau}$ of the particle at $\mathbf{x}$ at the time t depends on the whole history of the Lagrangian strain at that particle. To derive a linear model we proceed as follows. Let μ', λ' be independent of ρ. To the linear order in $\nabla \mathbf{u}$ we have

$$\boldsymbol{\tau}(t) = 2\int_0^\infty \mu'(s)\mathbf{E}_t^t(s)ds + \int_0^\infty \lambda'(s)[\mathrm{tr}\,\mathbf{E}_t^t(s)]ds\,\mathbf{1}.$$

Define

$$\tilde{\mu}(s) = -\int_s^\infty \mu'(w)dw, \qquad \tilde{\lambda}(s) = -\int_s^\infty \lambda'(w)dw$$

and require that

$$\lim_{s\to\infty} \tilde{\mu}(s), \tilde{\lambda}(s) = 0.$$

The observation that $d\mathbf{E}_t(\xi)/d\xi = \mathbf{D}(\xi)$ or

$$-\frac{d}{ds}\mathbf{E}_t^t(s) = \mathbf{D}^t(s) = \tfrac{1}{2}\left[\nabla\dot{\mathbf{u}}(t-s) + \nabla\dot{\mathbf{u}}^\dagger(t-s)\right], \qquad s \in \mathbb{R}^+,$$

and an integration by parts allow $\boldsymbol{\tau}$ to be expressed as

$$\boldsymbol{\tau}(t) = 2\int_0^\infty \tilde{\mu}(s)\mathbf{D}^t(s)ds + \int_0^\infty \tilde{\lambda}(s)[\mathrm{tr}\,\mathbf{D}^t(s)]ds\,\mathbf{1}.$$

Then we can write the (linear) constitutive functional for the stress $\mathbf{T}$ as

$$\mathbf{T}(t) = -p(\rho(t))\,\mathbf{1} + 2\int_0^\infty \tilde{\mu}(s)\,\overset{\circ}{\mathbf{D}}{}^t(s)ds + \int_0^\infty \tilde{\kappa}(s)[\mathrm{tr}\,\mathbf{D}^t(s)]ds\,\mathbf{1} \tag{5.4}$$

where $\tilde{\kappa}(s) = \tilde{\lambda}(s) + 2\tilde{\mu}(s)/3$ is the bulk relaxation function.

The viscous fluid model can be viewed as the limiting case when $\tilde{\mu}(s)$ and $\tilde{\lambda}(s)$ - or $\tilde{\kappa}(s)$ - are Dirac's delta-functions of the form $\mu\,\delta(s), \lambda\,\delta(s)$.

Even for fluids, thermodynamic restrictions provide inequalities which prove of fundamental importance in the analysis of wave propagation. Examine first the viscoelastic fluid and look for thermodynamic restrictions on the shear and bulk relaxation functions $\tilde{\mu}(s)$, $\tilde{\kappa}(s)$. Let $\hat{\mathbf{T}}(\rho, \mathbf{D}^t)$ denote the constitutive functional for the stress tensor $\mathbf{T}$. We express the second law by requiring that

$$\int_0^d \hat{\mathbf{T}}(\rho, \mathbf{D}^t)\cdot\mathbf{D}(t)\,dt > 0 \tag{5.5}$$

for any cycle starting at $t = 0$ and ending at $t = d > 0$ whereas $\mathbf{D}$ does not vanish identically in $[0, d)$. Here a cycle is a one-parameter family of pairs $(\rho(t), \mathbf{D}^t)$, the parameter t running on $[0, d)$, with $\mathbf{D}(\cdot)$ piecewise continuous in $[0, d)$, such that $\rho(d) = \rho(0)$, $\mathbf{D}^d = \mathbf{D}^0$. Really, by (2.2) we may write the continuity equation as

$$\dot{\rho} + \rho\,\mathrm{tr}\,\mathbf{D} = 0$$

whence it follows that

$$\rho(t) = \rho(0) \exp \big[- \int_0^t \operatorname{tr} \mathbf{D}(u)\, du \big].$$

Then $\rho(d) = \rho(0)$ if and only if the integral of $\operatorname{tr} \mathbf{D}$ on $[0, d)$ vanishes.

Any function $p(\rho)$ is compatible with the second law (5.5) in that, if H is the integral of p/ρ,

$$\int_0^d p(\rho(t)) \operatorname{tr} \mathbf{D}(t)\, dt = \int_0^d p\big(\rho(0) \exp \big[- \int_0^t \operatorname{tr} \mathbf{D}(u)\, du\big]\big) \operatorname{tr} \mathbf{D}(t)\, dt = -[H(\rho(d)) - H(\rho(0))]$$

and hence the integral of $p(\rho) \operatorname{tr} \mathbf{D}$ vanishes along any cycle. The same is trivially true for incompressible fluids because in that case $\operatorname{tr} \mathbf{D} = 0$. Now, let the function $\mathbf{D}(\cdot)$ be given by

$$\mathbf{D}(t) = \mathbf{D}_1 \cos \omega t + \mathbf{D}_2 \sin \omega t, \qquad \omega \in \mathbb{R}^{++}, \quad \mathbf{D}_1, \mathbf{D}_2 \in \mathrm{Sym},$$

and set $d = 2\pi/\omega$. Let the subscript c denote the half-range Fourier-cosine transform, e.g.,

$$\mu_c(\omega) = \int_0^\infty \tilde{\mu}(s) \cos \omega s\, ds.$$

By paralleling the procedure of §2.4, it is a routine matter to show that the functional (5.4) satisfies (5.5) only if

$$\mu_c(\omega) > 0, \qquad \kappa_c(\omega) > 0, \qquad \omega \in \mathbb{R}^{++}. \tag{5.6}$$

The inequalities (5.6) turn out to be also sufficient for the validity of the second law (5.5) - cf. [69].

The case when the fluid is Newtonian, namely $\boldsymbol{\tau}$ is given by (5.2) or (5.3), makes (5.5) into the form

$$\int_0^d [2\mu \stackrel{\circ}{\mathbf{D}} \cdot \stackrel{\circ}{\mathbf{D}} + \kappa (\operatorname{tr} \mathbf{D})^2]\, dt > 0$$

for any non-zero function $\mathbf{D}(t)$ with vanishing mean value of $\operatorname{tr} \mathbf{D}$. Hence the second law holds if and only if the viscosity coefficients satisfy

$$\mu > 0, \qquad \kappa > 0. \tag{5.7}$$

2.6 Equation of motion in prestressed dissipative bodies

The dynamics of a solid, when small displacements are superimposed on a finite deformation, is governed by the general equation (2.15). To make such an equation operative, though, we have to assign the stress perturbation $\hat{\mathbf{Y}}$ in terms of $\mathbf{u}$ or $\mathbf{H}$ and the predeformation gradient $\mathbf{F}$ or prestress $\mathbf{T}^0$. Both $\mathbf{F}$ and $\mathbf{T}^0$ enter explicitly the equation (2.15).

Moreover, they are likely to affect the value of the perturbation $\hat{\mathbf{Y}}$ in that tensor coefficients produced by linearization might depend on $\mathbf{F}$ and $\mathbf{T}^0$.

The assignment of $\hat{\mathbf{Y}}$ is a highly non-trivial task. Indeed, sometimes the question may be ill-posed, as we show in a moment. To fix ideas, take the simplest model of solid body, viz. the isotropic, linearly elastic, solid. In such a case the Cauchy stress $\mathbf{T}$, or the Piola-Kirchhoff stress $\mathbf{S}$, is given by

$$\mathbf{T} = 2\mu\,\mathbf{E} + \lambda(\operatorname{tr}\mathbf{E})\mathbf{1}, \tag{6.1}$$

$\mathbf{E}$ being the infinitesimal strain tensor. To determine $\hat{\mathbf{Y}}$ we should know the constitutive equation for $\mathbf{Y}$. But (6.1) models the stress-strain relation to within $o(\mathbf{H})$ and here only the superimposed displacement (gradient) is small and then (6.1) becomes meaningless if finite deformations are involved. As a first step it is natural to have recourse to non-linear elasticity and examine the corresponding linearization procedure.

Quite generally, let the stress $\mathbf{T}$ be a function of the deformation gradient $\mathbf{F}$. Indeed, as is well known, by objectivity arguments it follows that the stress $\mathbf{T}$ can depend on $\mathbf{F}$ only through the form

$$\mathbf{T} = \mathbf{F}\bar{\mathbf{T}}(\mathbf{C})\mathbf{F}^\dagger,$$

or equivalent ones, whence, by (2.10),

$$\mathbf{Y} = \bar{\mathbf{Y}}(\mathbf{C})$$

in that $J = (\det\mathbf{C})^{1/2}$. This indicates that insights into the question about $\hat{\mathbf{Y}}$ can be gained by examining $\mathbf{C}$ or $\boldsymbol{\mathcal{E}} = \frac{1}{2}(\mathbf{C} - \mathbf{1})$.

Following the notation of §2.2, let $\tilde{\mathbf{x}} = \mathbf{x} + \mathbf{u}$ be the current position, of a particle $\mathbf{X}$, obtained by superposition of the displacement $\mathbf{u}$ on the equilibrium position $\mathbf{x}$. Accordingly we set

$$\tilde{\mathbf{F}} = \nabla_{\mathbf{X}}\,\tilde{\mathbf{x}}^\dagger, \quad \tilde{\mathbf{C}} = \tilde{\mathbf{F}}^\dagger\tilde{\mathbf{F}}, \quad \mathbf{F} = \nabla_{\mathbf{X}}\,\mathbf{x}^\dagger, \quad \mathbf{C} = \mathbf{F}^\dagger\mathbf{F}, \quad \mathbf{H} = \nabla\mathbf{u}^\dagger.$$

By the definition of $\tilde{\mathbf{F}}$ and the chain rule we have

$$\tilde{\mathbf{F}} = \mathbf{F} + \nabla_{\mathbf{X}}\,\mathbf{u}^\dagger, \qquad \nabla_{\mathbf{X}}\,\mathbf{u}^\dagger = \mathbf{H}\mathbf{F}$$

whence

$$\tilde{\mathbf{F}} = (\mathbf{1} + \mathbf{H})\mathbf{F}.$$

As a consequence,

$$\tilde{\mathbf{C}} = \mathbf{C} + \mathbf{F}^\dagger(\mathbf{H} + \mathbf{H}^\dagger)\mathbf{F} + \mathbf{F}^\dagger\mathbf{H}^\dagger\mathbf{H}\mathbf{F}.$$

Since $\mathbf{H}$ is small we disregard $o(\mathbf{H})$ terms. This means that

$$\tilde{\mathbf{C}} \simeq \mathbf{C} + 2\mathbf{F}^\dagger\mathbf{E}\mathbf{F}$$

and this suggests that we regard the present value $\tilde{\mathbf{Y}}$, of the second Piola-Kirchhoff stress, as the first order approximation of $\bar{\mathbf{Y}}$ in the form

$$\tilde{\mathbf{Y}} = \bar{\mathbf{Y}}(\mathbf{C}) + 2\frac{\partial \bar{\mathbf{Y}}}{\partial \mathbf{C}}(\mathbf{F}^\dagger \mathbf{E} \mathbf{F}).$$

Accordingly we identify the perturbation $\hat{\mathbf{Y}}$ as

$$\hat{\mathbf{Y}} = 2\frac{\partial \bar{\mathbf{Y}}}{\partial \mathbf{C}}(\mathbf{F}^\dagger \mathbf{E} \mathbf{F}).$$

Let $\mathbf{K}$ be the fourth-order tensor $\partial \bar{\mathbf{Y}}/\partial \mathbf{C}$, namely $K_{MNPQ} = \partial \bar{Y}_{MN}/\partial C_{PQ}$. By definition

$$K_{MNPQ} = K_{NMPQ} = K_{NMQP}.$$

Substitution gives

$$\hat{Y}_{MN} = A_{MNij} E_{ij}$$

where $A_{MNij} = 2K_{MNPQ} F_{iP} F_{jQ}$ with

$$A_{MNij} = A_{NMij} = A_{NMji}.$$

As a consequence $\mathbf{AE} = \mathbf{AH}$. Finally, letting $\mathbf{b}$ be a constant vector, typically the gravity acceleration, and substituting into the equation of motion (2.15) we have

$$\rho \ddot{\mathbf{u}} = \nabla \cdot (\mathbf{H}\mathbf{T}^0 + \mathbf{B}\mathbf{H}) \tag{6.2}$$

where

$$B_{mnij} = \frac{1}{J} F_{mM} F_{nN} A_{MNij}$$

and hence

$$B_{mnij} = B_{nmij} = B_{nmji}. \tag{6.3}$$

If, further, we take into account compatibility with thermodynamics we conclude that the material is in fact hyperelastic (cf. [78]), namely there is a (free) energy function $\psi(\mathbf{C})$ such that $\mathbf{Y} = 2\rho_0 \partial\psi/\partial\mathbf{C}$. This implies that $\mathbf{K}$ and $\mathbf{B}$ are also symmetric

$$\mathbf{K} = \mathbf{K}^\dagger, \qquad \mathbf{B} = \mathbf{B}^\dagger.$$

Equation (6.2) coincides with the corresponding one, (1.4.3), of [95].

The corresponding analysis of the perturbation of the traction proceeds along similar lines. We omit such non-trivial analysis and are content with saying that

$$\mathbf{t} = (\mathbf{T}^0 + \mathbf{H}\mathbf{T}^0 + \mathbf{B}\mathbf{H})\mathbf{n}$$

is the traction to be considered when the first-order correction is incorporated.

The second equality in (6.3) allows us to say that, in general, $\mathbf{BH} = \mathbf{BE}$ and then the incremental part $\mathbf{BH}$ involves the displacement gradient $\mathbf{H}$ only through the infinitesimal strain tensor $\mathbf{E}$. Meanwhile observe that, although $\mathbf{T}^0$ is a (symmetric) Cauchy stress, the tensor $\mathbf{HT}^0$ involves both the symmetric and the skewsymmetric parts $\mathbf{E}, \mathbf{W}$ of $\mathbf{H}$,

$$\mathbf{HT}^0 = (\mathbf{E} + \mathbf{W})\mathbf{T}^0.$$

Once we know the tensor $\mathbf{B}$ we can apply (6.2) to any (elastic) prestressed solid. Of course to determine $\mathbf{B}$ we go back to the non-linear stress-strain relation. Yet, it may well happen that we have reasons for linearizing with respect to the superimposed motion while we have no (or not enough) information on the stress-strain relation. In such a case it is reasonable to assume that the incremental part $\mathbf{BH}$ is isotropic and this assumption is termed *incremental isotropy.* This means that we let

$$\mathbf{BE} = 2\,\mu\,\mathbf{E} + \lambda(\mathrm{tr}\,\mathbf{E})\mathbf{1}, \qquad \mathbf{E} = \mathrm{sym}\nabla\mathbf{u}.$$

By the incremental isotropy, for a viscoelastic solid we let

$$\mathbf{BE} \longrightarrow 2\mu_0\mathbf{E} + \lambda_0(\mathrm{tr}\,\mathbf{E})\mathbf{1} + \int_0^\infty \{2\mu'(s)\mathbf{E}^t(s) + \lambda'(s)[\mathrm{tr}\,\mathbf{E}^t(s)]\mathbf{1}\}ds$$

where $\mathbf{E} = \mathrm{sym}\nabla\mathbf{u}$ and $\mathbf{u}$ still represents the incremental displacement. Then (6.2) becomes

$$\rho\ddot{\mathbf{u}} = \nabla\cdot\left\{\mathbf{HT}^0 + 2\mu_0\mathbf{E} + \lambda_0(\mathrm{tr}\,\mathbf{E})\mathbf{1} + \int_0^\infty \{2\mu'(s)\mathbf{E}^t(s) + \lambda'(s)[\mathrm{tr}\,\mathbf{E}^t(s)]\mathbf{1}\}ds\right\}.$$

This scheme is applied in Ch. 9 in connection with wave propagation in the form of rays.

Consider the first term $\nabla\cdot(\mathbf{HT}^0)$, that is

$$\nabla\cdot(\mathbf{HT}^0) = \mathbf{H}\nabla\cdot\mathbf{T}^0 + (\mathbf{T}^0\cdot\nabla\nabla)\mathbf{u}.$$

By (2.11) we have $\nabla\cdot\mathbf{T}^0 + \rho\mathbf{b} = 0$ and hence, letting $\mathbf{b} = \mathbf{g}$, we can write

$$\mathbf{H}\nabla\cdot\mathbf{T}^0 = \rho(\mathbf{g}\cdot\nabla)\mathbf{u}.$$

Accordingly the equation of motion (6.2) becomes

$$\rho\ddot{\mathbf{u}} = (\mathbf{T}^0\cdot\nabla\nabla)\mathbf{u} + \rho(\mathbf{g}\cdot\nabla)\mathbf{u} + \mu\Delta\mathbf{u} + (\mu+\lambda)\nabla(\nabla\cdot\mathbf{u}). \tag{6.4}$$

For a more specific form of (6.4), suppose that $\mathbf{T}^0$ is diagonal and denote by T_1, T_2, T_3 the three entries. Further, use Cartesian coordinates with z the (upward) vertical axis and denote by u, v, w the components of $\mathbf{u}$. Then (6.4) yields

$$\rho\ddot{u} = (2\mu + \lambda + T_1)u_{,xx} + (\mu+\lambda)(v_{,xy} + w_{,xz}) + \mu(u_{,yy} + u_{,zz}) + T_2 v_{,yy} + T_3 w_{,zz} - \rho g u_{,z},$$

$$\rho\ddot{v} = (2\mu + \lambda + T_2)v_{,yy} + (\mu + \lambda)(u_{,xy} + w_{,yz}) + \mu(v_{,yy} + v_{,zz}) + T_1 v_{,xx} + T_3 v_{,zz} - \rho g v_{,z},$$

$$\rho\ddot{w} = (2\mu + \lambda + T_3)w_{,zz} + (\mu + \lambda)(u_{,xz} + v_{,yz}) + \mu(w_{,xx} + w_{,yy}) + T_1 w_{,xx} + T_2 w_{,yy} - \rho g w_{,z},$$

a comma denoting partial differentiation.

It is worth mentioning that a detailed investigation of the equation of motion for prestressed bodies is exhibited in [13]. The pertinent equations have been applied in [59] to study reflection and refraction of plane waves. Our equations are in partial agreement with those of [59] provided we identify their K with our $\mu + \lambda$.

Now we may ask about the equations which govern the dynamics of a fluid which is initially at equilibrium under a pressure field. The desired answer may be obtained by paralleling the procedure set up for solids. Yet for fluids an alternative procedure is available which proves more immediate and presumably more instructive. We follow this one.

Again we have in mind an intermediate placement $\mathcal{B}_i$ where the fluid is at equilibrium under the action of a body force and pressure fields at the boundary. Let $\mathcal{B}_t$ be the current placement and denote by $\mathbf{x}, \tilde{\mathbf{x}}$ the position of a particle in $\mathcal{B}_i, \mathcal{B}_t$, respectively. For convenience we regard $\mathcal{B}_i$ as reference. So, denoting by a superposed $\tilde{}$ the pertinent field corresponding to the motion $\tilde{\mathbf{x}}(\mathbf{x}, t)$, we write the continuity equation and the equation of motion in the (Eulerian) form

$$\dot{\tilde{\rho}} + \tilde{\rho}\nabla_{\tilde{\mathbf{x}}} \cdot \dot{\tilde{\mathbf{x}}} = 0, \tag{6.5}$$

$$\tilde{\rho}\ddot{\tilde{\mathbf{x}}} - \nabla_{\tilde{\mathbf{x}}} \cdot \tilde{\mathbf{T}} - \tilde{\rho}\mathbf{b} = 0; \tag{6.6}$$

to avoid inessential additional terms we regard $\mathbf{b}$ as constant. Now let $\mathbf{u} = \tilde{\mathbf{x}} - \mathbf{x}$ and look for the linearization of (6.5), (6.6) with respect to $\mathbf{u}$. Let ρ be the mass density in $\mathcal{B}_i$. The transformation $\mathbf{x} \to \tilde{\mathbf{x}}$ makes ρ into $\tilde{\rho}$. So letting $j = \det \nabla\tilde{\mathbf{x}}$, by (2.1) we have $j\tilde{\rho} = \rho$. Denote by ϱ the change of mass density,

$$\varrho = \tilde{\rho} - \rho = -\rho(j-1)/j.$$

By keeping only linear terms in $\mathbf{u}$,

$$j = 1 + \nabla \cdot \mathbf{u}, \qquad \varrho = -\rho\nabla \cdot \mathbf{u}. \tag{6.7}$$

Set $\mathbf{T} = -p(\tilde{\rho})\mathbf{1} + \boldsymbol{\tau}$, the extra-stress $\boldsymbol{\tau}$ being a functional $\boldsymbol{\mathcal{T}}(\tilde{\mathbf{D}}^t)$ of the history of $\tilde{\mathbf{D}} = \mathrm{sym}\nabla_{\tilde{\mathbf{x}}}\dot{\tilde{\mathbf{x}}}$. Multiply (6.6) by j and observe that $\ddot{\tilde{\mathbf{x}}} = \ddot{\mathbf{u}}$ to obtain

$$\rho\ddot{\mathbf{u}} + j[\nabla p(\rho + \varrho)]\nabla_{\tilde{\mathbf{x}}}\mathbf{x} - j\nabla_{\tilde{\mathbf{x}}} \cdot \boldsymbol{\mathcal{T}}(\tilde{\mathbf{D}}^t) - \rho\mathbf{b} = 0. \tag{6.8}$$

Now, linearly in ϱ,

$$p(\rho + \varrho) = p(\rho) + p_\rho\varrho,$$

where $p_\rho = dp/d\rho$ is evaluated at ρ, and hence

$$\nabla p(\rho + \varrho) = p_\rho \nabla \rho + p_{\rho\rho} \nabla \rho \, \varrho + p_\rho \nabla \varrho.$$

Substitution into (6.8), use of (6.7), and account of the equilibrium condition

$$p_\rho \nabla \rho - \rho \mathbf{b} = 0,$$

yield the desired equation

$$\rho \ddot{\mathbf{u}} - \rho(\nabla \mathbf{u})\mathbf{b} - \frac{p_{\rho\rho}\rho^2}{p_\rho} \mathbf{b}(\nabla \cdot \mathbf{u}) - \rho p_\rho \nabla(\nabla \cdot \mathbf{u}) - \nabla \cdot \boldsymbol{\mathcal{T}}(\mathbf{D}^t) = 0, \tag{6.9}$$

$\nabla \cdot \boldsymbol{\mathcal{T}}(\mathbf{D}^t)$ being the linear approximation of $j \nabla_{\tilde{\mathbf{x}}} \cdot \boldsymbol{\mathcal{T}}(\tilde{\mathbf{D}}^t)$. Obviously, if the fluid is non-dissipative, i.e. $\boldsymbol{\mathcal{T}}(\mathbf{D}^t) = 0$, then equation of motion reduces to

$$\rho \ddot{\mathbf{u}} - \rho(\nabla \mathbf{u})\mathbf{b} - \frac{p_{\rho\rho}\rho^2}{p_\rho} \mathbf{b}(\nabla \cdot \mathbf{u}) - \rho p_\rho \nabla(\nabla \cdot \mathbf{u}) = 0. \tag{6.10}$$

3 INHOMOGENEOUS WAVES IN UNBOUNDED MEDIA

The first step in the study of wave propagation is the analysis of waves in unbounded media where boundary effects are deliberately ruled out. The analysis is confined to monochromatic or time-harmonic waves for which the frequency ω is fixed and real. Solutions in the form of inhomogeneous waves are sought. Based on the equation $\mathbf{k} = \mathbf{k}(\omega)$ provided by the propagation condition, the propagation modes are examined by looking for the amplitude, or polarization, in terms of the wave vector $\mathbf{k}$. The use of the scalar and vector potentials for the displacement $\mathbf{u}$ proves advantageous in both (dissipative) solids and fluids.

The equation of energy may be viewed as the balance between the time rate of the energy density and the net contribution of the energy flux, along with the energy supply. This balance shows interesting features also because of non-uniqueness of the energy density. Within a general setting for this topic, as non-trivial applications the detailed expressions of the energy flux are determined for the pertinent waves in solids and fluids.

The propagation condition and the propagation modes are significantly affected by the occurrence of internal constraints and body forces. For internal constraints, such as inextensibility in one direction and incompressibility, results arise which generalize the analogous ones for non-dissipative media. A similar conclusion holds for the effect of body forces. They have a twofold role. First, they put the body in equilibrium in a placement where the stress cannot vanish. Wave propagation is then viewed as superimposed on a prestress and a predeformation. Second, body forces enter explicitly the propagation condition and then affect the relation $\mathbf{k} = \mathbf{k}(\omega)$ and the propagation modes. That is why prestressed media are examined in some detail.

3.1 Helmholtz representation and viscoelastic potentials

We are accustomed to the use of potentials in linear elasticity (cf. [160]). The displacement field as the gradient of a scalar potential (plus a solenoidal field) was introduced by Poisson in 1829. Later Lamé observed that every vector field of the form $\mathbf{u} = \nabla\phi + \nabla \times \boldsymbol{\psi}$ meets the equation of motion provided ϕ and $\boldsymbol{\psi}$ are solutions to appropriate wave equations. Next Clebsch proved that every solution of the equation of motion admits the aforementioned representation; a rigorous proof of this result was provided by Duhem.

For the dynamics of viscoelastic solids, the representation of the displacement field in terms of scalar and vector potentials is given by Edelstein & Gurtin [66] by adapting a method of Somigliana [157] (cf. [115], §38). A result and procedure, strictly analogous to those for elastic solids, can be exhibited when a time-harmonic dependence is involved. Here we give the detailed proof also with a view to applications to bulk waves, reflection and refraction at plane interfaces, surface waves.

As a preliminary step toward the introduction of viscoelastic potentials we recall the Helmholtz representation of any vector field $\mathbf{u}(\mathbf{x},t)$, $\mathbf{x} \in \mathcal{E}^3, t \in \mathbb{R}$. Let $\mathbf{w}(\mathbf{x},t)$ be the Newtonian potential defined by

$$\mathbf{w}(\mathbf{x},t) = -\frac{1}{4\pi}\int_\Omega \frac{\mathbf{u}(\mathbf{y},t)}{|\mathbf{x}-\mathbf{y}|}dy$$

where $\Omega \subset \mathcal{E}^3$. Throughout Ω we have

$$\Delta \mathbf{w} = \mathbf{u}.$$

Define $\eta(\mathbf{x},t)$ and $\mathbf{v}(\mathbf{x},t)$ by

$$\eta = \nabla \cdot \mathbf{w}, \qquad \mathbf{v} = -\nabla \times \mathbf{w}.$$

Then, by the identity $\Delta \mathbf{w} = \nabla(\nabla \cdot \mathbf{w}) - \nabla \times (\nabla \times \mathbf{w})$, we have

$$\mathbf{u} = \nabla\eta + \nabla \times \mathbf{v}, \quad \nabla \cdot \mathbf{v} = 0, \tag{1.1}$$

namely the Helmholtz representation of $\mathbf{u}$.

Consider a homogeneous viscoelastic solid (relative to a stress-free placement). The motion of the solid is described by the displacement vector $\mathbf{u}$ as a function of the position vector $\mathbf{x}$ and the time t. The material properties are summarized by the constitutive relation for the (Cauchy) stress tensor $\mathbf{T}$, at time t, in the form

$$\mathbf{T}(t) = 2\mu_0\mathbf{E}(t) + \lambda_0 \mathrm{tr}\,\mathbf{E}(t)\,\mathbf{1} + \int_0^\infty [2\mu'(s)\mathbf{E}(t-s) + \lambda'(s)\mathrm{tr}\,\mathbf{E}(t-s)\,\mathbf{1}]ds \tag{1.2}$$

where $\mathbf{E} = \mathrm{sym}\nabla\mathbf{u}$. We disregard the body force and write the equation of motion as

$$\rho\ddot{\mathbf{u}} = \nabla \cdot \mathbf{T}. \tag{1.3}$$

Assume a time-harmonic dependence for $\mathbf{u}$, namely $\mathbf{u} = \mathbf{U}(\mathbf{x})\exp(-i\omega t)$, where ω is real and, to fix ideas, positive; when convenient we write $\mathbf{U}(\mathbf{x};\omega)$ to emphasize that the dependence on $\mathbf{x}$ may be parameterized by the frequency ω. Upon substitution, (1.2) and (1.3) yield

$$\rho\omega^2\mathbf{U} + \mu\Delta\mathbf{U} + (\mu+\lambda)\nabla(\nabla \cdot \mathbf{U}) = 0 \tag{1.4}$$

where the moduli

$$\mu = \mu_0 + \int_0^\infty \mu'(s)\exp(i\omega s)\,ds, \qquad \lambda = \lambda_0 + \int_0^\infty \lambda'(s)\exp(i\omega s)\,ds$$

are usually complex-valued.

Specialize (1.1) to time-independent fields. Then we write for $\mathbf{U}$ the Helmholtz representation

$$\mathbf{U} = \nabla N + \nabla \times \mathbf{H}, \qquad \nabla \cdot \mathbf{H} = 0. \tag{1.5}$$

Substitution in (1.4) yields

$$\nabla[\rho\omega^2 + (2\mu + \lambda)\Delta]N + \nabla \times [\rho\omega^2 + \Delta]\mathbf{H} = 0. \tag{1.6}$$

Letting

$$a = [\rho\omega^2 + (2\mu + \lambda)\Delta]N, \qquad \mathbf{d} = [\rho\omega^2 + \mu\Delta]\mathbf{H}, \tag{1.7}$$

we can write (1.6) as

$$\nabla a + \nabla \times \mathbf{d} = 0. \tag{1.8}$$

Application to (1.8) of the operators ∇, $\nabla \times$ and account of (1.5) provide

$$\Delta a = 0, \qquad \Delta \mathbf{d} = 0, \quad \nabla \cdot \mathbf{d} = 0. \tag{1.9}$$

Let $N_1 = N - a/\rho\omega^2$, $\mathbf{W} = \mathbf{H} - \mathbf{d}/\rho\omega^2$. Along with (1.9), this changes (1.7) to

$$[\rho\omega^2 + (2\mu + \lambda)\Delta]N_1 = 0, \qquad [\rho\omega^2 + \mu\Delta]\mathbf{W} = 0. \tag{1.10}$$

Now we have to relate N_1 and $\mathbf{W}$ to $\mathbf{U}$. By (1.5) we can express $\mathbf{U}$ as

$$\mathbf{U} = \nabla N_1 + \nabla \times \mathbf{W} + \tilde{\mathbf{U}}$$

where $\tilde{\mathbf{U}} = (\nabla a + \nabla \times \mathbf{d})/\rho\omega^2$. In view of (1.9) we have

$$\nabla \cdot \tilde{\mathbf{U}} = \frac{1}{\rho\omega^2}\Delta a = 0, \qquad \nabla \times \tilde{\mathbf{U}} = -\frac{1}{\rho\omega^2}\Delta \mathbf{d} = 0$$

and then there exists a scalar function N_2 such that

$$\tilde{\mathbf{U}} = \nabla N_2, \quad \Delta N_2 = 0.$$

Hence

$$\mathbf{U} = \nabla N_1 + \nabla N_2 + \nabla \times \mathbf{W}$$

where N_1, $\mathbf{W}$ are solutions to (1.10) while N_2 is harmonic. To determine N_2 we have to require that $\mathbf{U}$ meet (1.4). Substitution in (1.4) gives

$$\nabla[\rho\omega^2 + (2\mu + \lambda)\Delta]N_1 + \nabla \times [\rho\omega^2 + \mu\Delta]\mathbf{W} + \nabla[\rho\omega^2 + (2\mu + \lambda)\Delta]N_2 = 0.$$

Since N_2 is harmonic, by (1.10) we have

$$\nabla N_2 = 0$$

and then N_2 is a (inessential) constant. Letting $M = N_1 + N_2$ we conclude that *for any solution* $\mathbf{U}$ *to (1.4) there exists a scalar function* $M(\mathbf{x})$ *and a vector function* $\mathbf{W}(\mathbf{x})$ *such that*

$$\mathbf{U} = \nabla M + \nabla \times \mathbf{W}, \qquad \nabla \cdot \mathbf{W} = 0 \tag{1.11}$$

and

$$[\rho\omega^2 + (2\mu + \lambda)\Delta]M = 0, \qquad [\rho\omega^2 + \mu\Delta]\mathbf{W} = 0. \tag{1.12}$$

As an aside, the analogous procedure for the equilibrium equation would lead to $\Delta N_2 = 0$ but not $\nabla N_2 = 0$. This means that N_2 may depend on $\mathbf{x}$, at least linearly. In a sense, equilibrium conditions seem to allow a more general representation for the potentials.

The representation for the potentials is non-unique. As a non-trivial example we mention that two equivalent representations are known for the equilibrium case. Papkovich's solution, which was independently rediscovered by Neuber, has the form

$$\mathbf{u} = \nabla(m + \mathbf{x} \cdot \mathbf{w}) - \frac{2(2\mu + \lambda)}{\mu + \lambda}\mathbf{w},$$

where m and $\mathbf{w}$ are harmonic. Lamé's solution for

$$\mathbf{u} = \nabla\varphi + \nabla \times \boldsymbol{\varepsilon}, \qquad \nabla \cdot \boldsymbol{\varepsilon} = 0$$

is given by the potentials

$$\varphi = m - \frac{2(\mu + \lambda)}{\mu + \lambda}\frac{\mu t^2}{\rho}\nabla \cdot \mathbf{w} - \frac{\mu}{\mu + \lambda}\mathbf{x} \cdot \mathbf{w}, \qquad \boldsymbol{\varepsilon} = \frac{2\mu + \lambda}{\mu + \lambda}\frac{\mu t^2}{\rho}\nabla \times \mathbf{w} + \nabla \times \mathbf{r}$$

where $\mathbf{r}$ is a particular solution of the pair of equations

$$\Delta\mathbf{r} = \frac{2(2\mu + \lambda)}{\mu + \lambda}\mathbf{m}, \qquad \nabla \cdot \mathbf{r} = \frac{2\mu + \lambda}{\mu + \lambda}\mathbf{x} \cdot \mathbf{m}.$$

The same result applies here with $\mathbf{u}$ replaced by $\mathbf{U}$.

3.2 Inhomogeneous waves in viscoelastic solids

Based on the result (1.11)-(1.12) for the time-harmonic dependence, we now look for solutions in terms of inhomogeneous waves. This means that we have to examine the existence of scalar and vector solutions to (1.11)-(1.12) of the form

$$M(\mathbf{x}) = \Phi \exp(i\mathbf{k} \cdot \mathbf{x}), \qquad \mathbf{W}(\mathbf{x}) = \Psi \exp(i\mathbf{k} \cdot \mathbf{x})$$

where $\mathbf{k}$ is a complex-valued vector; as usual, we let $\mathbf{k}_1, \mathbf{k}_2$ be the real and imaginary parts of $\mathbf{k}$; it is understood that $\Phi, \Psi, \mathbf{k}$ may be parameterized by the frequency ω. Substitution in (1.12) yields

$$[\rho\omega^2 - (2\mu + \lambda)\mathbf{k} \cdot \mathbf{k}]\Phi = 0, \qquad [\rho\omega^2 - \mu\mathbf{k} \cdot \mathbf{k}]\Psi = 0.$$

A non-zero Φ is allowed if and only if $\mathbf{k} = \mathbf{k}_L$ is such that

$$\kappa_L := \mathbf{k}_L \cdot \mathbf{k}_L = \frac{\rho\omega^2}{2\mu + \lambda} \tag{2.1}$$

and, by (1.11),

$$\mathbf{U} = i\mathbf{k}_L \Phi \exp(i\mathbf{k}_L \cdot \mathbf{x}). \tag{2.2}$$

Similarly, a non-zero Ψ is allowed if and only if $\mathbf{k} = \mathbf{k}_T$ is such that

$$\kappa_T := \mathbf{k}_T \cdot \mathbf{k}_T = \frac{\rho\omega^2}{\mu} \tag{2.3}$$

and, by (1.11),

$$\mathbf{U} = i\mathbf{k}_T \times \Psi \exp(i\mathbf{k}_T \cdot \mathbf{x}), \qquad \mathbf{k}_T \cdot \Psi = 0. \tag{2.4}$$

Since μ and λ are complex-valued, (2.1) and (2.3) show that $\mathbf{k}_L$ and $\mathbf{k}_T$ are complex-valued. Moreover, the amplitudes Φ and Ψ of the waves (2.2) and (2.4) are, respectively, parallel and orthogonal to the corresponding wave vectors. In connection with the observations in §1.2, and by analogy with the elastic case, the displacement fields (2.2) and (2.4) are regarded to describe longitudinal and transverse inhomogeneous waves. It is also natural to view $\Phi \exp(i\mathbf{k}_L \cdot \mathbf{x})$ as the scalar potential of the longitudinal wave and $\Psi \exp(i\mathbf{k}_T \cdot \mathbf{x})$ the vector potential of the transverse one. Here, though, the quantities Φ, Ψ, besides $\mathbf{k}_L, \mathbf{k}_T$, are complex-valued and then parallelism and orthogonality are meant in the sense of complex vectors. Some geometrical aspects are examined in a moment.

It is useful to examine some properties of the dependence of $\Phi, \Psi, \mathbf{k}$ on ω. Since the vector function $\mathbf{u}(\mathbf{x}, t)$ is real-valued, we can regard $\mathbf{U}(\mathbf{x};\omega)$ as complex-valued provided that

$$\mathbf{U}^*(\mathbf{x};\omega) = \mathbf{U}(\mathbf{x};-\omega). \tag{2.5}$$

where the superscript $*$ denotes the complex conjugate. This is so because a physical monochromatic wave of frequency ω is the superposition of two complex-valued waves of frequency ω and $-\omega$; the validity of (2.5) is necessary and sufficient for the resultant superposition to be real-valued. As a consequence of (2.5) we have

$$\Phi_1(\omega) = \Phi_1(-\omega), \quad \Phi_2(\omega) = -\Phi_2(-\omega), \qquad \Psi_1(\omega) = \Psi_1(-\omega), \quad \Psi_2(\omega) = -\Psi_2(-\omega), \tag{2.6}$$

$$\mathbf{k}_1(\omega) = -\mathbf{k}_1(-\omega), \quad \mathbf{k}_2(\omega) = \mathbf{k}_2(-\omega), \tag{2.7}$$

where the subscripts 1 and 2 denote the real part and the imaginary part, respectively.

It is worth emphasizing some consequences of thermodynamics. By (2.1) and (2.3) we have

$$\kappa_L = \frac{\rho\omega^2}{|2\mu+\lambda|^2}(2\mu^* + \lambda^*), \qquad \kappa_T = \frac{\rho\omega^2}{|\mu|^2}\mu^*.$$

In view of (2.4.7), whence $\operatorname{Im}\mu < 0$, $\operatorname{Im}(2\mu+\lambda) < 0$ for any strictly positive ω, we have

$$\operatorname{Im}\kappa_L > 0, \qquad \operatorname{Im}\kappa_T > 0, \qquad \omega \in \mathbb{R}^{++}. \tag{2.8}$$

Incidentally, the condition (2.8) is just the requirement $v > 0$ mentioned in connection with the general properties of inhomogeneous waves, cf. (1.3.1). Meanwhile thermodynamics does not place any conditions on $\operatorname{Re}\mu, \operatorname{Re}\lambda$. However, we know that

$$\operatorname{Re}\mu = \mu_0 + \mu_c'(\omega), \qquad \mu_0 > \mu_\infty > 0.$$

For any frequency ω, the cosine transform $\mu_c'(\omega)$ is likely to be much smaller than μ_0 and then $\operatorname{Re}\mu > 0$. By the same token we have $\operatorname{Re}(\lambda + 2\mu/3) > 0$. This in turn implies that

$$\operatorname{Re}\kappa_L > 0, \qquad \operatorname{Re}\kappa_T > 0, \qquad \omega \in \mathbb{R}. \tag{2.9}$$

In view of (2.8) we have

$$\omega \operatorname{Im}[\mathbf{k}(\omega)\cdot\mathbf{k}(\omega)] > 0, \qquad \omega \in \mathbb{R}\setminus\{0\},$$

for both $\mathbf{k}_L$ and $\mathbf{k}_T$. Then

$$\omega\, \mathbf{k}_1(\omega)\cdot\mathbf{k}_2(\omega) > 0, \qquad \omega \in \mathbb{R}\setminus\{0\}. \tag{2.10}$$

For both waves the space-time dependence has the form

$$\exp[i(\mathbf{k}_1\cdot\mathbf{x} - \omega t)]\exp(-\mathbf{k}_2\cdot\mathbf{x})$$

whereby the wave propagates in the direction of $\mathbf{k}_1$ and the amplitude decreases in the direction of $\mathbf{k}_2$. Because of (2.10), this means that by following the propagation of the wave (along $\mathbf{k}_1$) we see a decreasing amplitude. So, in essence, thermodynamics implies that the wave amplitude decays while the wave propagates.

The properties (2.6)-(2.8) are now exploited to derive the explicit form of $\mathbf{k}(\omega)$. Let k_1, k_2 be the moduli of $\mathbf{k}_1, \mathbf{k}_2$. We write any one of (2.1), (2.3) in the form

$$k_1^2 - k_2^2 = a(\omega), \tag{2.11}$$

$$2k_1k_2\alpha = b(\omega), \tag{2.12}$$

where a and b are the real and imaginary parts of the right-hand side of (2.1) and (2.3) while α is the cosine of the angle between $\mathbf{k}_1$ and $\mathbf{k}_2$. By thermodynamics we have

$$\omega b(\omega) > 0, \quad \omega \in \mathbb{R} \setminus \{0\} \tag{2.13}$$

while, by definition, $a(\omega) = a(-\omega)$, $b(\omega) = -b(-\omega)$. It follows at once from (2.12) and (2.13) that

$$\omega \alpha(\omega) > 0, \quad \omega \in \mathbb{R} \setminus \{0\};$$

the inversion of the cosine as $\omega \to -\omega$ follows also from (2.7). Upon obvious substitutions we obtain from (2.9) and (2.10) that

$$k_1 = \sqrt{[\sqrt{a^2 + b^2/\alpha^2} + a]/2}, \qquad k_2 = \sqrt{[\sqrt{a^2 + b^2/\alpha^2} - a]/2}. \tag{2.14}$$

This result shows a novel feature of genuinely inhomogeneous waves, relative to plane waves, namely that the wave vector $\mathbf{k}$ is not fully determined by the constitutive properties of the material. Rather, given the constitutive properties and the angle between $\mathbf{k}_1$ and $\mathbf{k}_2$ the wave mode is determined by (2.14). So any wave mode is parameterized by $\alpha(\neq 0)$ which is determined once we know how the wave has been produced.

As a particular case suppose that the material is "non-dissipative" in the sense that $b = 0$ and, quite reasonably, $a > 0$. Then by (2.12) two cases may occur. First, $\alpha = 0$ i.e. $\mathbf{k}_2$ is orthogonal to $\mathbf{k}_1$. Then we are left with a single equation, (2.11), in two unknowns, k_1, k_2. There are infinitely many solutions parameterized, e.g., by k_2, namely $k_1^2 = a + k_2^2$. Such solutions may be expressed, e.g., in the form

$$\exp(-k_2 y) \exp[i(k_1 x - \omega t)]. \tag{2.15}$$

On condition that we produce an amplitude which varies as $\exp(-k_2 y)$ we obtain a wave which propagates at the phase speed $\omega/\sqrt{a + k_2^2}$. For such wave solutions the energy flux velocity and the group velocity are different [85]. In the next chapter we show how such waves can be generated by refraction.

Second, $\alpha \neq 0$ and then k_1 or k_2 vanish. If $k_1 = 0$ no propagation occurs. So let $k_2 = 0$. Then $\mathbf{k}$ is a real vector and is determined, to within the direction, by (2.11) or (2.1) and (2.3). In such a case we have

$$k^2 = a,$$

i.e. $\mathbf{k}$ is any vector belonging to the spherical surface of radius $\sqrt{a}$. It is customary to define $\mathbf{q} = \mathbf{k}/\omega$ as the slowness vector in that its modulus is the inverse of the wave speed (and its direction is the direction of propagation). Denote by q_x, q_y, q_z the Cartesian components and, for convenience, choose the coordinate system so that $q_z = 0$. Then q_x and q_y are subject to

$$q_x^2 + q_y^2 = q^2 \tag{2.16}$$

where $q = \sqrt{a}/\omega$.

The relation (2.14) is often represented in a slowness diagram q_x, q_y (cf. [2]) through a circle of radius q. To each point of the circle corresponds a wave

$$\exp[i\omega(q_x x + q_y y - t)].$$

It is also observed that if $q_x > q$ then $q_y = \pm i\beta$, $\beta = \sqrt{q_x^2 - q^2}$ and

$$q_x^2 - \beta^2 = q^2$$

which represents an equilateral hyperbola. Accordingly, to each point of the equilateral hyperbola corresponds a wave

$$\exp(-\beta\omega y)\exp[i\omega(q_x x - t)].$$

Then we can say that, in a slowness diagram, the circle represents (particular) wave solutions, for non-dissipative materials, of the first type while the hyperbola represents wave solutions of the second type.

As already mentioned, waves of the form (2.15) are sometimes referred to as evanescent in that they are regarded as strongly attenuated by the propagation phenomenon ([76], §3.7). In the general scheme adopted here, evanescent waves are merely a particular example of inhomogeneous waves. Indeed, they correspond to $\mathbf{k}_2$ being orthogonal to $\mathbf{k}_1$, which is easily seen to describe waves of constant amplitude in the direction of propagation.

For later convenience we examine the dependence of the complex modulus

$$\mu(\omega) = \mu_0 + \int_0^\infty \mu'(s)\exp(i\omega s)ds$$

on the frequency ω. Since $\mu'(s)$ vanishes at $s = \infty$, an integration by parts yields

$$\int_0^\infty \mu'(s)\exp(i\omega s)ds = -\mu_0'(i\omega)^{-1} - (i\omega)^{-1}\int_0^\infty \mu''(s)\exp(i\omega s)ds.$$

Let $\mu_0^{(j)}$ denote the value at $s = 0$ of the j-th derivative of $\mu(s)$. Subsequent integration by parts, the standard assumption that the j-th derivative $\mu^{(j)}(s)$, $j = 1, 2, \ldots$, vanishes as $s \to \infty$, and application of Riemann-Lebesgue's lemma yield

$$\mu(\omega) = \sum_{h=0}^{n}(-1)^h \mu_0^{(h)}(i\omega)^{-h} + o(\omega^{-n}). \tag{2.17}$$

The same asymptotic behaviour holds for $\lambda(\omega)$.

3.3 Inhomogeneous waves in dissipative fluids

The simplest model of dissipative fluid is the viscous fluid (§2.6). Relative to an unstressed placement, with no body force, the linearized equation of motion takes the form

$$\rho\ddot{\mathbf{u}} = \rho p_\rho \nabla(\nabla\cdot\mathbf{u}) + \nabla\cdot[\mu_0(\nabla\dot{\mathbf{u}} + \nabla\dot{\mathbf{u}}^\dagger) + \lambda_0(\nabla\cdot\dot{\mathbf{u}})\mathbf{1}]$$

where μ_0, λ_0 are the viscosity coefficients. We observe that the result (2.6.7) is used to find that $\nabla p = -\rho\nabla(\nabla\cdot\mathbf{u})$. The viscoelastic fluid model accounts for dissipation also through memory effects (§2.6). Relative to an unstressed configuration the equation of motion reads

$$\rho\ddot{\mathbf{u}} = \rho p_\rho \nabla(\nabla\cdot\mathbf{u}) + \nabla\cdot\int_0^\infty \{\mu(s)[\nabla\mathbf{v}^t(s) + (\nabla\mathbf{v}^t)^\dagger(s)] + \lambda(s)\nabla\cdot\mathbf{v}^t(s)\,\mathbf{1}\}ds$$

where $\mu(s), \lambda(s)$ are relaxation functions and $\mathbf{v} = \dot{\mathbf{u}}$. For a time-harmonic field $\mathbf{u} = \mathbf{U}\exp(-i\omega t)$, both equations may be written in the form

$$\rho\omega^2\mathbf{U} - i\omega\mu\Delta\mathbf{U} + [\rho p_\rho - i\omega(\mu+\lambda)]\nabla(\nabla\cdot\mathbf{U}) = 0 \tag{3.1}$$

where $\mu = \mu_0, \lambda = \lambda_0$ for viscous fluids and

$$\mu = \int_0^\infty \tilde{\mu}(s)\exp(i\omega s)ds, \qquad \lambda = \int_0^\infty \tilde{\lambda}(s)\exp(i\omega s)ds$$

for viscoelastic fluids. The analysis developed in §3.1 for viscoelastic solids may be repeated step by step thus showing that for any solution $\mathbf{U}$ to (3.1) there exists a scalar function $M(\mathbf{x})$ and a vector function $\mathbf{W}(\mathbf{x})$ such that (1.11) holds and

$$\{\rho\omega^2 + [\rho p_\rho - i\omega(2\mu+\lambda)]\Delta\}M = 0, \qquad [\rho\omega^2 - i\omega\mu\Delta]\mathbf{W} = 0. \tag{3.2}$$

In view of (3.2) we look for inhomogeneous wave solutions by letting

$$M(\mathbf{x}) = \Phi\exp(i\mathbf{k}\cdot\mathbf{x}), \qquad \mathbf{W}(\mathbf{x}) = \mathbf{\Psi}\exp(i\mathbf{k}\cdot\mathbf{x}).$$

Upon substitution in (3.2) we have

$$\{\rho\omega^2 - [\rho p_\rho - i\omega(2\mu+\lambda)]\mathbf{k}\cdot\mathbf{k}\}\Phi = 0, \qquad [\rho\omega^2 + i\omega\mu\mathbf{k}\cdot\mathbf{k}]\mathbf{\Psi} = 0.$$

So a non-zero Φ is allowed if and only if $\mathbf{k} = \mathbf{k}_L$ is such that

$$\kappa_L := \mathbf{k}_L\cdot\mathbf{k}_L = \frac{\rho\omega^2}{\rho p_\rho - i\omega(2\mu+\lambda)} \tag{3.3}$$

and

$$\mathbf{U} = i\mathbf{k}_L \Phi \exp(i\mathbf{k}_L \cdot \mathbf{x}).$$

Similarly, a non-zero $\boldsymbol{\Psi}$ is allowed if and only if $\mathbf{k} = \mathbf{k}_T$ is such that

$$\kappa_T := \mathbf{k}_T \cdot \mathbf{k}_T = \frac{i\rho\omega}{\mu} \tag{3.4}$$

and

$$\mathbf{U} = i\mathbf{k}_T \times \boldsymbol{\Psi} \exp(i\mathbf{k}_T \cdot \mathbf{x}), \qquad \mathbf{k}_T \cdot \boldsymbol{\Psi} = 0.$$

Again, we can regard (3.3) and (3.4) as the characterization of longitudinal and transverse waves, respectively. The consequences of thermodynamics on the wave modes follow at once. Let $a = \operatorname{Re}\kappa_L, \operatorname{Re}\kappa_T$; $b = \operatorname{Im}\kappa_L, \operatorname{Im}\kappa_T$. For the viscoelastic fluid

$$a = \rho\omega^2 \frac{\rho p_\rho + \omega(2\mu_s + \lambda_s)}{[\rho p_\rho + \omega(2\mu_s + \lambda_s)]^2 + [\omega(2\mu_c + \lambda_c)]^2}, \quad \frac{\rho\omega\mu_s}{\mu_c^2 + \mu_s^2},$$

$$b = \rho\omega^3 \frac{2\mu_c + \lambda_c}{[\rho p_\rho + \omega(2\mu_s + \lambda_s)]^2 + [\omega(2\mu_c + \lambda_c)]^2}, \quad \frac{\rho\omega\mu_c}{\mu_c^2 + \mu_s^2},$$

depending on whether longitudinal or transverse waves are considered. Analogously, for the viscous fluid

$$a = \frac{\rho^2 p_\rho \omega^2}{(\rho p_\rho)^2 + \omega^2(2\mu_0 + \lambda_0)^2}, \quad 0, \tag{3.5}$$

$$b = \frac{\rho\omega^3(2\mu_0 + \lambda_0)}{(\rho p_\rho)^2 + \omega^2(2\mu_0 + \lambda_0)^2}, \quad \frac{\rho\omega}{\mu_0}. \tag{3.6}$$

By (2.5.6) it follows that $b > 0$ for both waves in the viscoelastic fluid. The analysis of the wave modes for the viscoelastic solid then applies, *mutatis mutandis*. As regards the viscous fluid, by (2.5.7) we still have $b > 0$ thus showing again the peculiar aspect of dissipativity on the wave propagation.

Observe that no restriction is placed by thermodynamics on the derivative p_ρ. Now, in perfect fluids ($k_2 = 0$, $k_1^2/\omega^2 = 1/p_\rho$) p_ρ is equal to the square of the wave speed. It is then reasonable to let $p_\rho > 0$ and to regard p_ρ as predominant on the viscoelastic analogue so that $a > 0$ for longitudinal waves. This is not necessarily so for transverse waves. In viscous fluids $a = 0$, which results in the condition that the real and imaginary parts of $\mathbf{k}_T$ are equal. Roughly speaking, propagation and attenuation have the same weight. In viscoelastic fluids no bound is given on μ_s and then we may have $a < 0$, whereby attenuation predominates on propagation. In conclusion, also for fluids

$$\operatorname{Im}\kappa_L > 0, \qquad \operatorname{Im}\kappa_T > 0, \qquad \omega \in \mathbb{R}^{++}$$

is true. Meanwhile $\operatorname{Re}\kappa_L > 0$ may be taken to be true but $\operatorname{Re}\kappa_T$ is zero for viscous fluids, not definite for viscoelastic fluids. It is often asserted that, in (linearly) viscous fluids, the attenuation of acoustic waves is proportional to the frequency squared ω^2. Indeed, on this basis numerical estimates are sometimes exhibited as, for example, in [22] where attenuation in water at 30° C is said to be $\epsilon\omega^2$ with $\epsilon = 4\ 10^{-15}$ dB m^{-1} Hz^{-1}. Of course, the knowledge of the attenuation is important in many respects; for example, the attenuation provides a (very sharp) upper cutoff frequency above which a given lens cannot be operated. To our knowledge, the standard argument for claiming that the attenuation is proportional to the frequency squared has been the one exhibited, e.g., in [112, 132]. For the benefit of the interested reader, here we follow very closely the pertinent part in [112].

Because of viscosity, the rate of energy dissipation is given by

$$\dot{e}_{mech} = -2\mu_0 \overset{\circ}{\mathbf{D}} \cdot \overset{\circ}{\mathbf{D}} - \kappa_0 (\mathrm{tr}\mathbf{D})^2$$

where e_{mech} is the mechanical energy per unit volume, $\overset{\circ}{\mathbf{D}}$ is the trace-free part of $\mathbf{D}$, and μ_0 and $\kappa_0 = \lambda_0 + 2\mu_0/3$ are the shear and bulk viscosity coefficients. Consider a wave propagating in the x-direction such that the (Cartesian) components v_x, v_y, v_z of the velocity are

$$v_x = v_0 \cos(kx - \omega t), \qquad v_y = 0, \qquad v_z = 0.$$

Denote by $\langle\cdot\rangle$ the mean value over a time period. Then the mean value $\langle \dot{e}_{mech}\rangle$ of $\dot{e}_{mech}$ turns out to be given by

$$\langle \dot{e}_{mech}\rangle = -\tfrac{1}{2}k^2(\tfrac{4}{3}\mu_0 + \kappa_0)v_0^2 \tag{3.7}$$

while the mean value $\langle e\rangle$ of the total energy of the wave e, per unit volume, is

$$\langle e\rangle = \tfrac{1}{2}\rho v_0^2. \tag{3.8}$$

The intensity of the waves decreases in time with damping coefficient $|\langle \dot{e}_{mech}\rangle|/2\langle e\rangle$. In the case of sound waves the intensity decreases with the distance x traversed and [112] "it is evident that this decrease will occur according to a law $\exp(-2\gamma x)$, and the amplitude will decrease as $\exp(-\gamma x)$, where the absorption coefficient γ is defined by

$$\gamma = \frac{|\langle \dot{e}_{mech}\rangle|}{2c\langle e\rangle},\text{"}$$

c being the phase speed. As a consequence, by (3.7) and (3.8) we obtain that the absorption coefficient is proportional to the frequency squared.

To our mind this conclusion is questionable. By means of the results (3.5)-(3.6) we can determine at once the absorption coefficient k_2 by applying the general solution (2.14). For transverse waves

$$k_2 = \sqrt{\frac{\rho\omega}{\mu_0\alpha}}.$$

Accordingly, for transverse waves the attenuation k_2 is proportional to the square root of ω. Consider longitudinal waves. Substitution of $(3.7)_1$ and $(3.8)_1$ in $(2.14)_2$ yields

$$k_2 = \sqrt{\frac{\omega^2[-1 + \sqrt{1 + \omega^2(2\mu_0 + \lambda_0)^2/p_\rho^2\alpha^2}]}{2p_\rho[1 + \omega^2(2\mu_0 + \lambda_0)^2/p_\rho^2]}}. \tag{3.9}$$

For small values of ω we have

$$k_2 \simeq \frac{2\mu_0 + \lambda_0}{2p_\rho^{3/2}\alpha}\omega^2.$$

Hence, the assertion that the attenuation of acoustic waves is proportional to the frequency squared should be related to longitudinal waves in the approximation of low frequencies, not generally to waves in (viscous) fluids.

For later reference we conclude this section by considering the limit case of the perfect fluid for which $\mu = 0$, $\lambda = 0$. By (3.2) we have $\mathbf{W} = 0$, and then $\Psi = 0$. So only longitudinal waves occur, $\Phi \neq 0$, with

$$\kappa_L = \frac{\omega^2}{p_\rho}, \tag{3.10}$$

or $k_l^2 = \omega^2/p_\rho$. For such waves the velocity $\mathbf{v} = \dot{\mathbf{u}}$ is expressed by

$$\mathbf{v} = \frac{\omega^2}{\sqrt{p_\rho}}\mathbf{n}\Phi \exp[i(\mathbf{k}_L \cdot \mathbf{x} - \omega t)], \tag{3.11}$$

$\mathbf{n}$ being the (real) unit vector of $\mathbf{k}_L$. The pressure p satisfies the equation of motion

$$\rho\dot{\mathbf{v}} = -\nabla p.$$

Then, letting $\tilde{p} = P\exp[i(\mathbf{k}_L \cdot \mathbf{x} - \omega t)]$ the deviation of the pressure from the equilibrium value, by (3.11) we have

$$P = \rho\omega^2\Phi. \tag{3.12}$$

3.4 Rate of energy and energy flux

Let $\mathbf{F}(\mathbf{x})\exp(-i\omega t)$ be the force acting on a particle which undergoes a motion with velocity $\mathbf{V}(\mathbf{x})\exp(-i\omega t)$. Since we have in mind that the real part is indeed the physical part of any quantity, we define the corresponding power on the particle as

$$\pi = \mathrm{Re}[\mathbf{F}(\mathbf{x})\exp(-i\omega t)] \cdot \mathrm{Re}[\mathbf{V}(\mathbf{x})\exp(-i\omega t)].$$

Hence we have

$$\pi = \tfrac{1}{4}(\mathbf{F} \cdot \mathbf{V}^* + \mathbf{F}^* \cdot \mathbf{V}) + \tfrac{1}{4}[\mathbf{F} \cdot \mathbf{V}\exp(-2i\omega t) + \mathbf{F}^* \cdot \mathbf{V}^*\exp(2i\omega t)].$$

If $\mathbf{F}$ represents a body force, we define the power $\mathcal{P}$ of $\mathbf{F}$ in a volume $\mathcal{V}$ of the body as

$$\mathcal{P}(t) = \int_{\mathcal{V}} \pi(\mathbf{x}, t) dx.$$

Owing to the oscillatory behaviour, a useful measure of the power is given by its average $\langle \mathcal{P} \rangle$ over a time period. Since $\mathbf{F} \cdot \mathbf{V}^* + \mathbf{F}^* \cdot \mathbf{V} = 2\,\mathrm{Re}(\mathbf{F} \cdot \mathbf{V}^*)$, we can write

$$\langle \mathcal{P} \rangle = \tfrac{1}{2}\mathrm{Re} \int_{\mathcal{V}} \mathbf{F} \cdot \mathbf{V}^* dx.$$

Consider a region Ω in the configuration of the body. By the same token we define the power $\mathcal{P}_t$ exerted by the traction force in the form

$$\begin{aligned} \langle \mathcal{P}_t \rangle &= \tfrac{1}{2}\mathrm{Re} \int_{\partial\Omega} (\mathbf{T}\mathbf{n}) \cdot \mathbf{v}^* da \\ &= \tfrac{1}{2}\mathrm{Re} \int_{\Omega} \nabla \cdot (\mathbf{T}\mathbf{v}^*) dx \end{aligned} \tag{4.1}$$

where $\mathbf{n}$ is the unit outward normal. By (1.3) we have

$$(\nabla \cdot \mathbf{T}) \cdot \mathbf{v}^* = -i\omega\rho\, \mathbf{v} \cdot \mathbf{v}^*$$

and then

$$\nabla \cdot (\mathbf{T}\mathbf{v}^*) = i\omega(\mathbf{T} \cdot \nabla\mathbf{u}^* - \rho\, \mathbf{v} \cdot \mathbf{v}^*).$$

Of course, $\rho\, \mathbf{v} \cdot \mathbf{v}^*$ is real. Meanwhile $\mathbf{T} \cdot \nabla\mathbf{u}^*$ turns out to be given by

$$\mathbf{T} \cdot \nabla\mathbf{u}^* = 2\mu\, \mathring{\mathbf{E}} \cdot \mathring{\mathbf{E}}^* + (\lambda + \tfrac{2}{3}\mu)\mathrm{tr}\,\mathbf{E}\; \mathrm{tr}\,\mathbf{E}^*.$$

Then we obtain

$$\langle \mathcal{P}_t \rangle = -\omega \int_{\Omega} [\mathrm{Im}\,\mu\; \mathring{\mathbf{E}} \cdot \mathring{\mathbf{E}}^* + \mathrm{Im}(\tfrac{1}{2}\lambda + \tfrac{1}{3}\mu)\mathrm{tr}\,\mathbf{E}\; \mathrm{tr}\,\mathbf{E}^*] dx. \tag{4.2}$$

If the body is elastic, namely μ and λ are real, then the right-hand side vanishes; this means that for elastic bodies the averaged power of the stress over a closed surface vanishes. For viscoelastic bodies, due to the thermodynamic inequalities (2.4.7), the right-hand side is strictly positive (for a nonrigid motion) and represents the power of the stress which is absorbed by the body and dissipated. Incidentally, this is a check of consistency of the definition of $\mathcal{P}_t$.

Owing to some discrepancy exhibited in the literature about energy density and flux (cf. [106]), it is worth showing how differences arose and may arise. Inner multiplication of the equation of motion (1.3) by $\dot{\mathbf{u}}$ and the identity

$$(\nabla \cdot \mathbf{T}) \cdot \dot{\mathbf{u}} = \nabla \cdot (\mathbf{T}\dot{\mathbf{u}}) - \mathbf{T} \cdot \dot{\mathbf{E}} \tag{4.3}$$

yield

$$\dot{\mathcal{K}} + \mathbf{T} \cdot \dot{\mathbf{E}} = \nabla \cdot (\mathbf{T}\dot{\mathbf{u}}) \tag{4.4}$$

where $\mathcal{K} = \frac{1}{2}\rho\dot{\mathbf{u}} \cdot \dot{\mathbf{u}}$. Examine separately (4.4) for elastic and viscoelastic bodies.

First regard the body as elastic. Then

$$\mathbf{T} \cdot \dot{\mathbf{E}} = \frac{d}{dt}(\tfrac{1}{2}\mathbf{T} \cdot \mathbf{E}).$$

So letting $\mathcal{U} = \frac{1}{2}\mathbf{T} \cdot \mathbf{E}$ and

$$\mathbf{J} = -\mathbf{T} \cdot \dot{\mathbf{u}},$$

we can write

$$\frac{d}{dt}(\mathcal{K} + \mathcal{U}) = -\nabla \cdot \mathbf{J}. \tag{4.5}$$

Equation (4.5) is usually interpreted as a balance of energy by saying that the derivative of the total energy (kinetic plus potential) $\mathcal{K} + \mathcal{U}$ is equal to (minus) the divergence of the *energy flux* vector $\mathbf{J}$. To our mind this interpretation is misleading. According to the general approach to balance equations, the balance of energy can be written as (cf. §2.2)

$$\frac{d}{dt}(\mathcal{K} + \mathcal{U}) = \nabla \cdot (\mathbf{T}\dot{\mathbf{u}}) - \nabla \cdot \mathbf{q} + \rho\mathbf{b} \cdot \dot{\mathbf{u}} + \rho r$$

where $\mathcal{U}$ is to be viewed as $\rho\epsilon$. This means that $\mathbf{J} = -\mathbf{T}\dot{\mathbf{u}}$ is at the outset the energy flux of mechanical character simply because the power of the traction force is $(\mathbf{T}\mathbf{n}) \cdot \dot{\mathbf{u}}$ and not because of the splitting (4.3). Otherwise no surprise should arise that some authors [17, 11], on the basis of the different splitting

$$(\nabla \cdot \mathbf{T}) \cdot \dot{\mathbf{u}} = -\frac{d}{dt}\tfrac{1}{2}\left[(\mu+\lambda)(\nabla \cdot \mathbf{u})^2 + \mu(\nabla\mathbf{u}) \cdot (\nabla\mathbf{u})\right] + \nabla \cdot \left[(\mu+\lambda)(\nabla \cdot \mathbf{u})\dot{\mathbf{u}} + \mu(\nabla\mathbf{u})\dot{\mathbf{u}}\right], \tag{4.6}$$

regard

$$\bar{\mathbf{J}} = -[(\mu + \lambda)(\nabla \cdot \mathbf{u})\dot{\mathbf{u}} + \mu(\nabla\mathbf{u})\dot{\mathbf{u}}]$$

as the energy flux.

Now let the body be viscoelastic. The possibility of inequivalent choices of energy density and flux is more apparent. By following a splitting procedure and having recourse to considerations about a spring-dashpot model, some authors (cf. [26]) have considered a "potential" function

$$\mathcal{U} = \int_{-\infty}^{t}\int_{-\infty}^{t} \mu(2t-\tau-\sigma)\dot{\mathbf{E}}(\tau) \cdot \dot{\mathbf{E}}(\sigma)d\tau\, d\sigma + \tfrac{1}{2}\int_{-\infty}^{t}\int_{-\infty}^{t} \lambda(2t-\tau-\sigma)\mathrm{tr}\dot{\mathbf{E}}(\tau)\,\mathrm{tr}\dot{\mathbf{E}}(\sigma)d\tau\, d\sigma.$$

and observed that, in terms of

$$\mathcal{D} = -2\int_{-\infty}^{t}\int_{-\infty}^{t} \mu'(2t-\tau-\sigma)\dot{\mathbf{E}}(\tau) \cdot \dot{\mathbf{E}}(\sigma)d\tau\, d\sigma - \int_{-\infty}^{t}\int_{-\infty}^{t} \lambda'(2t-\tau-\sigma)\mathrm{tr}\dot{\mathbf{E}}(\tau)\mathrm{tr}\dot{\mathbf{E}}(\sigma)d\tau\, d\sigma,$$

it follows that

$$\frac{d\mathcal{U}}{dt} = \mathbf{T} \cdot \dot{\mathbf{E}} - \mathcal{D}.$$

Accordingly they have written the energy equation as

$$\frac{d}{dt}(\mathcal{K} + \mathcal{U}) + \mathcal{D} = -\nabla \cdot \mathbf{J}, \tag{4.7}$$

where still $\mathbf{J} = -\mathbf{T}\dot{\mathbf{u}}$, and regarded $\mathcal{D}$ as the rate of dissipation. This splitting is highly subjective. For instance, consider

$$\begin{aligned} \tilde{\mathcal{U}} = {} & \mu_0 \mathbf{E}(t) \cdot \mathbf{E}(t) + \int_{-\infty}^{t} \Big(\int_0^\infty \mu'(s) \mathbf{E}(\tau - s) ds \Big) \cdot \dot{\mathbf{E}}(\tau) d\tau \\ & + \tfrac{1}{2}\lambda_0 (\mathrm{tr}\mathbf{E}(t))^2 + \tfrac{1}{2} \int_{-\infty}^{t} \Big(\int_0^\infty \lambda'(s) \mathrm{tr}\mathbf{E}(\tau - s) ds \Big) \mathrm{tr}\dot{\mathbf{E}}(\tau) d\tau. \end{aligned}$$

Time differentiation yields

$$\frac{d\tilde{\mathcal{U}}}{dt} = \mathbf{T} \cdot \dot{\mathbf{E}}.$$

Hence we can write (4.4) as

$$\frac{d}{dt}(\mathcal{K} + \tilde{\mathcal{U}}) = -\nabla \cdot \mathbf{J}. \tag{4.8}$$

Also, borrowing from the splitting (4.6) for the elastic case, we may consider

$$\begin{aligned} \hat{\mathcal{U}} = {} & \tfrac{1}{2}\mu_0 (\nabla \mathbf{u}(t)) \cdot (\nabla \mathbf{u}(t)) + \tfrac{1}{2}\mu_0 (\mathrm{tr}\mathbf{E}(t))^2 + 2 \int_{-\infty}^{t} \Big(\int_0^\infty \mu'(s) \mathbf{E}(\tau - s) ds \Big) \cdot \dot{\mathbf{E}}(\tau) d\tau \\ & + \tfrac{1}{2}\lambda_0 (\mathrm{tr}\mathbf{E}(t))^2 + \tfrac{1}{2} \int_{-\infty}^{t} \Big(\int_0^\infty \lambda'(s) \mathrm{tr}\mathbf{E}(\tau - s) ds \Big) \mathrm{tr}\dot{\mathbf{E}}(\tau) d\tau \end{aligned}$$

and

$$\begin{aligned} \hat{\mathbf{J}} = {} & - [(\mu_0 + \lambda_0)(\mathrm{tr}\mathbf{E})\dot{\mathbf{u}} + \mu_0 (\nabla \mathbf{u})\dot{\mathbf{u}}](t) \\ & - \Big[2 \int_0^\infty \mu'(s) \mathbf{E}(t - s) ds \Big] \dot{\mathbf{u}}(t) - \Big[\int_0^\infty \lambda'(s) \mathrm{tr}\mathbf{E}(t - s) ds \Big] \dot{\mathbf{u}}(t). \end{aligned}$$

Then (4.4) may be written as

$$\frac{d}{dt}(\mathcal{K} + \hat{\mathcal{U}}) = -\nabla \cdot \hat{\mathbf{J}}. \tag{4.9}$$

The triplets $(\mathcal{U}, \mathcal{D}, \mathbf{J}), (\tilde{\mathcal{U}}, 0, \mathbf{J})$, and $(\hat{\mathcal{U}}, 0, \hat{\mathbf{J}})$ show that the equation of motion, along with suitable splittings of $(\nabla \cdot \mathbf{T}) \cdot \dot{\mathbf{u}}$, leads to inequivalent expressions of potential energy, rate of dissipation, and energy flux (vector). Now, it is well known that energy for the viscoelastic stress is intrinsically non-unique even though compatibility with thermodynamics is required (cf. [131]). Further, by appropriate splittings the rate of dissipation is made to vanish, which might seem quite odd due to the dissipative character of the viscoelastic model. However, looking at $\mathcal{D}$ as the rate of dissipation is an arbitrary choice motivated by a subjective partition of the rate of working. The same is true for the energy flux which then is found to have different representations (and values).

A final remark seems to be in order. The literature has devoted noticeable attention (cf. [147, 118, 174, 85, 84, 165, 166]) to the energy flux velocity (or mean velocity of energy transport) as the ratio of the mean energy flux to the mean energy density. Needless to say, this quantity is well defined only when both the energy density and the energy flux are uniquely determined. As we have seen, sometimes such is not the case and the ambiguities remain even though we consider the mean values.

3.5 Energy flux at inhomogeneous waves in solids

We now consider the energy flux intensity associated with an inhomogeneous wave. We regard the energy flux intensity $\mathcal{I}$ as the energy transfer per unit area in the direction of wave propagation. For inhomogeneous waves the direction of propagation is given by $\mathbf{k}_1 = k_1\mathbf{n}_1$. Then we define the *energy flux intensity* $\mathcal{I}$ as $\mathbf{J}\cdot\mathbf{n}_1$ whence, by analogy with (4.1),

$$\langle\mathcal{I}\rangle = -\tfrac{1}{2}\mathrm{Re}[(\mathbf{T}\mathbf{n}_1)\cdot\mathbf{v}^*].$$

Because $\mathbf{v}^* = i\omega\mathbf{u}^*$ we can write

$$\langle\mathcal{I}\rangle = \tfrac{1}{2}\omega\mathrm{Im}[(\mathbf{T}\mathbf{n}_1)\cdot\mathbf{u}^*]. \tag{5.1}$$

Let $\mathbf{u}$ represent an inhomogeneous wave $\mathbf{U}_0\exp[i(\mathbf{k}\cdot\mathbf{x}-\omega t)]$. Substitution in (5.1) yields

$$\langle\mathcal{I}\rangle = \tfrac{1}{2}\omega\exp(-2\mathbf{k}_2\cdot\mathbf{x})\mathrm{Re}\{\mu[(\mathbf{k}\cdot\mathbf{U}_0^*)(\mathbf{U}_0\cdot\mathbf{n}_1)+(\mathbf{U}_0\cdot\mathbf{U}_0^*)(\mathbf{k}\cdot\mathbf{n}_1)]+\lambda(\mathbf{k}\cdot\mathbf{U}_0)(\mathbf{U}_0^*\cdot\mathbf{n}_1)\}.$$

If the material is elastic, and then $\mu, \lambda, \mathbf{k} = k\mathbf{n}_1$ are real, we have

$$\langle\mathcal{I}\rangle = \tfrac{1}{2}k\omega[(\mu+\lambda)(\mathbf{U}_0^*\cdot\mathbf{n}_1)(\mathbf{U}_0\cdot\mathbf{n}_1)+\mu\mathbf{U}_0\cdot\mathbf{U}_0^*].$$

For both longitudinal and transverse waves, letting $c = \omega/k$ be the pertinent phase speed we obtain

$$\langle\mathcal{I}\rangle = \tfrac{1}{2}\rho\omega^2 c|\mathbf{U}_0|^2,$$

namely a classical expression for the energy flux intensity [3].

Things are not that simple in viscoelasticity [39]. Also for an immediate connection with the literature on the subject we evaluate the mean energy flux

$$\langle\mathbf{J}\rangle = \tfrac{1}{2}\,\omega\,\mathrm{Im}(\mathbf{T}\mathbf{u}^*), \tag{5.2}$$

at the wave, and hence

$$\langle \mathcal{I} \rangle = \langle \mathbf{J} \rangle \cdot \mathbf{n}_1.$$

Letting $\mathbf{u} = \mathbf{U}_0(\omega)\exp[i(\mathbf{k}\cdot\mathbf{x} - \omega t)]$, $\mathbf{k} = \mathbf{k}_1 + i\mathbf{k}_2$, we have

$$\langle \mathbf{J} \rangle = \tfrac{1}{2}\omega \operatorname{Re}\{\exp(-2\mathbf{k}_2\cdot\mathbf{x})\,[\mu\mathbf{k}(\mathbf{U}_0\cdot\mathbf{U}_0^*) + \mu\mathbf{U}_0(\mathbf{k}\cdot\mathbf{U}_0^*) + \lambda(\mathbf{k}\cdot\mathbf{U}_0)\mathbf{U}_0^*]\}. \tag{5.3}$$

Let the wave be longitudinal, namely

$$\mathbf{k}\cdot\mathbf{k} = \frac{\rho\omega^2}{2\mu+\lambda}, \qquad \mathbf{U}_0 = i\mathbf{k}\Phi,$$

Φ being complex-valued. Upon substitution and some rearrangement we have

$$\mu(\mathbf{U}_0\cdot\mathbf{U}_0^*)\mathbf{k} + \mu(\mathbf{k}\cdot\mathbf{U}_0^*)\mathbf{U}_0 + \lambda(\mathbf{k}\cdot\mathbf{U}_0)\mathbf{U}_0^* = [(2\mu+\lambda)(\mathbf{k}\cdot\mathbf{k})\mathbf{k}^* + 2\mu\mathbf{k}\times(\mathbf{k}\times\mathbf{k}^*)]|\Phi|^2.$$

Replace $(2\mu+\lambda)\mathbf{k}\cdot\mathbf{k}$ with $\rho\omega^2$. By (5.3) we obtain

$$\langle \mathbf{J} \rangle = \tfrac{1}{2}\omega|\Phi|^2\exp(-2\mathbf{k}_2\cdot\mathbf{x})[\rho\omega^2\mathbf{k}_1 - 4(\mathbf{k}_1\times\mathbf{k}_2)\times(\mu_1\mathbf{k}_2 + \mu_2\mathbf{k}_1)] \tag{5.4}$$

where μ_1, μ_2 are the real and imaginary parts of μ, i.e. $\mu_1 = \mu_0 + \mu_c'$, $\mu_2 = \mu_s'$. The expression (5.4) coincides with the analogous one given by, e. g., Buchen [26]. Meanwhile the mean energy flux intensity takes the form

$$\langle \mathcal{I} \rangle = \tfrac{1}{2}\omega|\Phi|^2\exp(-2\mathbf{k}_2\cdot\mathbf{x})[\rho\omega^2 k_1 + 4\mu_1 k_1 k_2^2\sin^2\gamma]$$

where γ is the angle between $\mathbf{k}_1$ and $\mathbf{k}_2$. The peculiar aspect of dissipativity consists in the decay, here described by $\exp(-2\mathbf{k}_2\cdot\mathbf{x})$. Further, the inhomogeneity of the wave results in the direction of $\langle \mathbf{J} \rangle$ which is usually different than $\mathbf{n}_1$.

Now let the wave be transverse, namely

$$\mathbf{k}\cdot\mathbf{k} = \frac{\rho\omega^2}{\mu}, \qquad \mathbf{U}_0 = i\mathbf{k}\times\boldsymbol{\Psi}, \qquad \mathbf{k}\cdot\boldsymbol{\Psi} = 0.$$

Since $\mathbf{k}\cdot\mathbf{U}_0 = 0$ we have only to evaluate $(\mathbf{U}_0\cdot\mathbf{U}_0^*)\mathbf{k} + (\mathbf{k}\cdot\mathbf{U}_0^*)\mathbf{U}_0$. Substitution and a little algebra yield

$$\begin{aligned}(\mathbf{U}_0^*\cdot\mathbf{k})\mathbf{U}_0 &= -[(\mathbf{k}\times\boldsymbol{\Psi}^*)\cdot\mathbf{k}^*](\mathbf{k}\times\boldsymbol{\Psi}) \\ &= -\{[\mathbf{k}^*\times(\mathbf{k}\times\boldsymbol{\Psi})]\times(\mathbf{k}\times\boldsymbol{\Psi}^*) + [(\mathbf{k}\times\boldsymbol{\Psi}^*)\cdot(\mathbf{k}\times\boldsymbol{\Psi})]\mathbf{k}^*\}.\end{aligned}$$

Because $\mathbf{k}\cdot\boldsymbol{\Psi} = 0$ we have

$$(\mathbf{k}\times\boldsymbol{\Psi}^*)\cdot(\mathbf{k}\times\boldsymbol{\Psi}) = [\boldsymbol{\Psi}^*\times(\mathbf{k}\times\boldsymbol{\Psi})]\cdot\mathbf{k} = (\boldsymbol{\Psi}^*\cdot\boldsymbol{\Psi})(\mathbf{k}\cdot\mathbf{k}).$$

Substitution, use of the condition $\mathbf{k} \cdot \Psi = 0$, and some rearrangement yield

$$\begin{aligned}(\mathbf{U}_0^* \cdot \mathbf{k})\mathbf{U}_0 &= -\{[(\mathbf{k}^* \cdot \Psi)\mathbf{k} - (\mathbf{k}^* \cdot \mathbf{k})\Psi] \times (\mathbf{k} \times \Psi^*) + (\mathbf{k} \cdot \mathbf{k})(\Psi^* \cdot \Psi)\mathbf{k}^*\} \\ &= (\mathbf{k}^* \cdot \Psi)(\mathbf{k} \cdot \mathbf{k})\Psi^* - (\mathbf{k}^* \cdot \Psi)(\mathbf{k} \cdot \Psi^*)\mathbf{k} \\ &\quad + (\mathbf{k}^* \cdot \mathbf{k})(\Psi^* \cdot \Psi)\mathbf{k} - (\mathbf{k} \cdot \mathbf{k})(\Psi^* \cdot \Psi)\mathbf{k}^* \\ &= (\mathbf{k}^* \cdot \Psi)(\mathbf{k} \cdot \mathbf{k})\Psi^* - (\mathbf{k}^* \cdot \Psi)(\mathbf{k} \cdot \Psi^*)\mathbf{k} + (\Psi^* \cdot \Psi)[(\mathbf{k}^* \cdot \mathbf{k})\mathbf{k} - (\mathbf{k} \cdot \mathbf{k})\mathbf{k}^*].\end{aligned}$$

Meanwhile,

$$\begin{aligned}\mathbf{U}_0 \cdot \mathbf{U}_0^* &= (\Psi^* \times \Psi) \cdot (\Psi \times \mathbf{k}) = \Psi \cdot [\mathbf{k} \times (\Psi^* \times \mathbf{k}^*)] \\ &= (\mathbf{k} \cdot \mathbf{k}^*)(\Psi \cdot \Psi^*) - (\mathbf{k} \cdot \Psi^*)(\mathbf{k}^* \cdot \Psi).\end{aligned}$$

In conclusion we obtain

$$\begin{aligned}(\mathbf{U}_0 \cdot \mathbf{U}_0^*)\mathbf{k} + (\mathbf{k} \cdot \mathbf{U}_0^*)\mathbf{U}_0 =& (\Psi \cdot \Psi^*)[2(\mathbf{k} \cdot \mathbf{k}^*)\mathbf{k} - (\mathbf{k} \cdot \mathbf{k})\mathbf{k}^*] \\ & - 2(\mathbf{k} \cdot \Psi^*)(\mathbf{k}^* \cdot \Psi)\mathbf{k} + (\mathbf{k}^* \cdot \Psi)(\mathbf{k} \cdot \mathbf{k})\Psi^*\end{aligned}$$

and then, by (5.2) and (5.3),

$$\begin{aligned}\langle \mathbf{J} \rangle = \tfrac{1}{2}\omega \exp(-2\mathbf{k}_2 \cdot \mathbf{x})\mathrm{Re}\{\mu[(\Psi \cdot \Psi^*)(\mathbf{k} \times (\mathbf{k} \times \mathbf{k}^*) + (\mathbf{k}^* \cdot \mathbf{k})\mathbf{k}) \\ - 2(\mathbf{k} \cdot \Psi^*)(\mathbf{k}^* \cdot \Psi)\mathbf{k} + (\mathbf{k}^* \cdot \Psi)(\mathbf{k} \cdot \mathbf{k})\Psi^*]\}.\end{aligned} \tag{5.5}$$

We now evaluate explicitly the real part. Observe preliminarily that

$$\begin{aligned}\rho\omega^2 = \mu \mathbf{k} \cdot \mathbf{k} =& \mu_1(k_1^2 + k_2^2) - 2\mu_1 k_2^2 - 2\mu_2 \mathbf{k}_1 \cdot \mathbf{k}_2 \\ & + i[2\mu_2 k_1^2 - \mu_2(k_1^2 + k_2^2) + 2\mu_1 \mathbf{k}_1 \cdot \mathbf{k}_2].\end{aligned}$$

Hence we obtain the relations

$$\mu_1(k_1^2 + k_2^2) = \rho\omega^2 + 2\mu_1 k_2^2 + 2\mu_2 \mathbf{k}_1 \cdot \mathbf{k}_2, \tag{5.6}$$

$$\mu_2(k_1^2 + k_2^2) = 2\mu_2 k_1^2 + 2\mu_1 \mathbf{k}_1 \cdot \mathbf{k}_2. \tag{5.7}$$

By use of (5.6) and (5.7) we obtain

$$\begin{aligned}\mathrm{Re}\{\mu(\mathbf{k} \cdot \mathbf{k}^*)\mathbf{k}\} &= \rho\omega^2 \mathbf{k}_1 + 2(\mu_1 k_2^2 + \mu_2 \mathbf{k}_1 \cdot \mathbf{k}_2)\mathbf{k}_1 - 2(\mu_2 k_1^2 + \mu_1 \mathbf{k}_1 \cdot \mathbf{k}_2)\mathbf{k}_2 \\ &= \rho\omega^2 \mathbf{k}_1 - 2(\mathbf{k}_1 \times \mathbf{k}_2) \times (\mu_2 \mathbf{k}_1 + \mu_1 \mathbf{k}_2).\end{aligned}$$

Meanwhile,

$$\mathrm{Re}\{\mu(\mathbf{k} \cdot \mathbf{k})\mathbf{k}^*\} = \rho\omega^2 \mathbf{k}_1.$$

Then we have

$$\mathrm{Re}\{\mu[2(\mathbf{k}\cdot\mathbf{k}^*)\mathbf{k}-(\mathbf{k}\cdot\mathbf{k})\mathbf{k}^*]\}=\rho\omega^2\mathbf{k}_1-4(\mathbf{k}_1\times\mathbf{k}_2)\times(\mu_2\mathbf{k}_1+\mu_1\mathbf{k}_2).$$

Let $\boldsymbol{\Psi}_1, \boldsymbol{\Psi}_2$ be the real and imaginary parts of $\boldsymbol{\Psi}$. The "orthogonality" condition $\mathbf{k}\cdot\boldsymbol{\Psi}=0$ reads

$$\mathbf{k}_1\cdot\boldsymbol{\Psi}_1-\mathbf{k}_2\cdot\boldsymbol{\Psi}_2=0, \qquad \mathbf{k}_2\cdot\boldsymbol{\Psi}_1+\mathbf{k}_1\cdot\boldsymbol{\Psi}_2=0. \tag{5.8}$$

Then we have

$$\mathrm{Re}\{(\mathbf{k}^*\cdot\boldsymbol{\Psi})\boldsymbol{\Psi}^*\}=2[(\mathbf{k}_1\cdot\boldsymbol{\Psi}_1)\boldsymbol{\Psi}_1+(\mathbf{k}_1\cdot\boldsymbol{\Psi}_2)\boldsymbol{\Psi}_2],$$

$$(\mathbf{k}\cdot\boldsymbol{\Psi}^*)(\mathbf{k}^*\cdot\boldsymbol{\Psi})=4[(\mathbf{k}_1\cdot\boldsymbol{\Psi}_1)^2+(\mathbf{k}_1\cdot\boldsymbol{\Psi}_2)^2].$$

Letting $\mathbf{f}_1=\boldsymbol{\Psi}_1/|\boldsymbol{\Psi}|$, $\mathbf{f}_2=\boldsymbol{\Psi}_2/|\boldsymbol{\Psi}|$, upon substitution in (5.5) we obtain

$$\begin{aligned}\langle\mathbf{J}\rangle=\tfrac{1}{2}\omega\exp(-2\mathbf{k}_2\cdot\mathbf{x})|\boldsymbol{\Psi}|^2\{&\rho\omega^2\mathbf{k}_1-4(\mathbf{k}_1\times\mathbf{k}_2)\times(\mu_2\mathbf{k}_1+\mu_1\mathbf{k}_2)\\ &-8[(\mathbf{k}_1\cdot\mathbf{f}_1)^2+(\mathbf{k}_1\cdot\mathbf{f}_2)^2](\mu_1\mathbf{k}_1-\mu_2\mathbf{k}_2)+2\rho\omega^2[(\mathbf{k}_1\cdot\mathbf{f}_1)\mathbf{f}_1+(\mathbf{k}_1\cdot\mathbf{f}_2)\mathbf{f}_2]\}.\end{aligned} \tag{5.9}$$

Inner multiplication by $\mathbf{n}_1$ yields the energy flux intensity in the form

$$\begin{aligned}\langle\mathcal{I}\rangle=\tfrac{1}{2}\omega\exp(-2\mathbf{k}_2\cdot\mathbf{x})\frac{|\boldsymbol{\Psi}|^2}{k_1}\{&\rho\omega^2k_1^2+4\mu_1[k_1^2k_2^2-(\mathbf{k}_1\cdot\mathbf{k}_2)^2]\\ &+[(\mathbf{k}_1\cdot\mathbf{f}_1)^2+(\mathbf{k}_1\cdot\mathbf{f}_2)^2][2\rho\omega^2-8(\mu_1k_1^2-\mu_2\mathbf{k}_1\cdot\mathbf{k}_2)]\}.\end{aligned}$$

The first line of (5.9) coincides with the expression given, e.g., by Buchen [26]. The second line is new in the literature. The reason is due to the fact that people regard letting $\boldsymbol{\Psi}$ be real as no loss in generality. As we show in §§4.1 and 4.5, such is not the case. Here we are content with remarking that if $\boldsymbol{\Psi}$ is real, namely $\boldsymbol{\Psi}_2=0$, then by (5.8) also $\mathbf{k}_1\cdot\boldsymbol{\Psi}_1=0$ and the contribution of the second line vanishes. However in general this assumption is not allowed.

3.6 Energy flux at inhomogeneous waves in fluids

General considerations for energy flux in fluids parallel those developed for solids. Accordingly, to avoid prolixity, here we restrict attention to the peculiar aspects for fluids. Indeed, via suitable interpretations of the pertinent quantities, our developments apply to both viscous and viscoelastic fluids.

With reference to the equation of motion (2.6.8) and the equilibrium condition, the equivalent (incremental) stress tensor for a viscoelastic fluid can be written as

$$\mathbf{T}(t)=\rho p_\rho(\nabla\cdot\mathbf{u}(t))\mathbf{1}+\int_0^\infty\{\tilde{\mu}(s)[\nabla\dot{\mathbf{u}}(t-s)+\nabla\dot{\mathbf{u}}^\dagger(t-s)]+\tilde{\lambda}(s)(\nabla\cdot\dot{\mathbf{u}})(t-s)\mathbf{1}\}ds.$$

For a time-harmonic wave, $\mathbf{u} = \mathbf{U}_0 \exp[i(\mathbf{k}\cdot\mathbf{x} - \omega t)]$, we have

$$\mathbf{T} = \{i\rho p_\rho(\mathbf{k}\cdot\mathbf{U}_0)\mathbf{1} + \omega[\mu(\mathbf{k}\otimes\mathbf{U}_0 + \mathbf{U}_0\otimes\mathbf{k}) + \lambda(\mathbf{k}\cdot\mathbf{U}_0)\mathbf{1}]\}\exp[i(\mathbf{k}\cdot\mathbf{x} - \omega t)]$$

where, as usual, $\mu = \int_0^\infty \tilde{\mu}(s)\exp(i\omega s)ds$, $\lambda = \int_0^\infty \tilde{\lambda}(s)\exp(i\omega s)ds$. Viscous fluids are described by simply letting $\mu = \mu_0$, $\lambda = \lambda_0$, whereby μ, λ are real-valued.

In order to apply (5.2) for the mean energy flux we have to evaluate $\mathbf{Tu}^*$. Substitution and some rearrangement yield

$$\mathbf{Tu}^* = i\exp(-2\mathbf{k}_2\cdot\mathbf{x})\{\rho p_\rho(\mathbf{k}\cdot\mathbf{U}_0)\mathbf{U}_0^* - i\omega[\mu(\mathbf{U}_0\cdot\mathbf{U}_0^*)\mathbf{k} + \mu(\mathbf{k}\cdot\mathbf{U}_0^*)\mathbf{U}_0 + \lambda(\mathbf{k}\cdot\mathbf{U}_0)\mathbf{U}_0^*]\}. \quad (6.1)$$

As with solids, the evaluation of (6.1) changes accordingly as we are dealing with longitudinal or transverse waves.

Consider longitudinal waves, namely

$$\mathbf{k}\cdot\mathbf{k} = \frac{\rho\omega^2}{\rho p_\rho - i\omega(2\mu + \lambda)}, \qquad \mathbf{U}_0 = i\mathbf{k}\Phi.$$

Then (6.1) can be written as

$$\mathbf{Tu}^* = i\exp(-2\mathbf{k}_2\cdot\mathbf{x})|\Phi|^2[\rho\omega^2\mathbf{k}^* + 2i\omega\mu\mathbf{k}\times(\mathbf{k}^*\times\mathbf{k})]$$

and, by (5.2),

$$\langle\mathbf{J}\rangle = \tfrac{1}{2}\omega\exp(-2\mathbf{k}\cdot\mathbf{x})|\Phi|^2\mathrm{Re}\{\rho\omega^2\mathbf{k}^* + 2i\omega\mu\mathbf{k}\times(\mathbf{k}^*\times\mathbf{k})\}.$$

Direct evaluation of the real part allows $\langle\mathbf{J}\rangle$ to be written as

$$\langle\mathbf{J}\rangle = \tfrac{1}{2}\omega|\Phi|^2\exp(-2\mathbf{k}_2\cdot\mathbf{x})[\rho\omega^2\mathbf{k}_1 + 4\omega(\mu_2\mathbf{k}_2 - \mu_1\mathbf{k}_1)\times(\mathbf{k}_1\times\mathbf{k}_2)].$$

Correspondingly by (5.1) the mean energy flux intensity takes the form

$$\langle\mathcal{I}\rangle = \tfrac{1}{2}\omega|\Phi|^2\exp(-2\mathbf{k}_2\cdot\mathbf{x})[\rho\omega^2 k_1 + 4\omega\mu_2 k_1 k_2^2\sin^2\gamma] \quad (6.2)$$

where γ is the angle between $\mathbf{k}_1$ and $\mathbf{k}_2$.

If the fluid is viscoelastic, $\mu_1 > 0$, $\mu_2 \neq 0$, then the energy flux usually is not directed along $\mathbf{k}_1$ and the energy flux intensity depends on both k_1 and k_2. If the fluid is viscous, $\mu_1 = \mu_0, \mu_2 = 0$, then still the energy flux is not directed along $\mathbf{k}_1$, i.e.

$$\langle\mathbf{J}\rangle = \tfrac{1}{2}\omega|\Phi|^2\exp(-2\mathbf{k}_2\cdot\mathbf{x})[\rho\omega^2\mathbf{k}_1 - 4\omega\mu_0\mathbf{k}_1\times(\mathbf{k}_1\times\mathbf{k}_2)],$$

but the energy flux intensity is affected by the component along $\mathbf{k}_1$ only, that is

$$\langle\mathcal{I}\rangle = \tfrac{1}{2}\omega|\Phi|^2\exp(-2\mathbf{k}_2\cdot\mathbf{x})\rho\omega^2 k_1.$$

Further, if the fluid is non-dissipative, $\mu_1 = 0$, $\mu_2 = 0$, then we have

$$\langle \mathcal{I} \rangle = \tfrac{1}{2}\rho\omega^3|\Phi|^2 k_1 = \tfrac{1}{2}\rho c\omega^2|\mathbf{U}_0|^2$$

which formally is the same as for elastic solids, c being the sound speed, $c^2 = p_\rho$ (cf. [116]).

Consider transverse waves, namely

$$\mathbf{k}\cdot\mathbf{k} = \frac{i\rho\omega}{\mu}, \qquad \mathbf{U}_0 = i\mathbf{k}\times\Psi, \qquad \mathbf{k}\cdot\Psi = 0.$$

Obviously $\mathbf{k}\cdot\mathbf{U}_0 = 0$. Then we have

$$\mathbf{T}\mathbf{u}^* = \omega\exp(-2\mathbf{k}_2\cdot\mathbf{x})\mu[(\mathbf{U}_0\cdot\mathbf{U}_0^*)\mathbf{k} + (\mathbf{k}\cdot\mathbf{U}_0^*)\mathbf{U}_0].$$

Hence, by (5.2), on paralleling the procedure developed for solids we obtain

$$\begin{aligned}\langle \mathbf{J} \rangle = \tfrac{1}{2}\omega^2\exp(-2\mathbf{k}_2\cdot\mathbf{x})\mathrm{Im}\{\mu[(\Psi\cdot\Psi^*)(2(\mathbf{k}\cdot\mathbf{k}^*)\mathbf{k} - (\mathbf{k}\cdot\mathbf{k})\mathbf{k}^*)\\ - 2(\mathbf{k}\cdot\Psi^*)(\mathbf{k}^*\cdot\Psi)\mathbf{k} + (\mathbf{k}^*\cdot\Psi)(\mathbf{k}\cdot\mathbf{k})\Psi^*]\}.\end{aligned} \tag{6.3}$$

Observe that

$$\begin{aligned}i\rho\omega = \mu\mathbf{k}\cdot\mathbf{k} = & -\mu_1(k_1^2 + k_2^2) + 2\mu_1 k_1^2 - 2\mu_2\mathbf{k}_1\cdot\mathbf{k}_2\\ & + i[\mu_2(k_1^2 + k_2^2) - 2\mu_2 k_2^2 + 2\mu_1\mathbf{k}_1\cdot\mathbf{k}_2]\end{aligned}$$

whence

$$\mu_1(k_1^2 + k_2^2) = 2(\mu_1 k_1^2 - \mu_2\mathbf{k}_1\cdot\mathbf{k}_2), \tag{6.4}$$

$$\mu_2(k_1^2 + k_2^2) = \rho\omega + 2(\mu_2 k_2^2 - \mu_1\mathbf{k}_1\cdot\mathbf{k}_2). \tag{6.5}$$

By use of (6.4), (6.5) and some rearrangement we have

$$\begin{aligned}\mathrm{Im}[\mu(\mathbf{k}\cdot\mathbf{k}^*)\mathbf{k}] &= 2(\mu_1 k_1^2 - \mu_2\mathbf{k}_1\cdot\mathbf{k}_2)\mathbf{k}_2 + \rho\omega\mathbf{k}_1 + 2(\mu_2 k_2^2 - \mu_1\mathbf{k}_1\cdot\mathbf{k}_2)\mathbf{k}_1\\ &= \rho\omega\mathbf{k}_1 + 2(\mathbf{k}_1\times\mathbf{k}_2)(\mu_1\mathbf{k}_1 - \mu_2\mathbf{k}_2).\end{aligned}$$

Substitution in (6.3), replacing $\mu\mathbf{k}\cdot\mathbf{k}$ with $i\rho\omega$, introducing $\mathbf{f}_1 = \Psi_1/|\Psi|$, $\mathbf{f}_2 = \Psi_2/|\Psi|$, and some rearrangement yield

$$\begin{aligned}\langle \mathbf{J} \rangle = \tfrac{1}{2}\omega^2\exp(-2\mathbf{k}_2\cdot\mathbf{x})|\Psi|^2\{&\rho\omega\mathbf{k}_1 + 4(\mathbf{k}_1\times\mathbf{k}_2)\times(\mu_1\mathbf{k}_1 - \mu_2\mathbf{k}_2)\\ &- 8[(\mathbf{k}_1\cdot\mathbf{f}_1)^2 + (\mathbf{k}_1\cdot\mathbf{f}_2)^2](\mu_1\mathbf{k}_2 + \mu_2\mathbf{k}_1)\\ &+ 2\rho\omega[(\mathbf{k}_1\cdot\mathbf{f}_1)\mathbf{f}_1 + (\mathbf{k}_1\cdot\mathbf{f}_2)\mathbf{f}_2]\}\end{aligned}$$

where, of course, the condition $\mathbf{k}\cdot\Psi = 0$ has been taken into account. This result is formally similar to the analogous one for solids. Inner multiplication by $\mathbf{n}_1$ provides the mean energy flux intensity $\langle \mathcal{I} \rangle$.

For viscous fluids $\mu_2 = 0$ and the first line consists of a term parallel to $\mathbf{k}_1$ and a term orthogonal to $\mathbf{k}_1$. Then the contribution to $\langle \mathcal{I} \rangle$ is simply $\rho\omega k_1$ times the common factor $\frac{1}{2}\omega^2 \exp(-2\mathbf{k}_2 \cdot \mathbf{x})|\mathbf{\Psi}|^2$. The contribution of the terms in the second line is related to the component of $\mathbf{\Psi}$ in the direction of propagation $\mathbf{n}_1$ and vanishes if $\mathbf{\Psi} \cdot \mathbf{n}_1 = 0$.

3.7 Waves in constrained media

By constrained media we mean materials whose motion is subject to a priori constraints usually called internal constraints. Two common examples of internal constraints are inextensibility in one direction and incompressibility. If $\mathbf{e}$ is the direction of inextensibility in the reference placement then the relation $d\mathbf{x} = \mathbf{F}d\mathbf{X}$ provides the constraint of inextensibility in the form

$$\mathbf{Fe} \cdot \mathbf{Fe} - 1 = 0. \tag{7.1}$$

If instead the material is incompressible then by the continuity equation $\rho = \rho_0 / J$ we have the constraint

$$\det \mathbf{F} - 1 = 0. \tag{7.2}$$

More generally, regard the material as subject to a set of internal constraints

$$\Gamma^i(\mathbf{F}) = 0, \tag{7.3}$$

Γ^i denoting a, possibly nonlinear, scalar-valued function. By objectivity requirements, the value of any function (7.3) must be invariant under change of observer ([164], §30). This is easily shown to imply that Γ^i depends on $\mathbf{F}$ through $\mathbf{C} = \mathbf{F}^\dagger\mathbf{F}$ only. Then we can write (7.3) as

$$\gamma^i(\mathbf{C}) = 0. \tag{7.3'}$$

Incidentally, as it must be, (7.1) and (7.2) may be written in the form (7.3'), namely

$$\mathbf{e} \cdot \mathbf{Ce} - 1 = 0, \qquad \det \mathbf{C} - 1 = 0.$$

It is reasonable to expect that these restrictions on the motion shall not specify the strain. By (7.3') it is evident that this is the case if $i \leq 5$.

Whether the material is dissipative or not, the constraint (7.3'), or (7.3), gives rise to internal Piola-Kirchhoff stresses

$$\hat{\mathbf{S}}^i = \tfrac{1}{2}\alpha^i \frac{\partial \Gamma^i}{\partial \mathbf{F}} = \alpha^i \mathbf{F} \frac{\partial \gamma^i}{\partial \mathbf{C}}$$

which individually do no work in that, by (7.3), $(\partial \Gamma^i / \partial \mathbf{F}) \cdot \dot{\mathbf{F}} = 0$. The corresponding Cauchy stresses are

$$\hat{\mathbf{T}}^i = \alpha^i J^{-1} \mathbf{F} \frac{\partial \gamma^i}{\partial \mathbf{C}} \mathbf{F}^\dagger$$

which are apparently symmetric. The quantities α^i (Lagrange multipliers) are, so far, indeterminate. They are in fact functions of the position $\mathbf{x}$ and time t which are required to satisfy the equations of motion and the pertinent boundary conditions.

Again on disregarding the body force, we consider the equation of motion (1.3) which now has to be written in the form

$$\rho\ddot{\mathbf{u}} = \nabla\cdot\mathbf{T} + \sum_i \nabla\cdot\hat{\mathbf{T}}^i. \tag{7.4}$$

The reader interested in general properties of the solution $\mathbf{u}, \alpha^i$ to (7.4) is referred to, e.g., [88, 47, 153, 18]. Here, for the sake of definiteness, attention is focussed on a singly constrained body and it is assumed that the inextensibility constraint (7.1) holds. Observe that $\partial\gamma/\partial\mathbf{C} = \mathbf{e}\otimes\mathbf{e}$ and then

$$\hat{\mathbf{T}} = \alpha J^{-1}\mathbf{F}\mathbf{e}\otimes\mathbf{F}\mathbf{e}. \tag{7.5}$$

Take the body to be a viscoelastic solid and look for inhomogeneous wave solutions. While the constitutive stress is given by (1.2), we take the reactive stress (7.5) in the linear approximation as

$$\hat{\mathbf{T}} = \alpha[\mathbf{e}\otimes\mathbf{e} - (\nabla\cdot\mathbf{u})\mathbf{e}\otimes\mathbf{e} + \nabla\mathbf{u}^\dagger\mathbf{e}\otimes\mathbf{e} + \mathbf{e}\otimes\nabla\mathbf{u}^\dagger\mathbf{e}]$$

whence

$$\nabla\cdot\hat{\mathbf{T}} = (\mathbf{e}\cdot\nabla\alpha)\mathbf{e} + \alpha(\mathbf{e}\cdot\nabla)(\mathbf{e}\cdot\nabla)\mathbf{u}.$$

Upon substitution into the equation of motion (7.4) and letting $\mathbf{u}$ be in the form of inhomogeneous wave, $\mathbf{u} = \mathbf{U}_0\exp[i(\mathbf{k}\cdot\mathbf{x} - \omega t)]$, along with $\alpha = \alpha_0 + \hat{\alpha}_0\exp[i(\mathbf{k}\cdot\mathbf{x} - \omega t)]$ we obtain

$$\mathbf{Q}(\mathbf{k})\mathbf{U}_0 + \alpha_0(\mathbf{e}\cdot\mathbf{k})^2\mathbf{U}_0 - i(\mathbf{e}\cdot\mathbf{k})\mathbf{e}\hat{\alpha}_0 - \rho\omega^2\mathbf{U}_0 = 0. \tag{7.6}$$

Here

$$\mathbf{Q}(\mathbf{k}) = \mu(\mathbf{k}\cdot\mathbf{k})\mathbf{1} + (\mu+\lambda)\mathbf{k}\otimes\mathbf{k}$$

may be viewed as the analogue of the standard acoustic tensor. Observe that, because of the constraint, the propagation condition (7.6) results in the vanishing of a combination of the three vectors $\mathbf{U}_0, \mathbf{k}, \mathbf{e}$. Then, quite naturally, the internal constraint reduces the set of solutions. We examine two non-trivial cases.

I. $\mathbf{e}\cdot\mathbf{k} = 0$.

The wave vector is orthogonal to $\mathbf{e}$, namely to the direction of inextensibility. The transverse isotropy of the body suggests that we find the same result as for waves in unconstrained media. Such is really the case, i. e.,

$$[\mathbf{Q}(\mathbf{k}) - \rho\omega^2\,\mathbf{1}]\mathbf{U}_0 = 0,$$

whence we find the standard transverse and longitudinal wave solutions.

II. $\mathbf{e} \cdot \mathbf{U}_0 = 0$.

The polarization is orthogonal to the direction of inextensibility. It is convenient to take the inner product of (7.6) with $\mathbf{e}$,

$$\mathbf{e} \cdot \mathbf{Q}(\mathbf{k})\mathbf{U}_0 - i(\mathbf{e} \cdot \mathbf{k})\hat{\alpha}_0 = 0,$$

and determine $\hat{\alpha}_0$. Upon substitution we can write the propagation condition as

$$\{\mathbf{P}\mathbf{Q}(\mathbf{k}) - [\rho\omega^2 - \alpha_0(\mathbf{e} \cdot \mathbf{k})^2]\mathbf{1}\}\mathbf{U}_0 = 0$$

where $\mathbf{P} = \mathbf{1} - \mathbf{e} \otimes \mathbf{e}$ is the projection onto the plane perpendicular to $\mathbf{e}$.

In both cases we find formally the same results as for elastic bodies (cf. [50]), the difference being that now $\mathbf{Q}$ is complex-valued and so are $\mathbf{k}$ and $\mathbf{U}_0$.

3.8 Body force effects on waves

Wave propagation is usually investigated by disregarding the body force in the equation of motion. Really, from a qualitative standpoint, the body force has a twofold effect. First, it induces an equilibrium placement through the relation

$$\nabla \cdot \mathbf{T} + \rho \mathbf{b} = 0.$$

Then inevitably, whenever a body force occurs, we should consider a prestressed body (in the sense of §2.6). Second, even when $\mathbf{b}$ is constant, the body force enters the propagation condition thus affecting (in fact, reducing) the set of wave solutions. Both aspects are now emphasised by investigating wave propagation in dissipative fluids (cf. [34]).

Regard the body force $\mathbf{b}$ as constant and then apply the equation of motion in the form (2.6.9), namely

$$\rho\ddot{\mathbf{u}} - \rho\nabla\mathbf{u}\,\mathbf{b} - \frac{\rho^2 p_{\rho\rho}}{p_\rho}\mathbf{b}\nabla \cdot \mathbf{u} - \rho p_\rho \nabla(\nabla \cdot \mathbf{u}) - \nabla \cdot \boldsymbol{\mathcal{T}}(\mathbf{D}^t) = 0. \tag{8.1}$$

Take $\boldsymbol{\mathcal{T}}(\mathbf{E}^t)$ as the functional for viscoelastic fluids. Then, letting $\mathbf{u} = \mathbf{U}_0 \exp[i(\mathbf{k} \cdot \mathbf{x} - \omega t)]$, we have

$$\boldsymbol{\mathcal{T}}(\mathbf{D}^t) = \omega[\mu(\mathbf{k} \otimes \mathbf{U}_0 + \mathbf{U}_0 \otimes \mathbf{k}) + \lambda(\mathbf{k} \cdot \mathbf{U}_0)\mathbf{1}] \exp[i(\mathbf{k} \cdot \mathbf{x} - \omega t)]$$

where $\mu = \int_0^\infty \tilde{\mu}(s) \exp(i\omega s) ds$, $\lambda = \int_0^\infty \tilde{\lambda}(s) \exp(i\omega s) ds$, and

$$\nabla \cdot \boldsymbol{\mathcal{T}}(\mathbf{D}^t) = i\omega[\mu(\mathbf{k} \cdot \mathbf{k})\mathbf{U}_0 + (\mu + \lambda)(\mathbf{k} \cdot \mathbf{U}_0)\mathbf{k}] \exp[i(\mathbf{k} \cdot \mathbf{x} - \omega t)].$$

Then (8.1) yields the propagation condition in the form

$$[\rho\omega^2 + i\omega\mu\mathbf{k}\cdot\mathbf{k}]\mathbf{U}_0 + \{[i\omega(\mu+\lambda) - \rho p_\rho]\mathbf{k}\cdot\mathbf{U}_0 + i\rho\mathbf{U}_0\cdot\mathbf{b}\}\mathbf{k} + \frac{\rho^2 p_{\rho\rho}}{p_\rho}(\mathbf{k}\cdot\mathbf{U}_0)\mathbf{b} = 0. \quad (8.2)$$

Observe that if the fluid is viscous the propagation condition (8.2) still holds while μ and λ are the viscosity coefficients.

Incidentally, we expect that ρ and $p(\rho)$ depend on the position $\mathbf{x}$ in the fluid. By the equilibrium equation $p_\rho \nabla\rho = \rho\mathbf{b}$ we have $\rho/|\nabla\rho| = p_\rho/|\mathbf{b}|$. In water, for instance, $p_\rho = 2\,10^6$ m^2/s^2 and then $\rho/|\nabla\rho| \simeq 2\,10^5$m. For standard experimental conditions it is then reasonable to regard ρ in (8.2) as constant.

Investigate the possible solutions to (8.2) by considering first the case when $\mathbf{U}_0\cdot\mathbf{k}$ vanishes.

I. $\mathbf{k}\cdot\mathbf{U}_0 = 0$.

It follows from (8.2) that

$$[\rho\omega^2 + i\omega\mu\mathbf{k}\cdot\mathbf{k}]\mathbf{U}_0 + i\rho(\mathbf{b}\cdot\mathbf{U}_0)\mathbf{k} = 0$$

whence, by the orthogonality of $\mathbf{k}$ and $\mathbf{U}_0$,

$$\mathbf{k}\cdot\mathbf{k} = \frac{i\rho\omega}{\mu}, \qquad \mathbf{b}\cdot\mathbf{U}_0 = 0. \quad (8.3)$$

Quite naturally we regard this solution as a transverse wave in that any inhomogeneous wave satisfying (8.3) is necessarily transverse. In fact inner multiplication of (8.2) by $\mathbf{U}_0$ and use of $(8.3)_1$ yield

$$\big[-(\rho p_\rho - i\omega(\mu+\lambda))\mathbf{k}\cdot\mathbf{U}_0 + i\rho\big(1+\rho\frac{p_{\rho\rho}}{p_\rho}\big)\mathbf{b}\cdot\mathbf{U}_0\big]\mathbf{k}\cdot\mathbf{U}_0 = 0,$$

whence it follows that either

$$\mathbf{k}\cdot\mathbf{U}_0 = 0$$

or

$$\mathbf{b}\cdot\mathbf{U}_0 = \frac{\rho p_\rho - i\omega(\mu+\lambda)}{i\rho(1+\rho p_{\rho\rho}/p_\rho)}\mathbf{k}\cdot\mathbf{U}_0.$$

In the former case, substitution into (8.2) yields $(8.3)_2$. In the latter case, instead, (8.2) reduces to

$$[-\rho p_\rho + i\omega(\mu+\lambda)]\mathbf{k} + \rho\big(1+\rho\frac{p_{\rho\rho}}{p_\rho}\big)\mathbf{b} = 0.$$

Inner multiplication by $\mathbf{k}$ and $\mathbf{b}$, and comparison of the results leads to the requirement

$$\Big[\rho\frac{1+\rho p_{\rho\rho}/p_\rho}{\rho p_\rho - i\omega(\mu+\lambda)}\Big]^2\mathbf{b}^2 = i\frac{\rho\omega}{\mu},$$

which generally does not hold. Then we conclude that only $\mathbf{k}\cdot\mathbf{U}_0 = 0$ is possible and then the wave is transverse.

Two possibilities occur accordingly as $\mathbf{k}$ is parallel to $\mathbf{b}$ or is not. If $\mathbf{k}\times\mathbf{b} = 0$ then any $\mathbf{U}_0$ orthogonal to $\mathbf{b}$ satisfies $(8.3)_2$ - and (I) as well. If $\mathbf{k}\times\mathbf{b} \neq 0$ then $(8.3)_2$ and (I) yield

$$\mathbf{U}_0 = \zeta\mathbf{k}\times\mathbf{b}$$

where ζ is any complex number.

As a comment we can say that, whenever $\mathbf{k}\cdot\mathbf{U}_0 = 0$, the wave vector $\mathbf{k}$ is unaffected by $\mathbf{b}$. Yet, by $(8.3)_2$ the wave exists only if $\mathbf{U}_0$ and $\mathbf{b}$ are orthogonal.

II. $\mathbf{k}\cdot\mathbf{U}_0 \neq 0$.

Inner multiplication of (8.2) by $\mathbf{k}$ and $\mathbf{b}$ yields two equations which may be viewed as a linear homogeneous system in the unknowns $\mathbf{U}_0\cdot\mathbf{k}$ and $\mathbf{U}_0\cdot\mathbf{b}$. The determinantal equation, which allows non-trivial solutions, is

$$\det\begin{pmatrix} \rho\omega^2 - [\rho p_\rho - i\omega(2\mu+\lambda)]\mathbf{k}\cdot\mathbf{k} + i\rho^2(p_{\rho\rho}/p_\rho)\mathbf{b}\cdot\mathbf{k} & i\rho\mathbf{k}\cdot\mathbf{k} \\ i\rho^2(p_{\rho\rho}/p_\rho)\mathbf{b}^2 - [\rho p_\rho - i\omega(\mu+\lambda)]\mathbf{b}\cdot\mathbf{k} & \rho\omega^2 + i\omega\mu\mathbf{k}\cdot\mathbf{k} + i\rho\mathbf{b}\cdot\mathbf{k} \end{pmatrix} = 0. \tag{8.4}$$

Consistent with the physical framework, we regard the unit vectors $\mathbf{n}_1, \mathbf{n}_2$ of $\mathbf{k}_1, \mathbf{k}_2$ as given and then (8.4) becomes a system of two equations - the left-hand side of (8.4) is complex-valued - in the unknowns k_1, k_2. Once k_1 and k_2, and then $\mathbf{k}$, are determined we can find $\mathbf{U}_0\cdot\mathbf{b}$ in terms of $\mathbf{U}_0\cdot\mathbf{k}$ namely

$$\mathbf{U}_0\cdot\mathbf{b} = A\mathbf{U}_0\cdot\mathbf{k}$$

where

$$A = -\frac{\rho\omega^2 - [\rho p_\rho + i\omega(\lambda+2\mu)]\mathbf{k}\cdot\mathbf{k} + i\rho^2(p_{\rho\rho}/p_\rho)\mathbf{b}\cdot\mathbf{k}}{i\rho\mathbf{k}\cdot\mathbf{k}}.$$

Substitution in (8.2) yields

$$\mathbf{U}_0 = \frac{\mathbf{U}_0\cdot\mathbf{k}}{\rho\omega^2 + i\omega\mu\mathbf{k}\cdot\mathbf{k}}\{[\rho p_\rho - i\omega(\lambda+\mu) - i\rho A]\mathbf{k} - i\rho^2\frac{p_{\rho\rho}}{p_\rho}\mathbf{b}\}. \tag{8.5}$$

Since $\mathbf{U}_0$ is determined by (8.5) to within a scalar factor, we can always assume that $\mathbf{U}_0\cdot\mathbf{k} = 1$ and then (8.5) is the desired relation $\mathbf{U}_0 = \mathbf{U}_0(\mathbf{k})$.

In addition to these geometrical-analytical features for a dissipative fluid, it may be instructive to examine what happens in a perfect fluid for which $\mu, \lambda = 0$. A simple, immediate effect of the body force appears by considering the wave vector $\mathbf{k}$ in the plane orthogonal to $\mathbf{b}$, i.e. $\mathbf{k}\cdot\mathbf{b} = 0$. The determinantal equation (8.4) reduces to

$$\det\begin{pmatrix} \rho\omega^2 - \rho p_\rho\mathbf{k}\cdot\mathbf{k} & i\rho\mathbf{k}\cdot\mathbf{k} \\ i\rho^2(p_{\rho\rho}/p_\rho)\mathbf{b}^2 & \rho\omega^2 \end{pmatrix} = 0$$

whence

$$\mathbf{k}\cdot\mathbf{k} = \frac{\omega^2/p_\rho}{1 - \rho p_{\rho\rho}\mathbf{b}^2/p_\rho^2\omega^2}.$$

Neglecting the body force would have given $\mathbf{k}\cdot\mathbf{k} = \omega^2/p_\rho$. Then, if $p_{\rho\rho} > 0$, the body force results in a decreasing of the phase speed; the speed vanishes at the critical frequency ω_c such that $\omega_c^2 = \rho p_{\rho\rho}\mathbf{b}^2/p_\rho^2$. For water, if we consider the constitutive function $p(\rho)$ given by the Tait equation [103] along the adiabatic passing through 1 atmosphere and 20° C we have $\rho p_{\rho\rho}/p_\rho = 6$. Then $\omega_c \simeq 10^{-2}$ s^{-1}, a negligibly small value. If the fluid is allowed to be viscous, or viscoelastic, the determinantal equation becomes

$$(\rho\omega^2)^2 - \rho\omega^2[\rho p_\rho - i\omega(2\mu+\lambda) - \rho^2 p_{\rho\rho}\mathbf{b}^2/p_\rho\omega^2]\mathbf{k}\cdot\mathbf{k} - i\omega\mu[\rho p_\rho - i\omega(2\mu+\lambda)](\mathbf{k}\cdot\mathbf{k})^2 = 0$$

and the same conclusion for the critical frequency follows. Of course the quantitative effect on the phase speed may not be small in other fluids.

4 REFLECTION AND REFRACTION

Reflection and refraction constitute the fundamental phenomenon that is at the basis of the behaviour of waves at discontinuity interfaces. This motivates a detailed analysis of reflection and refraction of inhomogeneous waves at a plane interface. Of course, the general case occurs when the plane of the incident wave vector is not orthogonal to the interface. The behaviour of inhomogeneous waves at a plane interface is investigated by letting the interface be the boundary of a viscoelastic (solid) half-space, or the common boundary of two viscoelastic solids or two layers of a multilayered solid; this last case is of remarkable interest in seismology. Relative to the particular case when the plane of the incident wave vector is orthogonal to the interface, new effects are shown in connection with the polarization of the reflected and transmitted waves.

By generalizing a standard approach for wave propagation in elastic solids, the displacement field is represented in terms of complex potentials thus allowing a concise description of phenomena. To unify the treatment of incident longitudinal and transverse waves, the incident field is taken in the form of a conjugate pair, namely the superposition of a longitudinal and a transverse wave whose wave vectors have equal projections on the interface. In this framework the matrices describing the reflection and refraction can be derived in a straightforward way for any interface. Of course, owing to the complex nature of the polarizations of the pertinent waves, the determination of reflection and refraction coefficients deserves some attention.

As with the interface between elastic media, the refraction coefficients may become greater than unity for certain directions of the incident wave, usually around critical angles. This looks paradoxical also on the basis of an intuitive idea of energy conservation. In fact the seeming paradox is overcome by considering carefully the energy flux intensities of the pertinent waves; for definiteness this is performed in the case of a perfect fluid-viscoelastic solid interface. In essence, for any interface the energy flux intensity involves the cross section and the ratios of the cross sections approaches zero, or infinity, when the angle of the incident wave vector is around the critical value.

4.1 Coordinate representations for displacement and traction

As a preliminary step we derive expressions for displacement and traction that prove useful in the description of the behaviour of inhomogeneous waves at interfaces. Specifically, let

$\mathbf{u} = \mathbf{U}\exp(-i\omega t)$. By the existence of scalar and vector potentials in linear viscoelasticity we may represent $\mathbf{U}$ as

$$\mathbf{U} = \nabla\phi + \nabla \times \boldsymbol{\psi}, \qquad \nabla \cdot \boldsymbol{\psi} = 0,$$

where the potentials ϕ and $\boldsymbol{\psi}$, which are identified with M and $\mathbf{W}$ of the previous chapter, satisfy the complex Helmholtz equations

$$\Delta\phi + \frac{\rho\omega^2}{2\mu+\lambda}\phi = 0, \qquad \Delta\boldsymbol{\psi} + \frac{\rho\omega^2}{\mu}\boldsymbol{\psi} = 0. \tag{1.1}$$

Let x, y, z be Cartesian coordinates with unit vectors $\mathbf{e}_x$, $\mathbf{e}_y$, $\mathbf{e}_z$. The coordinate representation of $\mathbf{U}$ reads

$$\mathbf{U} = \Big(\frac{\partial\phi}{\partial x} + \frac{\partial\psi_z}{\partial y} - \frac{\partial\psi_y}{\partial z}\Big)\mathbf{e}_x + \Big(\frac{\partial\phi}{\partial y} + \frac{\partial\psi_x}{\partial z} - \frac{\partial\psi_z}{\partial x}\Big)\mathbf{e}_y + \Big(\frac{\partial\phi}{\partial z} + \frac{\partial\psi_y}{\partial x} - \frac{\partial\psi_x}{\partial y}\Big)\mathbf{e}_z. \tag{1.2}$$

In a moment we need the components of the traction $\mathbf{t} = \mathbf{T}\,\mathbf{n}$, $\mathbf{n} = -\mathbf{e}_z$. To within the exponential $\exp(-i\omega t)$ we have

$$\mathbf{T} = \mu(\nabla\mathbf{U} + \nabla\mathbf{U}^\dagger) + \lambda(\nabla\cdot\mathbf{U})\mathbf{1}.$$

Substitution of (1.2), comparison with (1.1), and the condition of vanishing divergence for $\boldsymbol{\psi}$ lead to the representation

$$\mathbf{t} = \Big[-\rho\omega^2\psi_y - 2\mu\Big(\frac{\partial^2\phi}{\partial x\partial z} - \frac{\partial^2\psi_x}{\partial x\partial y} + \frac{\partial^2\psi_y}{\partial x^2}\Big)\Big]\mathbf{e}_x + \Big[\rho\omega^2\psi_x - 2\mu\Big(\frac{\partial^2\phi}{\partial y\partial z} - \frac{\partial^2\psi_x}{\partial y^2} + \frac{\partial^2\psi_y}{\partial x\partial y}\Big)\Big]\mathbf{e}_y$$
$$+ \Big[\rho\omega^2\phi + 2\mu\Big(\frac{\partial^2\phi}{\partial x^2} + \frac{\partial^2\phi}{\partial y^2} + \frac{\partial^2\psi_x}{\partial y\partial z} - \frac{\partial^2\psi_y}{\partial x\partial z}\Big)\Big]\mathbf{e}_z. \tag{1.3}$$

For perfect fluids we have $\boldsymbol{\psi} = 0$ and $\mu = 0$, whence $t_x = t_y = 0$ and $t_z = \rho\omega^2\phi$.

Inhomogeneous waves correspond to special choices of the potentials. Longitudinal waves are generated by a scalar potential of the form $\phi = \Phi\exp(i\mathbf{k}_L\cdot\mathbf{x})$, Φ being the (complex) constant amplitude. As we know, the wave vector $\mathbf{k}_L$ is such that

$$\mathbf{k}_L\cdot\mathbf{k}_L = \frac{\rho\omega^2}{2\mu+\lambda} =: \kappa_L, \qquad \mathbf{U} = i\mathbf{k}_L\phi. \tag{1.4}$$

An inhomogeneous transverse wave comes from the vector potential $\boldsymbol{\psi} = \boldsymbol{\Psi}\exp(i\mathbf{k}_T\cdot\mathbf{x})$ where $\boldsymbol{\Psi}$ is the complex (constant) amplitude vector, and the complex wave vector $\mathbf{k}_T$ is such that

$$\mathbf{k}_T\cdot\mathbf{k}_T = \frac{\rho\omega^2}{\mu} =: \kappa_T, \qquad \mathbf{U} = i\mathbf{k}_T \times \boldsymbol{\psi}. \tag{1.5}$$

The divergence-free condition $\nabla\cdot\boldsymbol{\psi} = 0$ yields the constraint

$$\boldsymbol{\Psi}\cdot\mathbf{k}_T = 0. \tag{1.6}$$

For ease in writing we sometimes consider k_L and k_T such that

$$k_L = \sqrt{\kappa_L}, \qquad k_T = \sqrt{\kappa_T}$$

where the square root is meant with positive real part.

It might seem that letting Φ and Ψ be complex-valued is an unnecessary generality; there are references where the amplitude is complex-valued [89], others where it is real-valued [17, 26]. Now, regardless of the properties of the sources, we may consider incident waves with real-valued amplitudes. The trouble is that, as shown later, the real-valuedness is not invariant under reflection and refraction and this makes the real-valuedness assumption generally inconsistent.

By (1.4) and (1.5) the expressions (1.2) and (1.3) for $\mathbf{U}$ and $\mathbf{t}$ become

$$\begin{aligned} \mathbf{U} &= i\Phi(k_{Lx}\mathbf{e}_x + k_{Ly}\mathbf{e}_y + k_{Lz}\mathbf{e}_z)\exp(i\mathbf{k}_L\cdot\mathbf{x}) \\ &+ i\big[(k_{Ty}\Psi_z - k_{Tz}\Psi_y)\mathbf{e}_x + (k_{Tz}\Psi_x - k_{Tx}\Psi_z)\mathbf{e}_y + (k_{Tx}\Psi_y - k_{Ty}\Psi_x)\mathbf{e}_z\big]\exp(i\mathbf{k}_T\cdot\mathbf{x}), \end{aligned} \tag{1.7}$$

$$\begin{aligned} \mathbf{t} &= 2\mu\Phi\big[k_{Lz}(k_{Lx}\mathbf{e}_x + k_{Ly}\mathbf{e}_y) + (\tfrac{1}{2}\kappa_T - k_{Lx}^2 - k_{Ly}^2)\mathbf{e}_z\big]\exp(i\mathbf{k}_L\cdot\mathbf{x}) \\ &+ 2\mu\big[\tfrac{1}{2}\kappa_T(-\Psi_y\mathbf{e}_x + \Psi_x\mathbf{e}_y) - (k_{Ty}\Psi_x - k_{Tx}\Psi_y)(k_{Tx}\mathbf{e}_x + k_{Ty}\mathbf{e}_y + k_{Tz}\mathbf{e}_z\big]\exp(i\mathbf{k}_T\cdot\mathbf{x}), \end{aligned} \tag{1.8}$$

with obvious meaning of the symbols.

4.2 Generalized Snell's law

Consider a plane surface $\mathcal{P}$, with unit normal $\mathbf{n}$, that separates two homogeneous half-spaces. An inhomogeneous wave, incident on $\mathcal{P}$, has a wave vector

$$\mathbf{k}^i = \mathbf{k}_1^i + i\mathbf{k}_2^i = k_1^i\mathbf{n}_1^i + ik_2^i\mathbf{n}_2^i$$

where $\mathbf{n}_1^i, \mathbf{n}_2^i$ are the unit vectors of $\mathbf{k}_1^i, \mathbf{k}_2^i$; of course $\mathbf{k}_1^i$ and $\mathbf{k}_2^i$ are real. No choice is made about the polarization of the incident wave which may be longitudinal or transverse. Nor is the coplanarity of the vectors $\mathbf{k}_1^i, \mathbf{k}_2^i, \mathbf{n}$ assumed.

Because of the discontinuity surface, four inhomogeneous waves originate at (any point of) $\mathcal{P}$: a longitudinal and a transverse wave for each medium. A reflection-refraction problem consists in the determination of the reflected and transmitted (or refracted) waves in terms of the properties of the two media and the incident wave. To solve such a problem we first determine a priori geometrical restrictions on the pertinent wave vectors.

We assume that the two half-spaces are in welded contact. Then the boundary conditions governing the process of reflection and refraction are the continuity of displacement

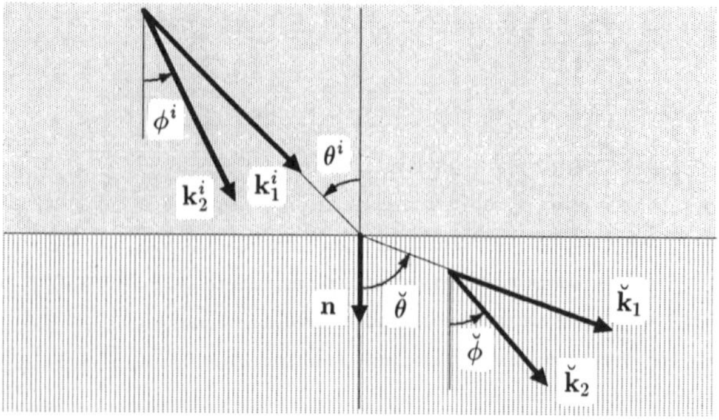

Fig. 4.1 The geometry of refraction of a wave at an interface when the incident wave vector is vertically polarized.

and traction across $\mathcal{P}$. Denote by the accent ˘ any quantity pertaining to the half-space opposite to that of the incident wave. The continuity of $\mathbf{U}$ and $\mathbf{t}$, as produced by the incident and reflected waves on one side and by the transmitted waves on the other, implies that the phase $\mathbf{k} \cdot \mathbf{x}$ at any point of $\mathcal{P}$ takes a common value for all waves. Letting $\mathbf{x}$ be the position vector of a current point of $\mathcal{P}$ relative to an origin in $\mathcal{P}$, we have

$$\mathbf{k}^i \cdot \mathbf{x} = \mathbf{k}_L^r \cdot \mathbf{x} = \mathbf{k}_T^r \cdot \mathbf{x} = \breve{\mathbf{k}}_L \cdot \mathbf{x} = \breve{\mathbf{k}}_T \cdot \mathbf{x}, \qquad \forall \mathbf{x} : \ \mathbf{n} \cdot \mathbf{x} = 0,$$

where the superscript r labels quantities pertaining to the reflected waves. For convenience represent any vector $\mathbf{x}$, such that $\mathbf{n} \cdot \mathbf{x} = 0$, in the form $\mathbf{x} = \mathbf{n} \times \mathbf{y}$; the set of vectors $\mathbf{x}$ is spanned by letting $\mathbf{y}$ be any vector in $\mathbb{R}^3$. The arbitrariness of $\mathbf{y}$ yields

$$\mathbf{k}^i \times \mathbf{n} = \mathbf{k}_L^r \times \mathbf{n} = \mathbf{k}_T^r \times \mathbf{n} = \breve{\mathbf{k}}_L \times \mathbf{n} = \breve{\mathbf{k}}_T \times \mathbf{n}. \tag{2.1}$$

We can view (2.1) as the general form of Snell's law. In particular it follows from the separate contributions of the real and imaginary parts of (2.1) that the planes $(\mathbf{n}_1, \mathbf{n})$ and $(\mathbf{n}_2, \mathbf{n})$ are invariant, namely all pairs $(\mathbf{n}_1, \mathbf{n})$ determine a common plane and so do the pairs $(\mathbf{n}_2, \mathbf{n})$.

The more standard form of Snell's law may be recovered by using Cartesian components. Identify $\mathcal{P}$ with the plane $z = 0$ and choose the half-space $z > 0$ as that supporting the incident wave. Accordingly the accent ˘ labels quantities relative to the half-space $z < 0$. No assumption is made about the choice of the x- and y-axes in the plane $\mathcal{P}$. Inner multiplication of (2.1) by $\mathbf{e}_x$ and $\mathbf{e}_y$ yields

$$k_x^i = k_{Lx}^r = k_{Tx}^r = \breve{k}_{Lx} = \breve{k}_{Tx} =: k_x, \tag{2.2}$$

$$k_y^i = k_{Ly}^r = k_{Ty}^r = \breve{k}_{Ly} = \breve{k}_{Ty} =: k_y. \tag{2.3}$$

The content of (2.1) may be phrased by saying that the projections on the plane $\mathcal{P}$ of the reflected and transmitted, longitudinal and transverse, waves coincide with those of the incident one.

Denote by θ and ϕ ($\leq \pi/2$) the angles between $\mathbf{n}$ and $\mathbf{n}_1$, $\mathbf{n}_2$, respectively. The real and imaginary parts of (2.1) result in

$$k_1^i \sin\theta^i = k_{L1}^r \sin\theta_L^r = k_{T1}^r \sin\theta_T^r = \breve{k}_{L1} \sin\breve{\theta}_L = \breve{k}_{T1} \sin\breve{\theta}_T, \tag{2.4}$$

$$k_2^i \sin\phi^i = k_{L2}^r \sin\phi_L^r = k_{T2}^r \sin\phi_T^r = \breve{k}_{L2} \sin\breve{\phi}_L = \breve{k}_{T2} \sin\breve{\phi}_T. \tag{2.5}$$

The peculiar features of the inhomogeneous waves motivate a closer inspection of some consequences of Snell's law. For instance, by (3.2.14) k_1 and k_2 are determined by the material parameters, along with the angle between $\mathbf{k}_1$ and $\mathbf{k}_2$, through (1.4) and (1.5); it is not immediately evident from (2.4) that the reflection angle θ_L^r equals the incidence angle θ^i. In this connection, owing to the invariance of the x- and y-components, it is essential to examine the behaviour of the z-component of $\mathbf{k}$.

For any wave vector $\mathbf{k}$, we can combine (3.2.11) and (3.2.12) to obtain

$$\kappa = k_x^2 + k_y^2 + k_z^2 = a + ib.$$

For convenience let $k_z = \zeta + i\sigma$, with real ζ and σ. Then we have

$$\zeta^2 - \sigma^2 = a - \mathrm{Re}\,(k_x^2 + k_y^2) =: A, \tag{2.6}$$

$$\zeta\sigma = b/2 - \mathrm{Im}\,(k_x^2 + k_y^2)/2 =: B/2, \tag{2.7}$$

where k_x and k_y are to be regarded as given. The solution of the algebraic system (2.6), (2.7) is unique, up to the sign. Indeed, we obtain

$$\zeta = \pm\sqrt{\frac{A + \sqrt{A^2 + B^2}}{2}}, \qquad \sigma = B/2\zeta = \pm\sqrt{\frac{\sqrt{A^2 + B^2} - A}{2}}.$$

The value (sign) of ζ is provided by the direction of propagation of the wave; the value of σ is then determined. Really, reflected waves correspond to the positive sign, while refracted waves require the negative sign. Notice also that the signs of ζ and σ are equal or opposite depending on the sign of the constant B.

Restrict attention to the reflected wave of the same kind of the incident one (i.e. longitudinal-longitudinal, transverse-transverse). The pairs ζ^i, σ^i and ζ^r, σ^r are solutions to the same algebraic system (2.6), (2.7) and then we find

$$\zeta^r = -\zeta^i, \qquad \sigma^r = -\sigma^i$$

the sign being determined by the observation that the z components of $\mathbf{k}$ are opposite. Indeed, this means that $k_z^r = -k_z^i$. Then it follows easily that $k_1^r = k_1^i$ and $k_2^r = k_2^i$. Because of (2.4) and (2.5) this yields

$$\theta^r = \theta^i, \qquad \phi^r = \phi^i,$$

as expected.

Further consequences follow if the two media are specified. Suppose that the upper half-space is filled by a perfect fluid and that the incident wave is longitudinal and homogeneous. It is not restrictive to let $\mathbf{k}^i$ belong to the (x, z)-plane; hence k_x is real and $k_y = 0$. Then (2.6) and (2.7) reduce to

$$\zeta^2 - \sigma^2 = a - k_x^2, \qquad \zeta\sigma = b/2. \tag{2.8}$$

If the lower medium is an elastic solid ($b = 0$) and we consider the transmitted transverse wave then we have $a = \breve{\kappa}_T$ and $\breve{\zeta}_T\breve{\sigma}_T = 0$. The following possibilities occur.

- If $\breve{\kappa}_T - k_x^2 > 0$ the condition of reality of ζ and σ yields $\breve{\beta}_T = \breve{\zeta}_T + i\breve{\sigma}_T = \sqrt{\breve{\kappa}_T - k_x^2}$; the transmitted transverse wave is homogeneous too.

- If $\breve{\kappa}_T - k_x^2 = 0$ then we find $\breve{\beta}_T = \breve{\zeta}_T + i\breve{\sigma}_T = 0$. In this case the wave vector is real and parallel to the interface, $\breve{\theta}_T = \pi/2$, and $k_1^i \sin\theta^i = \breve{k}_T$, which corresponds to a critical value of the incidence angle (cf. [2], Ch. 5).

- If $\breve{\kappa}_T - k_x^2 < 0$ then $\breve{\beta}_T = \breve{\zeta}_T + i\breve{\sigma}_T = -i\sqrt{\breve{\kappa}_T - k_x^2}$. The transmitted transverse wave is inhomogeneous, the real part $\breve{\mathbf{k}}_1$ of the wave vector is parallel to the interface, the amplitude of the wave decays with distance from the interface as $\exp(-\sqrt{\breve{\kappa}_T - k_x^2}\,|z|)$. Roughly speaking, when the incidence angle is greater than the critical value the transmitted wave is essentially confined to a neighbourhood of the interface.

If the lower medium is viscoelastic, then (2.8) shows that $\breve{\zeta}$ and $\breve{\sigma}$ have the same sign (because $b > 0$). Then they both are negative in view of the fact that we are considering transmitted waves. As in the elastic case, the amplitude decays with distance from the interface. If, however, also the upper medium is viscoelastic, then by (2.7) we may have $\breve{\zeta}\breve{\sigma} < 0$ which in turn allows for negative values of $\breve{\zeta}$ and positive values of $\breve{\sigma}$. In this circumstance the amplitude grows with distance from the interface. This, though, is in no contradiction to the wave decay as remarked in Ch. 1.

Observe that, in viscoelastic solids, $\zeta, \sigma = 0$ if and only if $A, B = 0$, namely $k_x^2 + k_y^2 = \kappa$. Suppose we have $B = 0$; then $\zeta = 0$ if $A < 0$ and $\sigma = 0$ if $A > 0$. In such a case, if $\zeta = 0$ we let $\sigma = \sqrt{-A}$, which provides the decay of the amplitude of the wave.

The previous scheme can be summarized as follows. According to Snell's law we define

$$\beta_L = \sqrt{\kappa_L - k_x^2 - k_y^2}, \qquad \beta_T = \sqrt{\kappa_T - k_x^2 - k_y^2}; \tag{2.9}$$

since the argument is generally complex, by (2.9) we mean the roots with positive real part; if the real part vanishes then the root is taken with positive imaginary part. The z-components of the incident and reflected waves coincide with the left-hand sides of (2.9), up to the sign, which is chosen as positive or negative according as the wave under consideration is upgoing (viz. reflected) or downgoing (viz. incident or transmitted). Represent the incident wave vector as $\mathbf{k}^i = k_x\mathbf{e}_x + k_y\mathbf{e}_y + k_z^i\mathbf{e}_z$, where k_z^i equals $-\beta_L$ or $-\beta_T$ according as the incident wave is longitudinal or transverse. Then we have

$$\mathbf{k}_L^r = k_x\mathbf{e}_x + k_y\mathbf{e}_y + \beta_L\mathbf{e}_z, \qquad \mathbf{k}_T^r = k_x\mathbf{e}_x + k_y\mathbf{e}_y + \beta_T\mathbf{e}_z,$$

$$\breve{\mathbf{k}}_L = k_x\mathbf{e}_x + k_y\mathbf{e}_y - \breve{\beta}_L\mathbf{e}_z, \qquad \breve{\mathbf{k}}_T = k_x\mathbf{e}_x + k_y\mathbf{e}_y - \breve{\beta}_T\mathbf{e}_z.$$

The wave vector is said to be *vertically polarized* if $\mathbf{k}_1$, $\mathbf{k}_2$ and $\mathbf{n}$ belong to a common (vertical) plane π. Whenever such is the case it is assumed that π coincides with the (x, z)-plane, which means that $k_y = 0$. In view of Snell's law, the property of being vertically polarized is inherited by the reflected and the transmitted wave vectors. Unlike the elastic case, in general the wave vector of the incident wave is not vertically polarized. Indeed, the vectors $\mathbf{k}_1$ and $\mathbf{k}_2$ depend on how the wave has been generated, and this is unrelated to the direction of $\mathbf{n}$. This feature is at the origin of the formal complications, connected with the behaviour at interfaces, which cannot be avoided unless restrictive assumptions are made.

Hereafter, the set of a longitudinal and a transverse wave that are propagating upwards (downwards) and whose wave vectors have the same x- and y-components are referred to as a *conjugate upgoing (downgoing) pair* [81]. Because of Snell's law, both pairs of reflected and transmitted waves still constitute conjugate pairs. To unify the treatment of longitudinal and transverse waves we will consider reflection and refraction of a conjugate pair of (downgoing) waves. Owing to conjugacy, there are only four, rather than eight, outgoing waves just as in the case of a single incident wave. By the linearity of the equations, the particular case of a longitudinal or transverse wave hitting the interface follows trivially.

4.3 Displacement and traction at interfaces

We proceed in the analysis of the continuity requirements which, besides providing the explicit form of Snell's law, allow the determination of the amplitudes of the reflected and transmitted conjugate pairs in terms of the amplitudes of the incident pair. It is understood that the z-components of the wave vectors involved are non-zero. The limit case when the z-components vanish will be examined separately.

In the half-space $z > 0$ a conjugate pair of incident waves is propagating downwards and a conjugate pair of reflected waves is propagating upwards; the overall scalar and vector potentials may be written as

$$\phi = [\Phi^+ \exp(i\beta_L z) + \Phi^- \exp(-i\beta_L z)] \exp[i(k_x x + k_y y)], \tag{3.1}$$

$$\boldsymbol{\psi} = [\boldsymbol{\Psi}^+ \exp(i\beta_T z) + \boldsymbol{\Psi}^- \exp(-i\beta_T z)] \exp[i(k_x x + k_y y)], \tag{3.2}$$

where the Φ's and the $\boldsymbol{\Psi}$'s are constant; the superscripts $+$ and $-$ label upgoing and downgoing waves, respectively. Quite naturally, for any conjugate pair the two constants $\Phi, \boldsymbol{\Psi}$ may be viewed as the amplitude of the pair. In the half-space $z < 0$ the scalar and vector potentials are given by

$$\breve{\phi} = \breve{\Phi}^- \exp(-i\breve{\beta}_L z) \exp[i(k_x x + k_y y)], \qquad \breve{\boldsymbol{\psi}} = \breve{\boldsymbol{\Psi}}^- \exp(-i\breve{\beta}_T z) \exp[i(k_x x + k_y y)]. \tag{3.3}$$

If either $\mathrm{Re}\,\breve{\beta}_L = 0$ or $\mathrm{Re}\,\breve{\beta}_T = 0$, the positive value of $\mathrm{Im}\,\breve{\beta}$ makes the amplitude of the transmitted waves decrease with distance from the interface.

For use in reflection-refraction problems we determine $\mathbf{U}$ and $\mathbf{t}$, at both sides of the interface, in terms of the amplitudes Φ, $\boldsymbol{\Psi}$ and the wave vectors $\mathbf{k}_L$, $\mathbf{k}_T$. Substitution of (3.1) and (3.2) into (1.7) and (1.8) yields the coordinate representations for displacement and traction in the upper half-space; the limiting values at the interface are obtained by simply letting $z \to 0_+$. By use of (1.5), the upper components of $\mathbf{U}$ and $\mathbf{t}$ at $\mathcal{P}$ are found to be

$$U_x = ik_x(\Phi^+ + \Phi^-) - i\beta_T(\Psi_y^+ - \Psi_y^-) + ik_y(\Psi_z^+ + \Psi_z^-), \tag{3.4}$$

$$U_y = ik_y(\Phi^+ + \Phi^-) + i\beta_T(\Psi_x^+ - \Psi_x^-) - ik_x(\Psi_z^+ + \Psi_z^-), \tag{3.5}$$

$$U_z = i\beta_L(\Phi^+ - \Phi^-) - ik_y(\Psi_x^+ + \Psi_x^-) + ik_x(\Psi_y^+ + \Psi_y^-), \tag{3.6}$$

$$t_x = -2\mu k_x[-\beta_L(\Phi^+ - \Phi^-) + k_y(\Psi_x^+ + \Psi_x^-) - q(\Psi_y^+ + \Psi_y^-)], \tag{3.7}$$

$$t_y = -2\mu k_y[-\beta_L(\Phi^+ - \Phi^-) + r(\Psi_x^+ + \Psi_x^-) - k_x(\Psi_y^+ + \Psi_y^-)], \tag{3.8}$$

$$t_z = 2\mu[A(\Phi^+ + \Phi^-) - k_y\beta_T(\Psi_x^+ - \Psi_x^-) + k_x\beta_T(\Psi_y^+ - \Psi_y^-)], \tag{3.9}$$

where

$$r = \frac{1}{k_y}(k_y^2 - \tfrac{1}{2}\kappa_T), \qquad q = \frac{1}{k_x}(k_x^2 - \tfrac{1}{2}\kappa_T), \qquad A = \tfrac{1}{2}\kappa_T - k_x^2 - k_y^2. \tag{3.10}$$

The common factor $\exp[i(k_x x + k_y y)]$ is understood and not written.

The analogous expressions for displacement and traction in the lower half space are simply recovered from (3.4)-(3.9) through cancellation of the contributions Φ^+, Ψ_x^+, Ψ_y^+, Ψ_z^+, and introduction of the superposed accent ˘ where appropriate.

The z-components of the amplitude vectors $\boldsymbol{\Psi}^+$ and $\boldsymbol{\Psi}^-$ can be eliminated from (3.4) and (3.5) through the use of the geometrical condition (1.6), $\boldsymbol{\Psi} \cdot \mathbf{k}_T = 0$. In fact, determination of Ψ_z^+ and Ψ_z^- from

$$k_x\Psi_x^+ + k_y\Psi_y^+ + \beta_T\Psi_z^+ = 0, \qquad k_x\Psi_x^- + k_y\Psi_y^- - \beta_T\Psi_z^- = 0$$

yields

$$\Psi_z^+ = -\frac{k_x\Psi_x^+ + k_y\Psi_y^+}{\beta_T}, \qquad \Psi_z^- = \frac{k_x\Psi_x^- + k_y\Psi_y^-}{\beta_T}. \tag{3.11}$$

This shows that the z-components of the vector $\mathbf{\Psi}$ can be eliminated from the expression of the continuity conditions at $\mathcal{P}$ thus reducing the number of unknowns. Really, addition of the left sides of (3.11) leads to the relation

$$\Psi_z^+ + \Psi_z^- = -[k_x(\Psi_x^+ - \Psi_x^-) + k_y(\Psi_y^+ - \Psi_y^-)]/\beta_T, \tag{3.12}$$

which allows (3.4) and (3.5) to be replaced by

$$U_x = ik_x(\Phi^+ + \Phi^-) - i\frac{k_x k_y}{\beta_T}(\Psi_x^+ - \Psi_x^-) - i\frac{k_y^2 + \beta_T^2}{\beta_T}(\Psi_y^+ - \Psi_y^-), \tag{3.4'}$$

$$U_y = ik_y(\Phi^+ + \Phi^-) + i\frac{k_x^2 + \beta_T^2}{\beta_T}(\Psi_x^+ - \Psi_x^-) + i\frac{k_x k_y}{\beta_T}(\Psi_y^+ - \Psi_y^-). \tag{3.5'}$$

Consider a conjugate pair with a vertically polarized wave vector $\mathbf{k}$ and then let $k_y = 0$. The component t_y does not vanish ($k_y r = k_y^2 - \frac{1}{2}\kappa_T \to -\frac{1}{2}\kappa_T$) and we find that

$$t_y = \rho\omega^2(\Psi_x^+ + \Psi_x^-) = \mu\kappa_T(\Psi_x^+ + \Psi_x^-), \tag{3.13}$$

while the remaining components of $\mathbf{t}$ and $\mathbf{U}$ are simply recovered from (3.4)-(3.7) and (3.9) through trivial substitutions of the condition $k_y = 0$.

Suppose now that the wave vectors of the incident pair are parallel to the z-axis, which means that $k_x = k_y = 0$. Of course, this condition is preserved for the reflected and transmitted pairs and, according to (2.9), we have

$$\beta_T = k_T, \qquad \beta_L = k_L.$$

The general condition $\mathbf{\Psi} \cdot \mathbf{k}_T = 0$ imposed on the vector potentials yields $\Psi_z^+ = \Psi_z^- = 0$. Substitution in (1.7) and (1.8) yields the desired expressions of displacement and traction at the upper side of the interface. In vector notation we have

$$\mathbf{U} = ik_T[-(\Psi_y^+ - \Psi_y^-)\mathbf{e}_x + (\Psi_x^+ - \Psi_x^-)\mathbf{e}_y] + ik_L(\Phi^+ - \Phi^-)\mathbf{e}_z, \tag{3.14}$$

$$\mathbf{t} = \rho\omega^2[-(\Psi_y^+ + \Psi_y^-)\mathbf{e}_x + (\Psi_x^+ + \Psi_x^-)\mathbf{e}_y + (\Phi^+ + \Phi^-)\mathbf{e}_z]. \tag{3.15}$$

The analogous result for the lower side follows by dropping Φ^+ and $\mathbf{\Psi}^+$.

4.4 Reflection at a free surface

Consider a viscoelastic medium occupying the half-space $z > 0$ and let the lower half-space $z < 0$ be empty. At the free surface $\mathcal{P}$ no condition is placed on displacement, while traction is required to vanish. A conjugate pair of incident waves produces a reflected (upgoing) conjugate pair. The components of the traction at the plane $z = 0$ assume the form (3.7)-(3.9); on setting them equal to zero we obtain a linear system in the unknown (reflected) amplitudes Φ^+, Ψ_x^+, Ψ_y^+ in terms of the known (incident) amplitudes Φ^-, Ψ_x^-, Ψ_y^-. Upon some rearrangement we can write

$$-\beta_L(\Phi^+ - \Phi^-) + k_y(\Psi_x^+ + \Psi_x^-) - q(\Psi_y^+ + \Psi_y^-) = 0, \tag{4.1}$$

$$-\beta_L(\Phi^+ - \Phi^-) + r(\Psi_x^+ + \Psi_x^-) - k_x(\Psi_y^+ + \Psi_y^-) = 0, \tag{4.2}$$

$$A(\Phi^+ - \Phi^-) - k_y\beta_T(\Psi_x^+ + \Psi_x^-) + k_x\beta_T(\Psi_y^+ + \Psi_y^-) = -2(A\Phi^- + k_y\beta_T\Psi_x^- - k_x\beta_T\Psi_y^-), \tag{4.3}$$

which shows the advantage of regarding $\Phi^+ - \Phi^-$, $\Psi_x^+ + \Psi_x^-$, $\Psi_y^+ + \Psi_y^-$ as intermediate unknowns. The solution is given by

$$\Phi^+ = \Phi^- + AE \tag{4.4}$$

$$\Psi_x^+ = -\Psi_x^- - k_y\beta_L E, \tag{4.5}$$

$$\Psi_y^+ = -\Psi_y^- + k_x\beta_L E, \tag{4.6}$$

where

$$E = -2(A\Phi^- + k_y\beta_T\Psi_x^- - k_x\beta_T\Psi_y^-)/D, \tag{4.7}$$

and

$$D = A^2 + \beta_L\beta_T(k_x^2 + k_y^2). \tag{4.8}$$

To determine the z-component of Ψ^+ we substitute (4.5) and (4.6) into (3.11) thus obtaining

$$\Psi_z^+ = \Psi_z^-. \tag{4.9}$$

This shows, in particular, that Ψ_z^+ vanishes if the incident wave is longitudinal.

To bring into evidence the origin of the various contributions to the amplitudes of the reflected pair we reformulate (4.4)-(4.6) in matrix form as

$$\begin{pmatrix} \Phi^+ \\ \Psi_x^+ \\ \Psi_y^+ \end{pmatrix} = R^h \begin{pmatrix} \Phi^- \\ \Psi_x^- \\ \Psi_y^- \end{pmatrix}, \tag{4.10}$$

where

$$R^h := \begin{pmatrix} 1 - 2A^2/D & -2Ak_y\beta_T/D & 2Ak_x\beta_T/D \\ 2Ak_y\beta_L/D & 2k_y^2\beta_L\beta_T/D - 1 & -2k_xk_y\beta_L\beta_T/D \\ -2Ak_x\beta_L/D & -2k_xk_y\beta_L\beta_T/D & 2k_x^2\beta_L\beta_T/D - 1 \end{pmatrix} \tag{4.11}$$

is called the *reflection matrix* for the viscoelastic half-space. The meaning of such a matrix is apparent. Suppose for definiteness that the incident wave is longitudinal. Then $\Psi_x^- = \Psi_y^- = 0$, and the first column of R^h yields the ratios Φ^+/Φ^-, Ψ_x^+/Φ^-, and Ψ_y^+/Φ^-, that is the reflection coefficients. Similarly, the second and third columns identify the reflection coefficients corresponding to the case of an incident transverse wave.

If $\mathbf{k}_1$, $\mathbf{k}_2$, and $\mathbf{n}$ are coplanar we may formally let $k_y = 0$, in which case the wave vector of the incident wave is vertically polarized. Then the system (4.1)-(4.3) decouples into a subsystem of two equations for the two unknowns Φ^+ and Ψ_y^+, plus a single equation for the unknown Ψ_x^+. The subsystem embodies the case of a longitudinal or a transverse vertically-polarized wave in the elastic framework, while the single equation corresponds to an incident transverse horizontally polarized wave. Precisely, we observe that if $k_y = 0$ then equation (4.2), which follows from the vanishing of t_y, is to be replaced by letting $t_y = 0$ in (3.13) whence $\Psi_x^+ = -\Psi_x^-$. Along with the limits of (4.1) and (4.3) - or (4.4) and (4.6) - we have

$$\begin{pmatrix} \Phi^+ \\ \Psi_x^+ \\ \Psi_y^+ \end{pmatrix} = \begin{pmatrix} 1 - 2A^2/D & 0 & 2Ak_x\beta_T/D \\ 0 & -1 & 0 \\ -2Ak_x\beta_L/D & 0 & 2k_x^2\beta_L\beta_T/D - 1 \end{pmatrix} \begin{pmatrix} \Phi^- \\ \Psi_x^- \\ \Psi_y^- \end{pmatrix}, \qquad (k_y = 0), \tag{4.12}$$

which is just the particular case of (4.10)-(4.11) as $k_y = 0$. A further simplification occurs in the case when the incident wave vector is vertical (i.e. parallel to the z-axis). Then the expression of $\mathbf{t}$ is given by (3.15) and the condition of vanishing traction at $\mathcal{P}$ yields

$$\begin{pmatrix} \Phi^+ \\ \Psi_x^+ \\ \Psi_y^+ \end{pmatrix} = \begin{pmatrix} -1 & 0 & 0 \\ 0 & -1 & 0 \\ 0 & 0 & -1 \end{pmatrix} \begin{pmatrix} \Phi^- \\ \Psi_x^- \\ \Psi_y^- \end{pmatrix}, \qquad (k_x = k_y = 0),$$

which corresponds to the limit of the general expressions (4.4)-(4.6) or (4.10) as $k_x, k_y \to 0$.

We are now in a position to consider some features, e.g. mode conversion, which are related to non-vertically polarized complex wave vectors incident on the free boundary of a half-space. For example, suppose that we ask under what conditions an incident longitudinal wave ($\Psi_x^- = \Psi_y^- = 0$) is transformed into a transverse one ($\Phi^+ = 0$). Inspection of (4.11) shows that this happens provided that the incident wave satisfies $2A^2 = D$. Comparison with the definitions of D and A, namely (4.8) and (3.10), reduces this condition to

$$\beta_L\beta_T(k_x^2 + k_y^2) = \left(\tfrac{1}{2}\kappa_T - k_x^2 - k_y^2\right)^2. \tag{4.13}$$

Equation (4.13) generalizes the relation that holds for vertically-polarized wave vectors [44] and the condition of mode conversion for elastic solids [22]. By (2.9) we obtain an equation for $k_x^2 + k_y^2$, that is a constraint on the incident wave vector. Hence (4.13) determines the x- and y-components of $\mathbf{k}^i$ only up to a parameter, thus implying that complete mode conversion is allowed for an infinite family of longitudinal incident waves. On the contrary, in the case of vertically polarized $\mathbf{k}^i$, only one wave allowing for such a complete mode conversion is allowed since $k_y = 0$. Furthermore, under the assumption that $\mathbf{k}^i$ is vertically polarized and $\Psi_x^- = 0$, no substantial difference between longitudinal and transverse incident waves is observed, in the sense that the same equation characterizes complete mode conversion. In fact if we assume $\Phi^- = \Psi_x^- = 0$ and require $\Psi_y^+ = \Psi_x^+ = 0$ we find the condition $2k_x^2\beta_L\beta_T = D$ that in general reduces to

$$\beta_L\beta_T(k_x^2 - k_y^2) = \left(\tfrac{1}{2}\kappa_T - k_x^2 - k_y^2\right)^2$$

and hence coincides with (4.13) if $k_y = 0$.

This result is not true in general. In the case of an incident transverse wave substitution of the condition $\Phi^- = 0$ into (4.10) and inspection of (4.11) show that Ψ_x^+ and Ψ_y^+ cannot vanish simultaneously if k_y and k_x are non-zero. The same behaviour also occurs if $\Psi_x^- = 0$. This shows that in general complete conversion of a transverse wave into a longitudinal one is not allowed.

We conclude this section by considering the two particular cases that correspond to an incident longitudinal (transverse) wave such that the related reflected transverse (longitudinal) wave has a vanishing z-component. Obviously, under these conditions we cannot consider an incident pair but it is appropriate to deal separately with the two allowable choices of the incident wave. In view of its inherent simplicity we study first the case when the incident wave is transverse.

Suppose that the incident and reflected transverse waves are described as in (3.2), namely

$$\boldsymbol{\psi} = [\boldsymbol{\Psi}^+ \exp(i\beta_T z) + \boldsymbol{\Psi}^- \exp(-i\beta_T z)] \exp[i(k_x x + k_y y)].$$

The incident wave is taken to satisfy

$$k_x^2 + k_y^2 = \kappa_L,$$

and then, as a consequence of Snell's law,

$$\beta_L = \sqrt{\kappa_L - k_x^2 - k_y^2} = 0. \tag{4.14}$$

Accordingly, we look for a reflected longitudinal wave described by

$$\phi = \Phi^+ \exp[i(k_x x + k_y y)], \tag{4.15}$$

where Φ^+ is constant. As is easily verified, the potential (4.15) satisfies the Helmholtz equation $(1.1)_1$, provided the definition $(1.4)_1$ of κ_L and the condition (4.14) are taken into account. Similarly, continuity of the phase is also ensured by (4.15). Therefore we have to consider the field obtained by superposition of the potentials (3.2) and (4.14), where $\boldsymbol{\Psi}^-$ is regarded as given while $\boldsymbol{\Psi}^+$ and Φ^+ are determined so as to ensure the vanishing of the traction at the interface $\mathcal{P}$.

The contribution $\mathbf{t}[\Phi^+]$ to the traction vector at $\mathcal{P}$ arising from the scalar potential ϕ follows from substitution of (4.15) into (1.3), and is given by

$$\mathbf{t}[\phi] = 2\mu A \kappa_T \Phi^+ \mathbf{e}_z. \tag{4.16}$$

The part of the traction that is originated by the vector potential $\boldsymbol{\psi}$ is easily recovered from (3.7)-(3.9) and reads

$$\begin{aligned}\mathbf{t}[\boldsymbol{\psi}] = 2\mu\{[-k_y(\Psi_x^+ + \Psi_x^-) - q(\Psi_y^+ + \Psi_y^-)]\mathbf{e}_x - [r(\Psi_x^+ + \Psi_x^-) - k_x(\Psi_y^+ + \Psi_y^-)]\mathbf{e}_y \\ + [-k_y\beta_T(\Psi_x^+ - \Psi_x^-) + k_x\beta_T(\Psi_y^+ - \Psi_y^-)]\mathbf{e}_z\}.\end{aligned} \tag{4.17}$$

The requirement that the traction $\mathbf{t}[\phi] + \mathbf{t}[\boldsymbol{\psi}]$ vanish at the boundary yields the linear system

$$k_y(\Psi_x^+ + \Psi_x^-) - q(\Psi_y^+ + \Psi_y^-) = 0,$$
$$r(\Psi_x^+ + \Psi_x^-) - k_x(\Psi_y^+ + \Psi_y^-) = 0,$$
$$A\Phi^+ - k_y\beta_T(\Psi_x^+ - \Psi_x^-) + k_x\beta_T(\Psi_y^+ - \Psi_y^-) = 0,$$

that may be regarded as obtained from (4.1)-(4.3) in the limit $\beta_L \to 0$, in the case of a vanishing Φ^-. The solution is then given in the form

$$\Psi_x^+ = -\Psi_x^-,$$
$$\Psi_y^+ = -\Psi_y^-,$$
$$\Phi^+ = -2\frac{\beta_T}{A}(k_y\Psi_x^- - k_x\Psi_y^-),$$

thus showing that the "reflected" longitudinal and transverse waves are uniquely determined. Actually, the wave vector of the longitudinal wave is purely horizontal, which means that a generic reference to a reflected wave of longitudinal type is, in a sense, a slight abuse.

The case of a longitudinal incident wave is developed along the same lines with interchange of the roles of scalar and vector potentials, although the analogy with the previous analysis of the expressions of displacement and traction is lost. Let us consider a scalar potential of the form (3.1), with Φ^- and Φ^+ representing the incident and reflected waves. As before we set

$$\boldsymbol{\psi} = \Psi^+ \exp\left[i(k_x x + k_y y)\right], \tag{4.18}$$

and assume that the incident wave is such that

$$\beta_T = \sqrt{\kappa_T - k_x^2 - k_y^2} = 0. \tag{4.19}$$

Again, the vector potential (4.18) is a solution to $(1.1)_2$ in view of $(1.5)_1$ and (4.19). The continuity condition for the phase at the interface is also satisfied. Unlike the previous procedure, however, the condition of vanishing divergence for $\boldsymbol{\psi}$ yields

$$\Psi_x^+ = -\frac{k_y}{k_x}\Psi_y^+, \tag{4.20}$$

instead of $(3.11)_1$. The traction $\mathbf{t}[\boldsymbol{\psi}]$ corresponding to the vector potential (4.18) follows from (1.3) through comparison with (4.20) in the form

$$\mathbf{t}[\boldsymbol{\psi}] = \mu\frac{\kappa_T}{k_x}\Psi_y^+(k_x\mathbf{e}_x + k_y\mathbf{e}_y). \tag{4.21}$$

The traction related to the scalar potential ϕ is found from (3.7)-(3.9) and (4.19), and reads

$$\mathbf{t}[\phi] = 2\mu\beta_L(\Phi^+ - \Phi^-)(k_x\mathbf{e}_x + k_y\mathbf{e}_y) - \mu\kappa_T(\Phi^+ + \Phi^-)\mathbf{e}_z.$$

Accordingly, we find the final expression for the traction at $\mathcal{P}$ in the form

$$\mathbf{t} = 2\mu\big[\beta_L(\Phi^+ - \Phi^-) + \tfrac{1}{2}\frac{\kappa_T}{k_x}\Psi_y^+\big](k_x\mathbf{e}_x + k_y\mathbf{e}_y) - \mu\kappa_T(\Phi^+ + \Phi^-)\mathbf{e}_z. \tag{4.22}$$

The requirement $\mathbf{t} = 0$ leads to

$$\Phi^+ = -\Phi^-,$$
$$\Psi_y^+ = 4\frac{\beta_L k_x}{\kappa_T}\Phi^-,$$

while Ψ_x^+ follows from substitution of Ψ_y^+ into (4.20). Notice that the wave vector of the transverse wave is horizontal. The z-component of the vector amplitude $\boldsymbol{\Psi}$ is left undetermined by the requirement that $\mathbf{t} = 0$. This is ultimately due to the fact that the constant parameter Ψ_z^+ does not enter the expression of $\mathbf{t}$.

4.5 Boundary between viscoelastic half-spaces

Let the half-spaces $z > 0$ and $z < 0$ be occupied by two isotropic, homogeneous viscoelastic solids with different material parameters. Suppose that a conjugate pair of inhomogeneous waves is incident on the boundary from the upper half-space. Owing to the presence of the discontinuity a reflected conjugate pair and a transmitted one are originated at $\mathcal{P}$. They are completely determined through the requirement of continuity for displacement and traction at the interface. The expressions for the wave vectors have already been given in §4.2. Here we determine the explicit form of the scalar and vector amplitudes. With reference to the representations (3.1) and (3.2) for the waves travelling in the upper medium, and to (3.3) for those in the lower one, we have to find the reduced reflected and transmitted amplitudes Φ^+, Ψ_x^+, Ψ_y^+, $\breve{\Phi}^-$, $\breve{\Psi}_x^-$, and $\breve{\Psi}_y^-$ in terms of the data Φ^-, Ψ_x^-, and Ψ_y^-.

It is convenient to introduce a matrix notation and to split the resulting linear system into two subsystems, so as to take advantage from the form of the resulting equations. The continuity of

$$U_x/i, \quad U_y/i, \quad t_z/2$$

leads to the matrix equation

$$C_1\begin{pmatrix}\Phi^+ + \Phi^- \\ \Psi_x^+ - \Psi_x^- \\ \Psi_y^+ - \Psi_y^-\end{pmatrix} = \breve{C}_1\begin{pmatrix}\breve{\Phi}^- \\ -\breve{\Psi}_x^- \\ -\breve{\Psi}_y^-\end{pmatrix}, \tag{5.1}$$

where the matrix C_1 is defined as

$$C_1 = \begin{pmatrix} k_x & -k_x k_y/\beta_T & -k_y^2/\beta_T - \beta_T \\ k_y & k_x^2/\beta_T + \beta_T & k_x k_y/\beta_T \\ \mu(\frac{1}{2}\kappa_T - k_x^2 - k_y^2) & -\mu k_y \beta_T & \mu k_x \beta_T \end{pmatrix}, \tag{5.2}$$

with inverse

$$(C_1)^{-1} = \frac{1}{\kappa_T \beta_T} \begin{pmatrix} 2k_x\beta_T & 2k_y\beta_T & 2\beta_T/\mu \\ -k_x k_y & \beta_T^2 - k_y^2 & -2k_y/\mu \\ k_x^2 - \beta_T^2 & k_x k_y & 2k_x/\mu \end{pmatrix}. \tag{5.3}$$

The definition of $\breve{C}_1$ follows directly through substitution into (5.2) of the quantities pertaining to the lower medium. Similarly, the continuity of

$$U_z/i, \quad t_x/2k_x, \quad t_y/2k_y,$$

yields the subsystem

$$C_2 \begin{pmatrix} \Phi^+ - \Phi^- \\ \Psi_x^+ + \Psi_x^- \\ \Psi_y^+ + \Psi_y^- \end{pmatrix} = -\breve{C}_2 \begin{pmatrix} \breve{\Phi}^- \\ -\breve{\Psi}_x^- \\ -\breve{\Psi}_y^- \end{pmatrix} \tag{5.4}$$

where C_2 denotes the matrix

$$C_2 = \begin{pmatrix} \beta_L & -k_y & k_x \\ \mu\beta_L & -\mu k_y & \mu(k_x - \frac{1}{2}\kappa_T/k_x) \\ \mu\beta_L & -\mu(k_y - \frac{1}{2}\kappa_T/k_y) & \mu k_x \end{pmatrix}, \tag{5.5}$$

with inverse

$$(C_2)^{-1} = \frac{2}{\mu\kappa_T\beta_L} \begin{pmatrix} \mu(\frac{1}{2}\kappa_T - k_x^2 - k_y^2) & k_x^2 & k_y^2 \\ -\mu k_y \beta_L & 0 & k_y\beta_L \\ \mu k_x \beta_L & k_x\beta_L & 0 \end{pmatrix}. \tag{5.6}$$

Of course the determinants of C_1 and C_2 are found to be non-zero in the present context.

Next we examine two different procedures for the determination of reflected and transmitted amplitudes. First we exhibit the solution in a matrix form, which is particularly suitable for numerical evaluation. Subsequently we find an explicit expression of the solution which is of great help in understanding qualitative aspects.

The subsystem (5.4) can be solved with respect to Φ^+, Ψ_x^+, and Ψ_y^+ to give

$$\begin{pmatrix} \Phi^+ \\ \Psi_x^+ \\ \Psi_y^+ \end{pmatrix} = \begin{pmatrix} \Phi^- \\ -\Psi_x^- \\ -\Psi_y^- \end{pmatrix} - (C_2)^{-1}\breve{C}_2 \begin{pmatrix} \breve{\Phi}^- \\ -\breve{\Psi}_x^- \\ -\breve{\Psi}_y^- \end{pmatrix}.$$

Substitution of this expression into (5.1) leads to a linear system for the scalar and vector amplitudes of the transmitted conjugate pair. This is written as

$$[\breve{C}_1 + C_1(C_2)^{-1}\breve{C}_2] \begin{pmatrix} \breve{\Phi}^- \\ -\breve{\Psi}_x^- \\ -\breve{\Psi}_y^- \end{pmatrix} = 2C_1 \begin{pmatrix} \Phi^- \\ -\Psi_x^- \\ -\Psi_y^- \end{pmatrix}.$$

If the determinant of $\breve{C}_1 + C_1(C_2)^{-1}\breve{C}_2$ is non-zero the matrix equation can be solved with respect to the reduced amplitudes of the conjugate transmitted pair to give the *transmission matrix* T. The result is

$$\begin{pmatrix} \breve{\Phi}^- \\ \breve{\Psi}_x^- \\ \breve{\Psi}_y^- \end{pmatrix} = T \begin{pmatrix} \Phi^- \\ \Psi_x^- \\ \Psi_y^- \end{pmatrix}$$

where we have set

$$T = 2\,\mathbb{I}\,[\breve{C}_1 + C_1(C_2)^{-1}\breve{C}_2]^{-1}C_1\,\mathbb{I},$$

and

$$\mathbb{I} := \mathrm{diag}\,(1,\,-1,\,-1).$$

As regards the reflected amplitudes, we take the explicit expression of the corresponding column vector and compare with the definition of T to deduce

$$\begin{pmatrix} \Phi^+ \\ \Psi_x^+ \\ \Psi_y^+ \end{pmatrix} = R \begin{pmatrix} \Phi^- \\ \Psi_x^- \\ \Psi_y^- \end{pmatrix},$$

where the *reflection matrix* is

$$R = \mathbb{I} - (C_2)^{-1}\breve{C}_2\mathbb{I}\,T.$$

We now determine the explicit solution to (5.1) and (5.4). In this regard it is convenient to describe the amplitudes in terms of the new set of variables Φ, V, W which are related to Φ, Ψ_x, Ψ_y by the trasformation

$$\begin{pmatrix} \Phi \\ V \\ W \end{pmatrix} = \begin{pmatrix} 1 & 0 & 0 \\ 0 & k_x & k_y \\ 0 & k_y & -k_x \end{pmatrix} \begin{pmatrix} \Phi \\ \Psi_x \\ \Psi_y \end{pmatrix} \tag{5.7}$$

with inverse

$$\begin{pmatrix} \Phi \\ \Psi_x \\ \Psi_y \end{pmatrix} = \frac{1}{k_x^2 + k_y^2} \begin{pmatrix} k_x^2 + k_y^2 & 0 & 0 \\ 0 & k_x & k_y \\ 0 & k_y & -k_x \end{pmatrix} \begin{pmatrix} \Phi \\ V \\ W \end{pmatrix}. \tag{5.7'}$$

As a consequence of Snell's law, the matrix entering (5.7) is independent of the medium under consideration. This allows the transformation (5.7) to apply to incident, reflected and transmitted amplitudes, provided the appropriate labels +, −, and ˇ are introduced. Left multiplication of (5.1) by $(C_1)^{-1}$ and application of the transformation (5.7) leads to

$$V^- - V^+ = \frac{\beta_T}{\breve{\beta}_T}\frac{\breve{\kappa}_T}{\kappa_T}\breve{V}^- \tag{5.8}$$

$$\Phi^+ + \Phi^- = (m - \Gamma)\breve{\Phi}^- + 2\frac{\breve{\beta}_T}{\kappa_T}(\frac{\breve{\mu}}{\mu} - 1)\breve{W}^-, \tag{5.9}$$

$$W^- - W^+ = \frac{k_x^2 + k_y^2}{\beta_T}(m - \Gamma - 1)\Phi^- + \frac{\breve{\beta}_T}{\beta_T}(\Gamma + 1)\breve{W}^-, \tag{5.10}$$

where

$$m = \frac{\breve{\rho}}{\rho}, \qquad \Gamma = 2\frac{k_x^2 + k_y^2}{\kappa_T}(\frac{\breve{\mu}}{\mu} - 1).$$

Similarly, multiplication of (5.4) by $(C_2)^{-1}$ and change of variables yields

$$\Phi^+ - \Phi^- = -\frac{\breve{\beta}_L}{\beta_L}(1 + \Gamma)\breve{\Phi}^- - \frac{1}{\beta_L}(1 + \Gamma - m)\breve{W}^-, \tag{5.11}$$

$$V^+ + V^- = m\breve{V}^- \tag{5.12}$$

$$W^+ + W^- = -\breve{\beta}_L\Gamma\breve{\Phi}^- + (m - \Gamma)\breve{W}^- \tag{5.13}$$

Now we solve the system (5.8)-(5.13) where Φ^-, V^-, and W^- are regarded as given data. Taking the difference of (5.9) and (5.11), the sum of (5.10) and (5.13), and the sum of (5.8) and (5.12) yields

$$\begin{aligned} \Phi^- &= a_1\breve{\Phi}^- + a_2\breve{W}^-, \\ W^- &= a_3\breve{\Phi}^- + a_4\breve{W}^-, \\ V^- &= \breve{V}^-/a_5, \end{aligned}$$

where the quantities a_i are defined by

$$\begin{aligned} a_1 &= \tfrac{1}{2}\Big[m - \Gamma + \frac{\breve{\beta}_L}{\beta_L}(1 + \Gamma)\Big] \\ a_2 &= \tfrac{1}{2}\Big[2\frac{\breve{\beta}_T}{\kappa_T}(\frac{\breve{\mu}}{\mu} - 1) + \frac{1}{\beta_L}(1 + \Gamma - m)\Big], \\ a_3 &= \tfrac{1}{2}\Big[\frac{1}{\beta_T}(k_x^2 + k_y^2)(m - 1 - \Gamma) - \breve{\beta}_L\Gamma\Big], \\ a_4 &= \tfrac{1}{2}\Big[m - \Gamma + \frac{\breve{\beta}_T}{\beta_T}(1 + \Gamma)\Big], \\ a_5 &= \frac{2}{m}\frac{\breve{\mu}\breve{\beta}_T}{\breve{\mu}\breve{\beta}_T + \mu\beta_T}. \end{aligned}$$

Now we write the solution in terms of the coefficients a_i, regarded as privileged parameters. The transmitted amplitudes can be expressed in terms of the incident ones through the matrix relation

$$\begin{pmatrix} \breve{\Phi}^- \\ \breve{V}^- \\ \breve{W}^- \end{pmatrix} = \begin{pmatrix} a_4/\mathcal{D} & 0 & -a_2/\mathcal{D} \\ 0 & a_5 & 0 \\ -a_3/\mathcal{D} & 0 & a_1/\mathcal{D} \end{pmatrix} \begin{pmatrix} \Phi^- \\ V^- \\ W^- \end{pmatrix} \tag{5.14}$$

where

$$\begin{aligned} \mathcal{D} = a_1a_4 - a_2a_3 = \tfrac{1}{4}\Big[&(m - \Gamma)^2 + \frac{\breve{\beta}_L}{\beta_L}\frac{\breve{\beta}_T}{\beta_T}(1 + \Gamma)^2 + m(\frac{\breve{\beta}_L}{\beta_L} + \frac{\breve{\beta}_T}{\beta_T}) \\ &+ \frac{k_x^2 + k_y^2}{\beta_L\beta_T}(1 + \Gamma - m)^2 + \frac{\breve{\beta}_L\breve{\beta}_T}{k_x^2 + k_y^2}\Gamma^2\Big]. \end{aligned}$$

Taking into account (5.7) and (5.7') for the change of variables we can write (5.14) in the equivalent form

$$\begin{pmatrix}\breve{\Phi}^-\\ \breve{\Psi}_x^-\\ \breve{\Psi}_y^-\end{pmatrix} = T\begin{pmatrix}\Phi^-\\ \Psi_x^-\\ \Psi_y^-\end{pmatrix},$$

where the explicit form of the transmission matrix T is now given as

$$T = \frac{1}{k_x^2+k_y^2}\begin{pmatrix}k_x^2+k_y^2 & 0 & 0\\ 0 & k_x & k_y\\ 0 & k_y & -k_x\end{pmatrix}\begin{pmatrix}a_4/\mathcal{D} & 0 & -a_2/\mathcal{D}\\ 0 & a_5 & 0\\ -a_3/\mathcal{D} & 0 & a_1/\mathcal{D}\end{pmatrix}\begin{pmatrix}1 & 0 & 0\\ 0 & k_x & k_y\\ 0 & k_y & -k_x\end{pmatrix}. \qquad (5.15)$$

Consider now the equations (5.11)-(5.13) and solve them with respect to the reflected amplitudes. The result is expressed as

$$\begin{pmatrix}\Phi^+\\ V^+\\ W^+\end{pmatrix} = \begin{pmatrix}\Phi^-\\ -V^-\\ -W^-\end{pmatrix} + \begin{pmatrix}-\breve{\beta}_L(1+\Gamma)/\beta_L & 0 & -(1+\Gamma-m)/\beta_L\\ 0 & m & 0\\ -\breve{\beta}_L\Gamma & 0 & m-\Gamma\end{pmatrix}\begin{pmatrix}\breve{\Phi}^-\\ \breve{V}^-\\ \breve{W}^-\end{pmatrix}.$$

In view of (5.15) and the transformation equation (5.7') we obtain

$$\begin{pmatrix}\Phi^+\\ \Psi_x^+\\ \Psi_y^+\end{pmatrix} = R\begin{pmatrix}\Phi^-\\ \Psi_x^-\\ \Psi_y^-\end{pmatrix}$$

where the explicit form of the reflection matrix R is obtained as

$$R = \mathbb{I} + \frac{1}{k_x^2+k_y^2}\begin{pmatrix}k_x^2+k_y^2 & 0 & 0\\ 0 & k_x & k_y\\ 0 & k_y & -k_x\end{pmatrix}\times$$
$$\begin{pmatrix}-\breve{\beta}_L(1+\Gamma)/\beta_L & 0 & -(1+\Gamma-m)/\beta_L\\ 0 & m & 0\\ -\breve{\beta}_L\Gamma & 0 & m-\Gamma\end{pmatrix}\begin{pmatrix}1 & 0 & 0\\ 0 & k_x & k_y\\ 0 & k_y & -k_x\end{pmatrix}T. \qquad (5.16)$$

The components Ψ_z^+ and Ψ_z^- follow from substitution into (3.11).

By way of example, consider an incident longitudinal wave of amplitude Φ^-; the transmitted and reflected pairs are obtained from the general expressions by letting $\Psi_x^- = \Psi_y^- = 0$. Notice that in general the coefficients a_i and the parameters k_x and k_y are complex-valued whence it follows that the amplitudes of the transmitted and reflected pairs turn out to be complex-valued too, even though the amplitudes of the incident pair are taken to be real. For this reason the scalar Φ and the vector Ψ entering the definition of longitudinal and transverse waves have been chosen as complex-valued.

For use in subsequent applications and comparison with known results, especially those concerning reflection and transmission at elastic plane interfaces [22], it is worth considering in some detail the particular case when the incident wave vector is vertically polarized, which means that

$$k_y = 0. \tag{5.17}$$

Actually, this condition can always be satisfied by homogeneous waves at the boundary between elastic half-spaces, if the axes are properly chosen, since the wave vector is real valued.

Insert (5.17) into the continuity conditions (5.1) and (5.4) and then into the definitions (5.2) and (5.5) of the matrices C_1 and C_2; in so doing the last line of C_2 should be 0, ρ, 0, as follows from comparison with (3.13). Because $k_x^2 + \beta_T^2 = \kappa_T - k_y^2 = \kappa_T$ and in view of (5.17) we find that the continuity requirement for displacement and traction results in

$$\begin{pmatrix} k_x & 0 & -\beta_T \\ 0 & \kappa_T/\beta_T & 0 \\ \mu(\frac{1}{2}\kappa_T - k_x^2) & 0 & \mu k_x \beta_T \end{pmatrix} \begin{pmatrix} \Phi^+ + \Phi^- \\ \Psi_x^+ - \Psi_x^- \\ \Psi_y^+ - \Psi_y^- \end{pmatrix} = \begin{pmatrix} k_x & 0 & -\breve{\beta}_T \\ 0 & \breve{\kappa}_T/\breve{\beta}_T & 0 \\ \breve{\mu}(\frac{1}{2}\breve{\kappa}_T - k_x^2) & 0 & \breve{\mu} k_x \breve{\beta}_T \end{pmatrix} \begin{pmatrix} \breve{\Phi}^- \\ -\breve{\Psi}_x^- \\ -\breve{\Psi}_y^- \end{pmatrix}, \tag{5.18}$$

$$\begin{pmatrix} \beta_L & 0 & k_x \\ \mu\beta_L & 0 & \mu(k_x - \frac{1}{2}\kappa_T/k_x) \\ 0 & \rho & 0 \end{pmatrix} \begin{pmatrix} \Phi^+ - \Phi^- \\ \Psi_x^+ + \Psi_x^- \\ \Psi_y^+ + \Psi_y^- \end{pmatrix} = -\begin{pmatrix} \breve{\beta}_L & 0 & k_x \\ \breve{\mu}\breve{\beta}_L & 0 & \breve{\mu}(k_x - \frac{1}{2}\breve{\kappa}_T/k_x) \\ 0 & \breve{\rho} & 0 \end{pmatrix} \begin{pmatrix} \breve{\Phi}^- \\ -\breve{\Psi}_x^- \\ -\breve{\Psi}_y^- \end{pmatrix}. \tag{5.19}$$

Before we proceed to the evaluation of the solution, we observe that the linear system decouples into a subsystem of two equations for the two unknowns Ψ_x^+ and $\breve{\Psi}_x^-$ in terms of Ψ_x^-, and another subsystem of four equations for the four unknowns Φ^+, Ψ_y^+, $\breve{\Phi}^-$, $\breve{\Psi}_y^-$, in terms of Φ^- and Ψ_y^-. The former case corresponds to an incident wave, with *horizontally polarized amplitude*, which is generated by a vector potential with vanishing y-component. The latter case is related to an incident wave with *vertically polarized amplitude*, either generated by a scalar potential or a vector potential with vanishing x-component. The decoupling of the linear system corresponds to the conservation of vertical and horizontal polarization at a plane interface [22].

To solve the system we follow the procedure of the general case: first we solve with respect to the column vector in the left sides, then we change variables through (5.7) and (5.7'), which hold even though $k_y = 0$. We observe that the inverse of the matrix in the left side of (5.18) is obtained from the inverse (5.3) of C_1, simply by letting k_y be zero. As regards (5.19), the inverse of the square matrix in the left side is

$$\frac{2}{\mu\kappa_T\beta_L} \begin{pmatrix} \mu(\frac{1}{2}\kappa_T - k_x^2) & k_x^2 & 0 \\ 0 & 0 & \mu\kappa_T\beta_L/2\rho \\ \mu k_x \beta_L & -k_x\beta_L & 0 \end{pmatrix}.$$

Once the required matrix products are performed we obtain the equations (5.8)-(5.10) as a consequence of (5.18), and (5.11)-(5.13) as the counterpart of (5.19). It is not necessary to examine here these equations; we only observe that the requirement $k_y = 0$ has to be taken into account.

As in the general case we arrive at (5.14), namely

$$\begin{pmatrix} \breve{\Phi}^- \\ \breve{V}^- \\ \breve{W}^- \end{pmatrix} = \begin{pmatrix} a_4/\mathcal{D} & 0 & -a_2/\mathcal{D} \\ 0 & a_5 & 0 \\ -a_3/\mathcal{D} & 0 & a_1/\mathcal{D} \end{pmatrix} \begin{pmatrix} \Phi^- \\ V^- \\ W^- \end{pmatrix}.$$

Since $k_y = 0$, by (5.7) and (5.7') we have

$$\Psi_x = V/k_x, \qquad \Psi_y = -W/k_x.$$

Accordingly, the last matrix equation yields

$$\begin{pmatrix} \breve{\Phi}^- \\ \breve{\Psi}_x^- \\ \breve{\Psi}_y^- \end{pmatrix} = T_V \begin{pmatrix} \Phi^- \\ \Psi_x^- \\ \Psi_y^- \end{pmatrix}, \qquad (k_y = 0) \tag{5.20}$$

where

$$T_V = \begin{pmatrix} a_4/\mathcal{D} & 0 & a_2 k_x/\mathcal{D} \\ 0 & a_5 & 0 \\ a_3/(\mathcal{D} k_x) & 0 & a_1/\mathcal{D} \end{pmatrix}$$

denotes the transmission matrix in the case of vertically polarized incident wave vector. Similarly we find that

$$\begin{pmatrix} \Phi^+ \\ \Psi_x^+ \\ \Psi_y^+ \end{pmatrix} = R_V \begin{pmatrix} \Phi^- \\ \Psi_x^- \\ \Psi_y^- \end{pmatrix}, \qquad (k_y = 0) \tag{5.21}$$

where the reflection matrix for vertically polarized incident waves is

$$R_V = \mathbb{I} + \begin{pmatrix} -(\breve{\beta}_L/\beta_L)(1+\Gamma) & 0 & (k_x/\beta_L)(1+\Gamma-m) \\ 0 & m & 0 \\ \breve{\beta}_L \Gamma/k_x & 0 & m-\Gamma \end{pmatrix} T_V.$$

We conclude this point with the analysis of reflection and refraction when the wave vector of the incident pair is parallel to the normal to the interface, that is $k_x = k_y = 0$. Comparison with (3.14) and (3.15) shows that the continuity of displacement and traction results in the two vector equations

$$k_T\big[-(\Psi_y^+ - \Psi_y^-)\mathbf{e}_x + (\Psi_x^+ - \Psi_x^-)\mathbf{e}_y\big] + k_L(\Phi^+ - \Phi^-)\mathbf{e}_z = \breve{k}_T(\breve{\Psi}_y^- \mathbf{e}_x - \breve{\Psi}_x^- \mathbf{e}_y) - \breve{k}_L \breve{\Phi}^- \mathbf{e}_z,$$

$$\rho\big[-(\Psi_y^+ + \Psi_y^-)\mathbf{e}_x + (\Psi_x^+ + \Psi_x^-)\mathbf{e}_y + (\Phi^+ + \Phi^-)\mathbf{e}_z\big] = \breve{\rho}(-\breve{\Psi}_y^- \mathbf{e}_x + \breve{\Psi}_x^- \mathbf{e}_y + \breve{\Phi}^- \mathbf{e}_z).$$

By solving the resulting three linear systems of two equations in two unknowns we have the result

$$\breve{\Phi}^- = \frac{2\rho k_L}{\breve{\rho}k_L + \rho\breve{k}_L}\Phi^-, \qquad \Phi^+ = \frac{\breve{\rho}k_L - \rho\breve{k}_L}{\breve{\rho}k_L + \rho\breve{k}_L}\Phi^-,$$
$$\breve{\Psi}_x^- = \frac{2\rho k_T}{\breve{\rho}k_T + \rho\breve{k}_T}\Psi_x^-, \qquad \Psi_x^+ = \frac{\breve{\rho}k_T - \rho\breve{k}_T}{\breve{\rho}k_T + \rho\breve{k}_T}\Psi_x^-,$$
$$\breve{\Psi}_y^- = \frac{2\rho k_T}{\breve{\rho}k_T + \rho\breve{k}_T}\Psi_y^-, \qquad \Psi_y^+ = \frac{\breve{\rho}k_T - \rho\breve{k}_T}{\breve{\rho}k_T + \rho\breve{k}_T}\Psi_y^-. \tag{5.22}$$

Degenerate cases. Now we examine some degenerate situations where the z-components of reflected or transmitted wave vectors vanish. In principle there are four possibilities. By analogy with the analysis given in §4.4, if the z-components of the reflected longitudinal or transverse wave vectors vanish then we cannot consider an incident pair, but rather an incident tranverse or longitudinal mode, respectively. This case is very simple and then is not considered. Rather we examine transmitted wave vectors with a vanishing z-component in which case the incident waves may still constitute a pair.

Let the incident pair satisfy

$$k_x^2 + k_y^2 = \breve{\kappa}_L \quad \Longleftrightarrow \quad \breve{\beta}_L = 0, \tag{5.23}$$

and consider a scalar potential of the form

$$\breve{\phi} = \breve{\Phi}^- \exp\left[i(k_x x + k_y y)\right].$$

By (5.23), $\breve{\phi}$ is a solution to the Helmholtz scalar equation in the lower half-space. The related displacement $\mathbf{U}[\breve{\phi}]$ and traction $\mathbf{t}[\breve{\phi}]$ are found through substitution into (1.2) and (1.3) and take the form

$$\mathbf{U}[\breve{\phi}] = i\breve{\Phi}^-(k_x\mathbf{e}_x + k_y\mathbf{e}_y),$$
$$\mathbf{t}[\breve{\phi}] = 2\breve{\mu}\breve{A}\breve{\Phi}^-\mathbf{e}_z.$$

The vector potential is of the general form

$$\breve{\boldsymbol{\psi}} = \breve{\Psi}^- \exp(-i\beta_T z)\exp\left[i(k_x x + k_y y)\right]$$

and the corresponding expressions for displacement and traction follow by comparison with (1.7), (1.8) or their component formulations (3.4)-(3.9). Setting aside inessential details, we say that when these results are replaced into the continuity conditions (5.1) and (5.4), the matrices $\breve{C}_1$ and $\breve{C}_2$ take the forms obtained from (5.2) and (5.5), provided the accent ˇ is introduced where appropriate. We only observe that the first column of the matrix $\breve{C}_2$ vanishes because $\breve{\beta}_L = 0$ in view of the assumption (5.23). Nevertheless, we can still solve

(5.4) for the reflected amplitudes Φ^+, Ψ_x^+, Ψ_y^+, to find (5.11)-(5.13), provided we recall that $\breve{\beta}_L$ vanishes. So, apart from this substitution, everything goes as in the general case, as far as the determination of reflected and transmitted amplitudes is concerned.

Other cases occur when the x- and y-components of the incident wave vectors satisfy the condition

$$k_x^2 + k_y^2 = \breve{\kappa}_T \quad \Longleftrightarrow \quad \breve{\beta}_T = 0. \tag{5.24}$$

Things are not simple as in the previous case (5.23), since we cannot refer to the results holding for the general case. Actually, a vector potential of the form

$$\breve{\boldsymbol{\psi}} = \breve{\boldsymbol{\Psi}}^- \exp[i(k_x x + k_y y)], \tag{5.25}$$

with constant vector amplitude $\breve{\boldsymbol{\Psi}}^-$ is a solution for the vector Helmholtz equation $(1.1)_2$, as a consequence of the constraint (5.24). With reference to (5.25), the condition of vanishing divergence for $\breve{\boldsymbol{\psi}}$ can be expressed as

$$\breve{\Psi}_x^- = -\frac{k_y}{k_x}\breve{\Psi}_y^-, \tag{5.26}$$

in analogy with (4.20). If $k_x \neq 0$, this means that we are regarding $\breve{\Psi}_y^-$ and $\breve{\Psi}_z^-$ as independent components of $\breve{\boldsymbol{\Psi}}^-$, whereas $\breve{\Psi}_x^-$ is found through (5.26). This is the reason why the conclusions of the previous analysis of the general case, where we found it convenient to regard $\breve{\Psi}_x^-$ and $\breve{\Psi}_y^-$ as independent parameters, cannot be extended to the present framework.

Substitution of (5.25) into the general representations (1.2) and (1.3) and account of (5.24) and (5.26) yields the contributions of the vector potential $\breve{\boldsymbol{\psi}}$ to the lower limits at the interface for displacement and traction in the form

$$\mathbf{U}[\breve{\boldsymbol{\psi}}] = i\big[k_y\breve{\Psi}_z^-\mathbf{e}_x - k_x\breve{\Psi}_z^-\mathbf{e}_y + (k_x\breve{\Psi}_y^- - k_y\breve{\Psi}_x^-)\mathbf{e}_z\big],$$

$$\mathbf{t}[\breve{\boldsymbol{\psi}}] = \breve{\mu}\frac{\breve{\kappa}_T}{k_x}\breve{\Psi}_y^-(k_x\mathbf{e}_x + k_y\mathbf{e}_y).$$

The pertinent contributions arising from the lower scalar potential

$$\breve{\phi} = \breve{\Phi}^- \exp(-i\breve{\beta}_L z)\exp[i(k_x x + k_y y)],$$

follow from comparison with (3.4)-(3.9). Accordingly, when we consider the interface $\mathcal{P}$, we find the following limit values from below

$$\mathbf{U} = i\big[(k_x\breve{\Phi}^- + k_y\breve{\Psi}_z^-)\mathbf{e}_x + (k_y\breve{\Phi}^- - k_x\breve{\Psi}_z^-)\mathbf{e}_y + (-\breve{\beta}_L\breve{\Phi}^- + \frac{\breve{\kappa}_T}{k_x}\breve{\Psi}_y^-)\mathbf{e}_z\big],$$

$$\mathbf{t} = 2\breve{\mu}\big(-\breve{\beta}_L\breve{\Phi}^- + \tfrac{1}{2}\frac{\breve{\kappa}_T}{k_x}\breve{\Psi}_y^-\big)(k_x\mathbf{e}_x + k_y\mathbf{e}_y) - \breve{\mu}\breve{\kappa}_T\breve{\Phi}^-\mathbf{e}_z.$$

These relations can be used to construct the analogues of the continuity conditions (5.1) and (5.4). Specifically, the continuity of U_x/i, U_y/i, and $t_z/2$ yields

$$C_1 \begin{pmatrix} \Phi^+ + \Phi^- \\ \Psi_x^+ - \Psi_x^- \\ \Psi_y^+ - \Psi_y^- \end{pmatrix} = \begin{pmatrix} k_x & 0 & k_y \\ k_y & 0 & -k_x \\ -\breve{\mu}\breve{\kappa}_T/2 & 0 & 0 \end{pmatrix} \begin{pmatrix} \breve{\Phi}^- \\ \breve{\Psi}_y^- \\ \breve{\Psi}_z^- \end{pmatrix}, \tag{5.27}$$

with C_1 as defined in (5.2). Similarly, imposing the continuity requirement for U_z/i, $t_x/2k_x$, and $t_y/2k_y$ leads to the correspondent of (5.4), namely,

$$C_2 \begin{pmatrix} \Phi^+ - \Phi^- \\ \Psi_x^+ + \Psi_x^- \\ \Psi_y^+ + \Psi_y^- \end{pmatrix} = \begin{pmatrix} -\breve{\beta}_L & \breve{\kappa}_T/k_x & 0 \\ -\breve{\mu}\breve{\beta}_L & \breve{\mu}\breve{\kappa}_T/2k_x & 0 \\ -\breve{\mu}\breve{\beta}_L & \breve{\mu}\breve{\kappa}_T/2k_x & 0 \end{pmatrix} \begin{pmatrix} \breve{\Phi}^- \\ \breve{\Psi}_y^- \\ \breve{\Psi}_z^- \end{pmatrix}, \tag{5.28}$$

with C_2 given by (5.5). Upon left multiplication by the matrix

$$\begin{pmatrix} k_y & -k_x & 0 \\ k_x & k_y & 1/\mu \\ k_x & k_y & 0 \end{pmatrix},$$

equation (5.27) gives rise to the system

$$k_x(\Psi_x^+ - \Psi_x^-) + k_y(\Psi_y^+ - \Psi_y^-) = -\beta_T \frac{\breve{\kappa}_T}{\kappa_T} \breve{\Psi}_z^-, \tag{5.29}$$

$$\Phi^+ + \Phi^- = 2\frac{\breve{\kappa}_T}{\kappa_T}(1 - \frac{\breve{\mu}}{2\mu})\breve{\Phi}^-, \tag{5.30}$$

$$k_y(\Psi_x^+ - \Psi_x^-) - k_x(\Psi_y^+ - \Psi_y^-) = \frac{\breve{\kappa}_T}{\beta_T}\Big[1 - 2\frac{\breve{\kappa}_T}{\kappa_T}(1 - \frac{\breve{\mu}}{2\mu})\Big]\breve{\Phi}^-. \tag{5.31}$$

Similarly, multiplication of both sides of (5.28) by $(C_2)^{-1}$ leads to

$$\Phi^+ - \Phi^- = \frac{2}{\mu\kappa_T\beta_L}\big[-\breve{\beta}_L(\mu A + \breve{\mu}\breve{\kappa}_T)\breve{\Phi}^- + \frac{\breve{\kappa}_T}{k_x}(\mu A + \frac{1}{2}\breve{\kappa}_T\breve{\mu})\breve{\Psi}_y^-\big], \tag{5.32}$$

$$\Psi_x^+ + \Psi_x^- = \frac{2k_y}{\mu\kappa_T}\big[\breve{\beta}_L(\mu - \breve{\mu})\breve{\Phi}^- - \frac{\breve{\kappa}_T}{k_x}(\mu - \frac{1}{2}\breve{\mu})\breve{\Psi}_y^-\big], \tag{5.33}$$

$$\Psi_y^+ + \Psi_y^- = -\frac{2k_x}{\mu\kappa_T}\big[\breve{\beta}_L(\mu - \breve{\mu})\breve{\Phi}^- - \frac{\breve{\kappa}_T}{k_x}(\mu - \frac{1}{2}\breve{\mu})\breve{\Psi}_y^-\big]. \tag{5.34}$$

We first determine the amplitudes of transmitted fields. We multiply (5.33) by k_x and (5.34) by k_y. Then we sum the results and contrast with (5.29) to find

$$\breve{\Psi}_z^- = \frac{2}{\beta_T}\frac{\kappa_T}{\breve{\kappa}_T}(k_x\Psi_x^- + k_y\Psi_y^-);$$

this gives one of the required transmission coefficients. Next we multiply (5.33) by k_y, subtract the product of k_x and (5.34), and contrast the result with (5.31) to find

$$k_y\Psi_x^- - k_x\Psi_y^- = \{\breve{\beta}_L\frac{\breve{\kappa}_T}{\kappa_T}(1-\frac{\breve{\mu}}{\mu}) - \frac{\breve{\kappa}_T}{\beta_T}[\frac{1}{2} - \frac{\breve{\kappa}_T}{\kappa_T}(1-\frac{\breve{\mu}}{2\mu})]\}\breve{\Phi}^- - \frac{1}{k_x}\frac{\breve{\kappa}_T^2}{\kappa_T}(1-\frac{\breve{\mu}}{2\mu})\breve{\Psi}_y^-. \quad (5.35)$$

Meanwhile, the difference of (5.30) and (5.32) yields

$$\Phi^- = [\frac{\breve{\kappa}_T}{\kappa_T}(1-\frac{\breve{\mu}}{2\mu}) + \frac{\breve{\beta}_L}{\beta_L}\frac{\breve{\kappa}_T}{\kappa_T}(\frac{1}{2}\frac{\kappa_T}{\breve{\kappa}_T} + \frac{\breve{\mu}}{\mu} - 1)]\breve{\Phi}^- - \frac{1}{k_x\beta_L}[\frac{\breve{\kappa}_T}{2} - \frac{\breve{\kappa}_T^2}{\kappa_T}(1-\frac{\breve{\mu}}{2\mu})]\breve{\Psi}_y^-. \quad (5.36)$$

Now we multiply (5.36) by β_L, add the result to (5.35) and solve with respect to $\breve{\Psi}_y^-$. The result reads

$$\begin{aligned}\breve{\Psi}_y^- &= -2\frac{k_x}{\breve{\kappa}_T}(k_y\Psi_x^- - k_x\Psi_y^- + \beta_L\Phi^-)\\ &\quad + k_x\{2\frac{\beta_L}{\kappa_T}(1-\frac{\breve{\mu}}{2\mu}) + \frac{\breve{\beta}_L}{\breve{\kappa}_T} - \frac{1}{\beta_T}[1 - 2\frac{\breve{\kappa}_T}{\kappa_T}(1-\frac{\breve{\mu}}{2\mu})]\}\breve{\Phi}^-.\end{aligned} \quad (5.37)$$

Insertion of (5.37) into (5.35) yields

$$\breve{\Phi}^- = \frac{1}{\breve{\kappa}_T}\frac{[\kappa_T - 2\breve{\kappa}_T(1-\frac{\breve{\mu}}{\mu})](k_y\breve{\Psi}_x^- - k_x\breve{\Psi}_y^-) - 2\beta_L\breve{\kappa}_T(1-\frac{\breve{\mu}}{\mu})\breve{\Phi}^-}{-\frac{1}{2}\frac{\breve{\mu}}{\mu}\breve{\beta}_L - 2\beta_L\frac{\breve{\kappa}_T}{\kappa_T}(1-\frac{\breve{\mu}}{2\mu})^2 - \frac{1}{2\beta_T\kappa_T}[\kappa_T - 2\breve{\kappa}_T(1-\frac{\breve{\mu}}{2\mu})]^2}. \quad (5.38)$$

Replacement of this expression for $\breve{\Phi}^-$ into (5.37) gives the component $\breve{\Psi}_y^-$ in terms of the amplitudes of the incident pair. Then Φ^+, Ψ_x^+ and Ψ_y^+ follow from (5.32)-(5.34).

4.6 Reflection and refraction coefficients

While the previous section provides a scheme of reflection and refraction in terms of potentials, now the connection is examined between the amplitudes of the displacement fields pertaining to the various waves. Incidentally, this allows a detailed account of the amplitude vectors (or polarizations). Moreover, the more familiar description in terms of angles is established.

For definiteness we consider the relationships between the transmitted, reflected, and incident pairs in the form

$$\begin{cases}\breve{\Phi}^- = T_{11}\Phi^- + T_{12}\Psi_x^- + T_{13}\Psi_y^-\\ \breve{\Psi}_x^- = T_{21}\Phi^- + T_{22}\Psi_x^- + T_{23}\Psi_y^-\\ \breve{\Psi}_y^- = T_{31}\Phi^- + T_{32}\Psi_x^- + T_{33}\Psi_y^-\end{cases} \qquad \begin{cases}\Phi^+ = R_{11}\Phi^- + R_{12}\Psi_x^- + R_{13}\Psi_y^-\\ \Psi_x^+ = R_{21}\Phi^- + R_{22}\Psi_x^- + R_{23}\Psi_y^-\\ \Psi_y^+ = R_{31}\Phi^- + R_{32}\Psi_x^- + R_{33}\Psi_y^-\end{cases} \quad (6.1)$$

where the entries of the matrices T and R are specified by (5.15) and (5.16). The columns of T and R may be interpreted as refraction and reflection coefficients. For instance, if we take an incident longitudinal wave so that both Ψ_x^- and Ψ_y^- vanish, while only Φ^- is non-zero, we may regard T_{11} and R_{11} as the refraction and reflection coefficients of the longitudinal inhomogeneous wave; T_{21}, T_{31} and R_{21}, R_{31} yield the transformation coefficients of the longitudinal wave into a transverse in refraction and reflection, respectively. As opposed to the elastic case, two pairs of coefficients, instead of two coefficients, are required to recover the transmitted and reflected transverse waves originated by an incident longitudinal one [22]. Of course, similar remarks also apply to the case of an incident transverse wave.

Actually, the form of the coefficients of reflection and refraction is rather involved, in that they depend on the characteristics of the incident pair, via k_x and k_y, and the material parameters of the media via κ_L, κ_T, $\breve{\kappa}_L$, $\breve{\kappa}_T$. By analogy with elasticity, one might think that a qualitative information is obtained by regarding the coefficients as functions of the geometric characteristics of the incident waves, which is obtained by looking at k_x as the product $k^i \sin\theta^i$, with variable θ^i. Seemingly, such is not the case in the present context in that we have to vary the two components k_x and k_y, both of them being complex-valued; furthermore, also the reflection and refraction coefficients are complex-valued. A rather special case where this can be done is treated in the next section.

The analysis of elastic waves at an interface is greatly simplified by viewing any transverse wave with arbitrary direction of the displacement vector as the superposition of a shear wave of horizontal polarization and of a wave of vertical polarization. At the origin of these distinctions is the fact that vertically and horizontally polarized waves do not interact with each other at a plane interface [22] and hence they can be studied separately. Owing to the characteristic features of inhomogeneous waves, we will see that the distinction between vertically and horizontally polarized waves at the plane boundary between viscoelastic bodies does not hold in general, but applies if the wave vectors of the incident pair are vertically polarized.

To avoid any ambiguities it is worth recalling that if the real and imaginary parts of the incident wave vectors belong to a common vertical plane, then this plane is invariant under reflection and refraction. That is why special attention has been drawn to wave vectors in the vertical plane, which allows considerable simplifications in the expressions of the reflection and transmission matrices. The analogy with elastic waves suggests that we consider also amplitude vectors that are vertically or horizontally polarized. Therefore, in dealing with the polarization we have to specify whether we are referring to amplitudes or wave vectors, unless this is unambiguously clear through the context. In any case the vertical plane is systematically identified with the (x, z)-plane.

A further remark is in order. The behaviour at a plane boundary has been examined through the representation of incident reflected and transmitted pairs in terms of the corresponding (complex) scalar and vector potentials. This approach results in much simpler calculations and brings immediately into evidence mode conversion phenomena at interfaces, the only drawback being that the relations between the (complex) amplitudes pertaining to longitudinal and transverse waves are left aside. Of course, these relations must be considered before polarization effects are investigated. On the contrary, in elasticity one may avoid finding the explicit relations between amplitudes (see e.g. [22]). This feature is ultimately due to the fact that, as is shown later, the wave vectors of homogeneous (plane) waves in non-dissipative media are vertically polarized.

With reference to the representation (1.5) of the amplitude of transverse waves, define the amplitude vector $\mathbf{A}$ through

$$\mathbf{A} = \mathbf{k}_T \times \mathbf{\Psi}. \tag{6.2}$$

By a vertically, or horizontally, polarized amplitude we mean that $\mathbf{A}$ belongs to a plane which is perpendicular, or parallel, to the interface. Accordingly as the transverse wave belongs to an upgoing or downgoing pair, the Cartesian representation of $\mathbf{A}$ reads

$$\mathbf{A}^+ = (k_y\Psi_z^+ - \beta_T\Psi_y^+)\mathbf{e}_x + (\beta_T\Psi_x^+ - k_x\Psi_z^+)\mathbf{e}_y + (k_x\Psi_y^+ - k_y\Psi_x^+)\mathbf{e}_z,$$

$$\mathbf{A}^- = (k_y\Psi_z^- + \beta_T\Psi_y^-)\mathbf{e}_x - (\beta_T\Psi_x^- + k_x\Psi_z^-)\mathbf{e}_y + (k_x\Psi_y^- - k_y\Psi_x^-)\mathbf{e}_z.$$

Incidentally, by (6.2) we find the condition $\mathbf{A}\cdot\mathbf{k}_T = 0$ which is characteristic of transverse waves. Once A_x and A_y are given, the third component A_z follows from the vanishing of the inner product with $\mathbf{k}_T$, whence

$$A_z^+ = -(k_xA_x^+ + k_yA_y^+)/\beta_T \quad \text{or} \quad A_z^- = (k_xA_x^- + k_yA_y^-)/\beta_T, \tag{6.3}$$

depending on whether the transverse wave is upgoing or downgoing. We can then say that the problem in study consists in the determination of Φ^+, A_x^+, A_y^+ and $\breve{\Phi}^-$, $\breve{A}_x^-$, $\breve{A}_y^-$ in terms of the given quantities Φ^-, A_x^-, A_y^-.

To benefit from the results of the previous section, it is worth observing that, upon substitution of (3.11) and comparison with the definitions (5.7) of V and W, the x- and y-components of $\mathbf{A}$ can be related to V and W through the transformations laws

$$\begin{pmatrix}\Phi^-\\ A_x^-\\ A_y^-\end{pmatrix} = \frac{1}{k_x^2+k_y^2}\begin{pmatrix}k_x^2+k_y^2 & 0 & 0\\ 0 & k_y\kappa_T/\beta_T & -k_x\beta_T\\ 0 & -k_x\kappa_T/\beta_T & -k_y\beta_T\end{pmatrix}\begin{pmatrix}\Phi^-\\ V^-\\ W^-\end{pmatrix}, \tag{6.4a}$$

$$\begin{pmatrix}\Phi^-\\ V^-\\ W^-\end{pmatrix} = \begin{pmatrix}1 & 0 & 0\\ 0 & k_y\beta_T/\kappa_T & -k_x\beta_T/\kappa_T\\ 0 & -k_x/\beta_T & -k_y/\beta_T\end{pmatrix}\begin{pmatrix}\Phi^-\\ A_x^-\\ A_y^-\end{pmatrix}, \tag{6.4b}$$

for downgoing waves and

$$\begin{pmatrix}\Phi^+\\ A_x^+\\ A_y^+\end{pmatrix} = \frac{1}{k_x^2+k_y^2}\begin{pmatrix}k_x^2+k_y^2 & 0 & 0\\ 0 & -k_y\kappa_T/\beta_T & k_x\beta_T\\ 0 & k_x\kappa_T/\beta_T & k_y\beta_T\end{pmatrix}\begin{pmatrix}\Phi^+\\ V^+\\ W^+\end{pmatrix}, \tag{6.5a}$$

$$\begin{pmatrix} \Phi^+ \\ V^+ \\ W^+ \end{pmatrix} = \begin{pmatrix} 1 & 0 & 0 \\ 0 & -k_y\beta_T/\kappa_T & k_x\beta_T/\kappa_T \\ 0 & k_x/\beta_T & k_y/\beta_T \end{pmatrix} \begin{pmatrix} \Phi^+ \\ A_x^+ \\ A_y^+ \end{pmatrix}, \tag{6.5b}$$

for upgoing waves; the scalar amplitudes Φ are considered here for future convenience.

Comparison with (6.4) and the analogous relations holding for the lower medium allows (5.14) to be written in the equivalent form

$$\begin{pmatrix} \breve{\Phi}^- \\ \breve{A}_x^- \\ \breve{A}_y^- \end{pmatrix} = T_A \begin{pmatrix} \Phi^- \\ A_x^- \\ A_y^- \end{pmatrix}, \tag{6.6}$$

where the transformation matrix for transmitted amplitudes is defined as

$$T_A = \frac{1}{k_x^2+k_y^2} \begin{pmatrix} k_x^2+k_y^2 & 0 & 0 \\ 0 & k_y\breve{\kappa}_T/\breve{\beta}_T & -k_x\breve{\beta}_T \\ 0 & -k_x\breve{\kappa}_T/\breve{\beta}_T & -k_y\breve{\beta}_T \end{pmatrix} \begin{pmatrix} a_4/\mathcal{D} & 0 & -a_2/\mathcal{D} \\ 0 & a_5 & 0 \\ -a_3/\mathcal{D} & 0 & a_1/\mathcal{D} \end{pmatrix} \times$$
$$\begin{pmatrix} 1 & 0 & 0 \\ 0 & k_y\beta_T/\kappa_T & -k_x\beta_T/\kappa_T \\ 0 & -k_x/\beta_T & -k_y/\beta_T \end{pmatrix}. \tag{6.7}$$

In a similar way we can exploit (5.16) to obtain

$$\begin{pmatrix} \Phi^+ \\ A_x^+ \\ A_y^+ \end{pmatrix} = R_A \begin{pmatrix} \Phi^- \\ A_x^- \\ A_y^- \end{pmatrix}, \tag{6.8}$$

where the reflection matrix for amplitudes is defined as

$$R_A = \begin{pmatrix} 1 & 0 & 0 \\ 0 & 1 & 0 \\ 0 & 0 & 1 \end{pmatrix} + \frac{1}{k_x^2+k_y^2} \begin{pmatrix} k_x^2+k_y^2 & 0 & 0 \\ 0 & -k_y\kappa_T/\beta_T & k_x\beta_T \\ 0 & k_x\kappa_T/\beta_T & k_y\beta_T \end{pmatrix} \times$$
$$\begin{pmatrix} -\breve{\beta}_L(1+\Gamma)/\beta_L & 0 & (m-\Gamma-1)/\beta_L \\ 0 & m & 0 \\ -\breve{\beta}_L\Gamma & 0 & m-\Gamma \end{pmatrix} \begin{pmatrix} 1 & 0 & 0 \\ 0 & k_y\breve{\beta}_T/\breve{\kappa}_T & -k_x\breve{\beta}_T/\breve{\kappa}_T \\ 0 & -k_x/\breve{\beta}_T & -k_y/\breve{\beta}_T \end{pmatrix} T_A, \tag{6.9}$$

where $m = \breve{\rho}/\rho$. In view of (6.6) and (6.8) the entries of the matrices T_A and R_A may be interpreted as the refraction and reflection coefficients for the amplitudes.

As an application of these transformation laws for amplitudes we show that the condition of vertical polarization for amplitudes is not preserved at the boundary between viscoelastic media. Consider an incident transverse wave, suppose that the (complex) amplitude vector is vertically polarized, and choose the (x,z)-plane so that $A_y^- = 0$. On observing that $\Phi^- = 0$, by (6.6) and (6.7) we have

$$\breve{A}_y^- = \frac{k_x k_y}{k_x^2+k_y^2}\Big(-a_5\frac{\breve{\kappa}_T}{\kappa_T}\frac{\beta_T}{\breve{\beta}_T} + \frac{a_1}{\mathcal{D}}\frac{\breve{\beta}_T}{\beta_T}\Big)A_x^-.$$

This proves that in general $\breve{A}_y^- \neq 0$, which means that the amplitude of the refracted transmitted wave has a non-vanishing y-component and hence is not contained in the vertical plane of the incident wave. Accordingly, the vertical polarization for amplitudes is not preserved under refraction.

Assume that the wave vector is vertically polarized and let $k_y = 0$. From the vector representation of the amplitudes we obtain

$$A_x^- = \beta_T \Psi_y^-, \qquad A_y^- = -\kappa_T \Psi_x^- / \beta_T, \tag{6.10}$$

for downgoing waves, and

$$A_x^+ = -\beta_T \Psi_y^+, \qquad A_y^+ = \kappa_T \Psi_x^+ / \beta_T, \tag{6.11}$$

for upgoing ones. By (6.10) and (6.11), the results (5.20) and (5.21), which provide the potentials of reflected and transmitted waves, may be expressed in terms of amplitudes. In particular, (5.20) can be written in the equivalent form

$$\begin{pmatrix} \breve{\Phi}^- \\ \breve{A}_x^- \\ \breve{A}_y^- \end{pmatrix} = \begin{pmatrix} 1 & 0 & 0 \\ 0 & 0 & \breve{\beta}_T \\ 0 & -\breve{\kappa}_T/\breve{\beta}_T & 0 \end{pmatrix} \begin{pmatrix} a_4/\mathcal{D} & 0 & a_2 k_x/\mathcal{D} \\ 0 & a_5 & 0 \\ a_3/(\mathcal{D}k_x) & 0 & a_1/\mathcal{D} \end{pmatrix} \times$$
$$\begin{pmatrix} 1 & 0 & 0 \\ 0 & 0 & -\beta_T/\kappa_T \\ 0 & 1/\beta_T & 0 \end{pmatrix} \begin{pmatrix} \Phi^- \\ A_x^- \\ A_y^- \end{pmatrix},$$

that is

$$\begin{pmatrix} \breve{\Phi}^- \\ \breve{A}_x^- \\ \breve{A}_y^- \end{pmatrix} = \begin{pmatrix} a_4/\mathcal{D} & a_2 k_x/(\mathcal{D}\beta_T) & 0 \\ a_3\breve{\beta}_T/(\mathcal{D}k_x) & a_1\breve{\beta}_T/(\mathcal{D}\beta_T) & 0 \\ 0 & 0 & a_5\beta_T\breve{\kappa}_T/(\breve{\beta}_T\kappa_T) \end{pmatrix} \begin{pmatrix} \Phi^- \\ A_x^- \\ A_y^- \end{pmatrix}. \tag{6.12}$$

As regards (5.21) we find that

$$\begin{pmatrix} \Phi^+ \\ A_x^+ \\ A_y^+ \end{pmatrix} = \begin{pmatrix} c_{11} & c_{12} & 0 \\ c_{21} & c_{22} & 0 \\ 0 & 0 & 1 - m a_5 \end{pmatrix} \begin{pmatrix} \Phi^- \\ A_x^- \\ A_y^- \end{pmatrix}, \tag{6.13}$$

where the entries c are defined as

$$\begin{aligned} c_{11} &= 1 + [a_3(1+\Gamma-m) - a_4\breve{\beta}_L(1+\Gamma)]/(\mathcal{D}\beta_L) \\ c_{12} &= k_x[a_1(1+\Gamma-m) - a_2\breve{\beta}_L(1+\Gamma)]/(\mathcal{D}\beta_L\beta_T) \\ c_{21} &= \beta_T[a_3(\Gamma-m) - a_4\breve{\beta}_L\Gamma]/(\mathcal{D}k_x) \\ c_{22} &= 1 + [a_1(\Gamma-m) - a_2\breve{\beta}_L\Gamma]/\mathcal{D}. \end{aligned}$$

Consider a transverse wave with vertically polarized amplitude and assume $A_y^- = 0$. As a consequence of (6.12) and (6.13) we find $A_y^+ = \breve{A}_y^- = 0$, which means that the reflected and transmitted waves are vertically polarized. Notice that this holds independently of the value of Φ^-. In addition, according to (6.10) and (6.11) we find that $\Psi_x^+ = \breve{\Psi}_x^- = 0$. Substitution into the expressions (3.11) of the z-component of Ψ shows that

$$\Psi_z^+ = \breve{\Psi}_z^- = 0.$$

It follows that only the y-component of the vector potential is non-zero, and this is consistent with the description of elastic vertically polarized waves (cf. [22]).

Suppose that a transverse wave with horizontally polarized amplitude is incident at the boundary, which means $A_z^- = 0$ and $\Phi^- = 0$. Comparison with (6.3) and account of the condition $k_y = 0$ shows that also $A_x^- = 0$. Then application of (6.11) and (6.12) shows that $\Phi^+ = \breve{\Phi}^- = 0$ and $A_x^+ = \breve{A}_x^- = 0$, thus implying that no longitudinal wave is originated at $\mathcal{P}$, and that both reflected and transmitted transverse waves are horizontally polarized. Again, we have the same formal results as for the shear waves of horizontal polarization in linear elasticity [22].

Results on the behaviour of inhomogeneous waves at a plane interface have been described in terms of Cartesian components of wave vectors. Most often, though, incidence, reflection, and transmission angles are the parameters used in the investigation of plane waves within the framework of linear elasticity and linearized fluid dynamics [2, 3]. It is then worth establishing a connection between the two descriptions. According to the notation of §4.2 we represent wave vectors in the form

$$\mathbf{k} = \mathbf{k}_1 + i\mathbf{k}_2 = k_1\mathbf{n}_1 + ik_2\mathbf{n}_2,$$

with $\mathbf{n}_1$ and $\mathbf{n}_2$ unit vectors. The wave vector of the incident wave is regarded as given; those pertaining to the reflected and transmitted waves are determined through the use of Snell's law and the material properties of the medium where propagation takes place. In terms of the Cartesian components of $\mathbf{k}$ we find

$$k_1 = \sqrt{(\mathrm{Re}\,k_x)^2 + (\mathrm{Re}\,k_y)^2 + (\mathrm{Re}\,k_z)^2}, \qquad k_2 = \sqrt{(\mathrm{Im}\,k_x)^2 + (\mathrm{Im}\,k_y)^2 + (\mathrm{Im}\,k_z)^2}.$$

Following a standard approach, we denote by $\theta \in [0, \pi/2]$ the angle between the directions of $\mathbf{n}_1$ and $\mathbf{n}$ (or $\mathbf{e}_z = -\mathbf{n}$). Moreover we denote by ϕ the angle between the unit vectors $\mathbf{n}_2$ and $\mathbf{e}_z$, with $\phi \in [0, \pi)$, and by α and ψ the angles between the x-axis and the projection of $\mathbf{n}_1$ and $\mathbf{n}_2$ on the (x, y)-plane. It follows that the unit vectors $\mathbf{n}_1$ and $\mathbf{n}_2$ may be represented as

$$\mathbf{n}_1 = \sin\theta\,(\cos\alpha\,\mathbf{e}_x + \sin\alpha\,\mathbf{e}_y) \pm \cos\theta\,\mathbf{e}_z,$$

$$\mathbf{n}_2 = \sin\phi(\cos\psi\,\mathbf{e}_x + \sin\psi\,\mathbf{e}_y) + \cos\phi\,\mathbf{e}_z.$$

The $+(-)$ sign for $\mathbf{n}_1$ is relative to upgoing (downgoing) waves. As a consequence of Snell's law, the angles α and ψ are the same for the various kinds of waves involved in the reflection-refraction phenomena at the interface. The relationships between the Cartesian components of $\mathbf{k}_1$ and $\mathbf{k}_2$ and the above angles follow from straightforward geometric considerations; they are expressed as

$$\begin{aligned} k_x &= k_1 \sin\theta \cos\alpha + ik_2 \sin\phi \cos\psi, \\ k_y &= k_1 \sin\theta \sin\alpha + ik_2 \sin\phi \sin\psi, \\ k_z &= \pm k_1 \cos\theta + ik_2 \cos\phi. \end{aligned} \tag{6.14}$$

Of course, the relations (6.14) can be solved for the angles in terms of Cartesian components, provided the definitions of k_1 and k_2 are taken into account. Further, the analogues of the relations (6.14) for the longitudinal and transverse waves belonging to the incident, reflected, and transmitted pairs can be used to obtain the reflection and transmission matrices in terms of angles.

In general, the better understanding allowed by the use of angles has an unpleasant counterpart due to the fact that the expressions of the matrices T and R become more and more involved. This fact is essentially the motivation for our preference to the use of Cartesian components. In special cases, though, the use of angles may be more profitable. A case in this sense occurs when the wave vectors of the incident pair are vertically polarized. By Snell's law this holds for the reflected and the transmitted pairs as well. Accordingly we let all wave vectors belong to the (x, z)-plane; this corresponds to setting $\alpha = 0$ and $\psi = 0$. Then we find

$$\mathbf{k} = k_1(\sin\theta\, \mathbf{e}_x \pm \cos\theta\, \mathbf{e}_z) + ik_2(\sin\phi\, \mathbf{e}_x + \cos\phi\, \mathbf{e}_z), \tag{6.15}$$

where

$$k_1 = \sqrt{(\operatorname{Re} k_x)^2 + (\operatorname{Re} k_z)^2}, \qquad k_2 = \sqrt{(\operatorname{Im} k_x)^2 + (\operatorname{Im} k_z)^2}. \tag{6.16}$$

In a more familiar notation we may write

$$\begin{aligned} k_x = k_1^i \sin\theta^i + ik_2^i \sin\phi^i &= \\ k_{L1}^r \sin\theta_L^r + ik_{L2}^r \sin\phi_L^r &= k_{T1}^r \sin\theta_T^r + ik_{T2}^r \sin\phi_T^r = \\ \breve{k}_{L1} \sin\breve{\theta}_L + i\breve{k}_{L2} \sin\breve{\phi}_L &= \breve{k}_{T1} \sin\breve{\theta}_T + i\breve{k}_{T2} \sin\breve{\phi}_T. \end{aligned} \tag{6.17}$$

Meanwhile the value of β is expressed as

$$\begin{aligned} \beta_L &= -\mathbf{k}_L^i \cdot \mathbf{e}_z = k_{L1}^i \cos\theta_L^i - ik_{L2}^i \cos\phi_L^i = \mathbf{k}_L^r \cdot \mathbf{e}_z = k_{L1}^r \cos\theta_L^r + ik_{L2}^r \cos\phi_L^r, \\ \beta_T &= -\mathbf{k}_T^i \cdot \mathbf{e}_z = k_{T1}^i \cos\theta_T^i - ik_{T2}^i \cos\phi_T^i = \mathbf{k}_T^r \cdot \mathbf{e}_z = k_{T1}^r \cos\theta_T^r + ik_{T2}^r \cos\phi_T^r, \\ \breve{\beta}_L &= -\breve{\mathbf{k}}_L \cdot \mathbf{e}_z = \breve{k}_{L1} \cos\breve{\theta}_L - i\breve{k}_{L2} \cos\breve{\phi}_L, \\ \breve{\beta}_T &= -\breve{\mathbf{k}}_T \cdot \mathbf{e}_z = \breve{k}_{T1} \cos\breve{\theta}_T - i\breve{k}_{T2} \cos\breve{\phi}_T. \end{aligned}$$

Substitution into the expressions of T_V and R_V obtained from (5.20) and (5.21) yields the expressions of the reflection and transmission matrices in terms of the moduli of the real and the imaginary parts of the wave vectors and of the angles with the unit vector $\mathbf{n}$.

4.7 Applications and numerical results

Now we examine quantitative aspects of reflection and refraction. For definiteness and for simplicity we restrict attention to the case when the interface is the common boundary of a viscoelastic solid and an inviscid fluid. No essential effect is lost, while considerable simplifications of the expressions involved are achieved. In addition we take this opportunity to complete our analysis of reflection and refraction phenomena, since the behaviour of the perfect fluid does not fit our general scheme in that now we require the continuity of the traction and the normal component (only) of the displacement. Via straightforward changes the procedure is applicable to the other cases already examined. In our numerical calculations the fluid is identified with water and the solid with annealed copper. It is assumed that the incident wave is coming from the fluid in the upper half-space. The reflected and transmitted wave vectors, reflection and refraction coefficients, and suitable ratios between the various kinds of energy densities are regarded as functions of the incidence angle.

The displacement field of the inviscid fluid occupying the upper half-space is described through the scalar potential ϕ. The incident wave is taken as plane and homogeneous. Hence the wave vector $\mathbf{k}_L^i$ is real, as well as its specular image $\mathbf{k}_L^r$. We choose the (x,z)-plane as the plane of $\mathbf{k}_L^i$ and $\mathbf{n}$, so that $k_y = 0$; it is assumed that the x-axis is oriented in the direction induced by $\mathbf{k}_L^i$. In view of the generalized Snell's law the wave vectors of incident and reflected waves admit the representations

$$\mathbf{k}_L^i = k_L(\sin\theta^i\,\mathbf{e}_x - \cos\theta^i\,\mathbf{e}_z), \qquad \mathbf{k}_L^r = k_L(\sin\theta^i\,\mathbf{e}_x + \cos\theta^i\,\mathbf{e}_z) \tag{7.1}$$

where θ^i denotes the angle between $\mathbf{n}$ and $\mathbf{k}_L^i$. As an immediate consequence of (7.1) it follows that

$$k_x = k_L \sin\theta^i, \qquad \beta_L = k_L\cos\theta^i.$$

Similarly, for the transmitted pair we find

$$\breve{\mathbf{k}}_T = k_L\sin\theta^i\,\mathbf{e}_x - \sqrt{\breve{\kappa}_T - (k_L\sin\theta^i)^2}\,\mathbf{e}_z, \tag{7.2}$$

$$\breve{\mathbf{k}}_L = k_L\sin\theta^i\,\mathbf{e}_x - \sqrt{\breve{\kappa}_L - (k_L\sin\theta^i)^2}\,\mathbf{e}_z, \tag{7.3}$$

showing in particular that the imaginary parts of $\breve{\mathbf{k}}_T$ and $\breve{\mathbf{k}}_L$ are vertical (downgoing) vectors, which corresponds to $\breve{\phi}_L = \breve{\phi}_T = \pi$. Consequently, the angles between the real and imaginary parts of $\breve{\mathbf{k}}_L$ and $\breve{\mathbf{k}}_T$ equal $\breve{\theta}_L$ and $\breve{\theta}_T$, respectively. So the significant part of the dependence of the wave vectors, on the incidence angle and the frequency ω, is given by the z-components of $\breve{\mathbf{k}}_L$ and $\breve{\mathbf{k}}_T$ through (7.2) and (7.3) in the form

$$\breve{\beta}_L = \sqrt{\breve{\kappa}_L - (k_L \sin\theta^i)^2} \qquad \breve{\beta}_T = \sqrt{\breve{\kappa}_T - (k_L \sin\theta^i)^2}. \tag{7.4}$$

An alternative, more intuitive formulation is obtained on observing that

$$\breve{\beta}_L = \breve{k}_{L1}\cos\breve{\theta}_L + i\breve{k}_{L2}, \qquad \breve{\beta}_T = \breve{k}_{T1}\cos\breve{\theta}_T + i\breve{k}_{T2}. \tag{7.5}$$

The moduli $\breve{k}_{L1}, \breve{k}_{L2}, \breve{k}_{T1}, \breve{k}_{T2}$, which, along with the two angles $\breve{\theta}_L, \breve{\theta}_T$, enter (7.5) depend on the incidence angle θ^i. We examine such dependence for both longitudinal and transverse waves at the same time. Omit the subscripts L or T and consider (3.2.11) and (3.2.12) with $\alpha = \cos\gamma = \cos\breve{\theta}$. Letting $c = k^i \sin\theta^i$ and making use of Snell's law we are left with the system

$$\breve{k}_1^2 - \breve{k}_2^2 = a,$$

$$2\breve{k}_1\breve{k}_2\cos\breve{\theta} = b,$$

$$\breve{k}_1 \sin\breve{\theta} = c,$$

in the unknowns $\breve{k}_1$, $\breve{k}_2$, $\breve{\theta}$; obviously $\breve{k}_1 \geq c$. Observe that in such a case the solution (3.2.14) is not operative in that $\alpha = \cos\breve{\theta}$ has to be determined. Now, it follows at once that only one solution holds, namely

$$\breve{k}_1 = \tfrac{1}{2}\sqrt{2(a+c^2) + 2\sqrt{(a-c^2)^2 + b^2}}, \tag{7.6}$$

$$\breve{k}_2 = \tfrac{1}{2}\sqrt{2(-a+c^2) + 2\sqrt{(a-c^2)^2 + b^2}}, \tag{7.7}$$

$$\tan\breve{\theta} = \frac{2c}{\sqrt{2(a-c^2) + 2\sqrt{(a-c^2)^2 + b^2}}}. \tag{7.8}$$

It may be useful to observe that the identity

$$\frac{2}{\sqrt{2(a-c^2) + 2\sqrt{(a-c^2)^2 + b^2}}} = \frac{1}{b}\sqrt{2(-a+c^2) + 2\sqrt{(a-c^2)^2 + b^2}}$$

holds. Incidentally, if the material is non-dissipative in that $b = 0$, and moreover $a > 0$, then we have $\breve{k}_2 = 0$. In such a case $\breve{k}_1 = \sqrt{a}$ and then a critical value θ_c^i for θ^i is such that $\theta = \pi/2$, namely

$$\sin\theta_c^i = \frac{\sqrt{a}}{k^i}.$$

In order to determine the numerical values of $\breve{\kappa}_L$ and $\breve{\kappa}_T$ we identify the solid with a material such as annealed copper. On adopting cgs units the mass density is taken as

$$\breve{\rho} = 8.9.$$

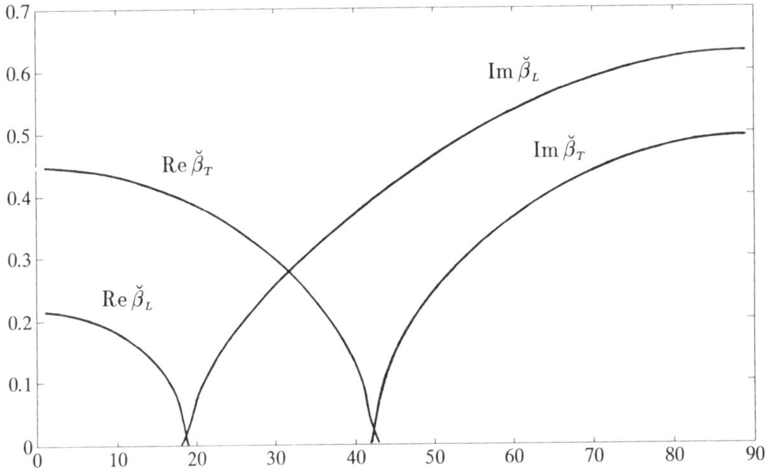

Fig 4.2 Behaviour of $\operatorname{Re}\breve{\beta}_L$, $\operatorname{Im}\breve{\beta}_L$, $\operatorname{Re}\breve{\beta}_T$, $\operatorname{Im}\breve{\beta}_T$ against the incidence angle θ^i at the fixed frequency $\omega = 10^5$ Hz.

As regards the relaxation functions, upon an analysis of creep tests [120] and the passage from Kelvin and Maxwell models to the stress functional [35] we have

$$\breve{\mu}' = -\breve{\mu}_0[a\exp(-b\tau) + c\exp(-d\tau)]$$

$$\breve{\lambda}' = -\breve{\lambda}_0[a\exp(-b\tau) + c\exp(-d\tau)],$$

where the constants a, b, c, d are given by

$$a = 0.1567\ 10^{-5}, \quad b = 0.1616\ 10^{-5}, \quad c = 0.1565\ 10^{-1}, \quad d = 0.5157$$

and

$$\breve{\lambda}_0 = 103.70\ 10^{10} \qquad \breve{\mu}_0 = 44.44\ 10^{10}.$$

Substitution into the expressions of λ and μ (cf. §3.1) leads to

$$\frac{\breve{\mu}}{\breve{\mu}_0} = \frac{\breve{\lambda}}{\breve{\lambda}_0} = 1 - \frac{ab}{b^2 + \omega^2} - \frac{cd}{d^2 + \omega^2} - i\omega\Big(\frac{a}{b^2 + \omega^2} + \frac{c}{d^2 + \omega^2}\Big).$$

Then (3.2.1) and (3.2.3) yield the explicit expressions of $\breve{\kappa}_L$ and $\breve{\kappa}_T$. As far as the fluid (water) is concerned we let

$$\rho = 1, \qquad dp/d\rho = 2.25\ 10^{10}.$$

As shown in Fig. 4.2, in correspondence with small incidence angles the attenuation (measured by Im $\breve{\beta}_L$ and Im $\breve{\beta}_L$) is negligible for both waves; this behaviour closely resembles that of elastic bodies. However the effect of dissipation becomes more and more influential for growing angles of incidence. Strictly speaking, no critical angle occurs though the real parts of $\breve{\beta}_L$ and $\breve{\beta}_T$ may be regarded as vanishing beyond suitable angles. Similar considerations also hold when the solid body is elastic; in that case the angles giving vanishing values for $\breve{\beta}_L$ and $\breve{\beta}_T$ coincide with the critical angles ([2], Ch. 5). When the incidence angle is greater than the critical one Re $\breve{\beta}_L$ and Re $\breve{\beta}_T$ vanish, which means that the transmitted waves propagate parallel to the interface and decay with distance from the interface. As shown also by the vanishing of the energy flux intensity, this behaviour corresponds to total reflection.

To determine the emerging waves we need the reflection and transmission coefficients as functions of the incidence angle. Now, the results of the previous section follow from the requirement of continuity for displacement and traction at the boundary. If one of the media is an inviscid fluid then, in addition to the continuity of the traction, we require only the continuity of the normal component of the displacement, in that the fluid can freely slip on the boundary without formation of cavitations.

To find the appropriate formulation of these conditions we recall that the scalar potential within the fluid may be represented as

$$\phi = \Phi^+ \exp(i\beta_L z) + \Phi^- \exp(-i\beta_L z),$$

where Φ^- is the real amplitude of the incident wave, Φ^+ is the (possibly complex) unknown amplitude of the reflected wave, and the common factor $\exp(ik_x x)$ is understood. By use of (1.7) and (1.8) we find that the expressions of displacement and traction at the boundary are given by

$$\mathbf{U} = ik_x(\Phi^+ + \Phi^-)\,\mathbf{e}_x + i\beta_L(\Phi^+ - \Phi^-)\,\mathbf{e}_z$$

$$\mathbf{t} = \rho\omega^2(\Phi^+ + \Phi^-)\,\mathbf{e}_z.$$

In the viscoelastic solid, according to (3.3) we have

$$\breve{\phi} = \breve{\Phi}^- \exp(-i\breve{\beta}_L z), \qquad \breve{\boldsymbol{\psi}} = \breve{\boldsymbol{\Psi}}^- \exp(-i\breve{\beta}_T z)$$

$\breve{\Phi}^-, \breve{\boldsymbol{\Psi}}^-$ being the transmitted amplitudes. Accordingly, displacement and traction follow from (3.6)-(3.9). Therefore the continuity of the normal component of $\mathbf{U}$, U_z, yields

$$\beta_L(\Phi^+ - \Phi^-) - \breve{\kappa}_T \breve{\Psi}_y^- /(2k_x) = 0, \tag{7.9}$$

whereas the continuity of t_x, t_y and t_z gives

$$\breve{\beta}_L \breve{\Phi}^- - \left[k_x - \breve{\kappa}_T/(2k_x)\right] \breve{\Psi}_y^- = 0, \tag{7.10}$$

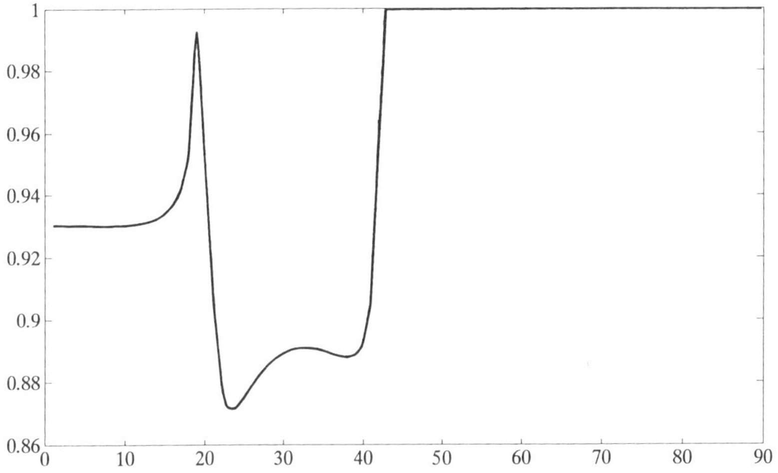

Fig. 4.3 Modulus of the reflection coefficient $\mathcal{R}_L$ versus the incidence angle θ^i.

$$\breve{\Psi}_x^- = 0, \tag{7.11}$$

$$\rho\omega^2(\Phi^+ + \Phi^-) = 2\breve{\mu}\big[-(k_x^2 - \breve{\kappa}_T/2)\breve{\Phi}^- - k_x\breve{\beta}_T\breve{\Psi}_y^-\big]. \tag{7.12}$$

Substitution of $\breve{\Psi}_x^- = 0$ into (3.11) and account of $k_y = 0$ shows that $\breve{\Psi}_z^- = 0$. The remaining amplitudes Φ^+, $\breve{\Phi}^-$ and $\breve{\Psi}_y^-$ are determined by solving the linear system (7.9), (7.10) and (7.12). The result is

$$\Phi^+ = \mathcal{R}_L\Phi^-, \qquad \breve{\Phi}^- = \mathcal{T}_L\Phi^-, \qquad \breve{\Psi}_y^- = \mathcal{T}_T\Phi^-, \tag{7.13}$$

where the reflection coefficient $\mathcal{R}_L$ and the refraction coefficients $\mathcal{T}_L$ and $\mathcal{T}_T$ are defined by

$$D\mathcal{R}_L = 4mk_x^2\beta_L[\breve{\beta}_L\breve{\beta}_T + (k_x^2 - \tfrac{1}{2}\breve{\kappa}_T)^2/k_x^2] - \breve{\beta}_L\breve{\kappa}_T^2, \tag{7.14}$$

$$D\mathcal{T}_L = -4\beta_L(k_x^2 - \tfrac{1}{2}\breve{\kappa}_T)\breve{\kappa}_T, \tag{7.15}$$

$$D\mathcal{T}_T = -4k_x\beta_L\breve{\beta}_L\breve{\kappa}_T, \tag{7.16}$$

with

$$D = 4mk_x^2\beta_L[\breve{\beta}_L\breve{\beta}_T + (k_x^2 - \tfrac{1}{2}\breve{\kappa}_T)^2/k_x^2] + \breve{\beta}_L\breve{\kappa}_T^2. \tag{7.17}$$

Substitution from (7.1)-(7.4) yields the reflection and refraction coefficients in terms of the incidence angle.

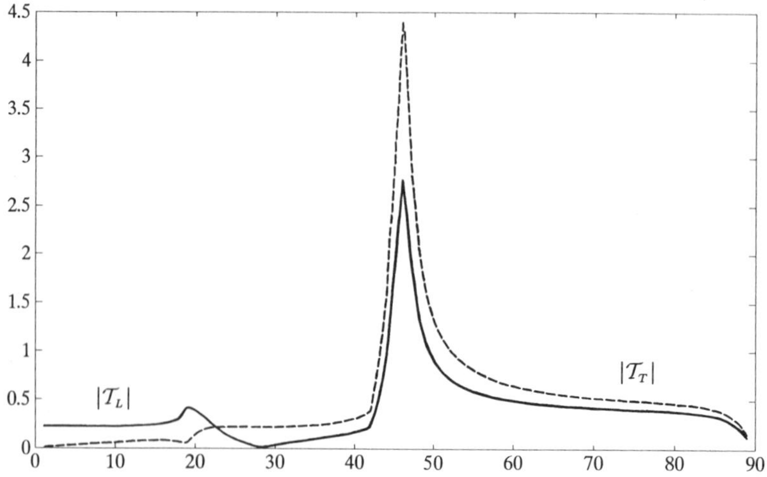

Fig. 4.4 Modulus of the longitudinal and transverse refraction coefficients versus the incidence angle.

The behaviour in Fig. 4.3 shows the existence of two minima for $|\mathcal{R}_L|$ corresponding to suitable angles. In particular $|\mathcal{R}_L|$ attains a remarkable absolute minimum at a critical angle of about 24°. Moreover $|\mathcal{R}_L| = 1$ for incidence angles greater than 41°, which means total reflection. These results parallel those obtained by Mott [133] in the analysis of incidence at a water-stainless steel interface and those of [44] under the influence of dissipation.

Figure 4.4 yields the behaviour of $|\mathcal{T}_L|$ and $|\mathcal{T}_T|$ for angles of incidence varying between 0° and 90°. It is apparent that they are greater than unity at certain values of the angle of incidence. This looks quite paradoxical in that we have in mind that the energy of the incident wave is partitioned among the reflected and transmitted waves. The paradox is solved by the following analysis of the energy flux intensity associated with the pertinent waves.

We recall that, according to §3.5, the (mean) energy flux intensity for the longitudinal transmitted wave in the viscoelastic half-space is

$$\langle \breve{\mathcal{I}}_L \rangle = \tfrac{1}{2}\breve{\rho}\omega^3 |\mathcal{T}_L|^2 |\Phi^-|^2 \exp(-2\breve{\mathbf{k}}_{L2}\cdot\mathbf{x})\breve{k}_{L1}\big(1 + \frac{4\breve{\mu}_1 \breve{k}_{L2}^2 \sin^2\breve{\gamma}_L}{\breve{\rho}\omega^2}\big), \tag{7.18}$$

where $\breve{\gamma}_L$ denotes the angle between the real part $\breve{\mathbf{k}}_{L1}$ and the imaginary part $\breve{\mathbf{k}}_{L2}$ of the longitudinal wave vector $\breve{\mathbf{k}}_L$, and the representation (7.13) for the amplitude $\breve{\Phi}^-$ of the transmitted longitudinal wave has been considered. As regards the transverse wave, the wave vector is vertically polarized while the vector amplitude $\breve{\Psi}^-$ is along the y-axis.

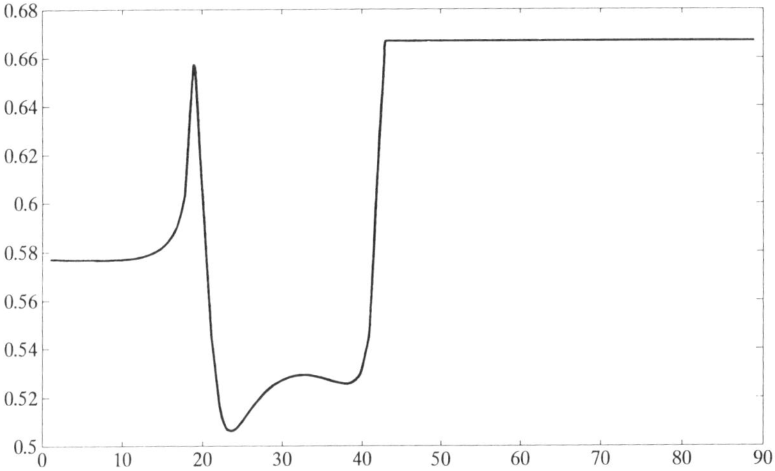

Fig. 4.5 Intensity $\mathcal{I}^r$ of the reflected wave versus the incidence angle.

Therefore (3.5.9) simplifies to

$$\langle \breve{\mathbf{J}}_T \rangle = \tfrac{1}{2}\omega \exp(-2\breve{\mathbf{k}}_{T2} \cdot \mathbf{x})|\breve{\Psi}^-|^2[\breve{\rho}\omega^2 \breve{\mathbf{k}}_{T1} - 4(\breve{\mathbf{k}}_1 \times \breve{\mathbf{k}}_2) \times (\breve{\mu}_2 \breve{\mathbf{k}}_{T1} + \breve{\mu}_1 \breve{\mathbf{k}}_{T2})].$$

Then the energy flux intensity for the transverse wave is given by

$$\langle \breve{\mathcal{I}}_T \rangle = \tfrac{1}{2}\breve{\rho}\omega^3 |\mathcal{T}_T|^2 |\Phi^-|^2 \exp(-2\breve{\mathbf{k}}_{T2} \cdot \mathbf{x})\breve{k}_{T1}\big(1 + \frac{4\breve{\mu}_1 \breve{k}_{T2} \sin^2 \breve{\gamma}_T}{\breve{\rho}\omega^2}\big), \tag{7.19}$$

with obvious meaning of the symbols, on observing that

$$|\breve{\Psi}^-|^2 = |\breve{\Psi}_y^-|^2 = |\mathcal{T}_T|^2 |\Phi^-|^2.$$

In the half-space occupied by the inviscid fluid the incident energy flux intensity is given by

$$\langle \mathcal{I}^i \rangle = \tfrac{1}{2}\rho\omega^3 |\Phi^-|^2 k_L, \tag{7.20}$$

whereas, for the reflected wave, we find

$$\langle \mathcal{I}^r \rangle = \tfrac{1}{2}\rho\omega^3 |\mathcal{R}_L|^2 |\Phi^-|^2 k_L, \tag{7.21}$$

where Φ^+ has been replaced by $\mathcal{R}_L \Phi^-$.

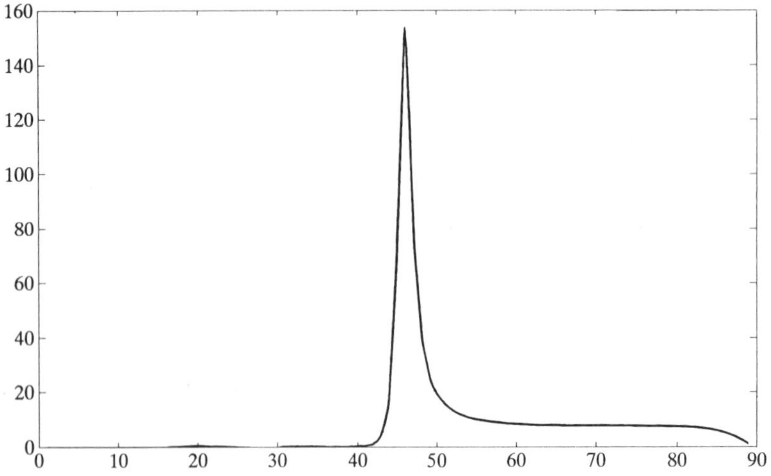

Fig. 4.6 Intensity $\breve{\mathcal{I}}_L$ of a transmitted, longitudinal wave versus the incidence angle.

The dependence of the intensities of reflected and transmitted, longitudinal waves on the incidence angle θ^i is represented in Figs. 4.5, 4.6. The transmitted, transverse wave shows a behaviour very close to that of the transmitted, longitudinal wave. It is henceforth assumed that $|\Phi^-| = 1$, and it is understood that the intensities are evaluated up to a common factor $\omega^3/2$. According to (7.20) and (7.21) the ratio of the reflected to the incident intensity is $|\mathcal{R}_L|^2$. Since $|\mathcal{R}_L| = 1$ for incidence angles greater than 40°, then the constant incident energy flux intensity equals the (constant) value of that reflected at high incidence angles. As a second remark we observe that the longitudinal and transverse transmitted energy flux intensities become very large at certain angles greater than 42°. This seems to contradict the energy conservation, especially because the ratio between reflected and incident densities is equal to unity.

To solve this seeming paradox we observe that, since $\langle\mathcal{I}\rangle$ represents the flow of energy, per unit time, per unit area in a plane orthogonal to $\mathbf{k}_1$, then the energy flow through a tube with generatrix $\mathbf{k}_1$, is $\langle\mathcal{I}\rangle$ times the cross section.

Figure 4.7 represents an incident wave and two transmitted waves. The tubes have a common intersection with the interface and, for technical convenience, we regard the common area da as infinitesimal. The cross sections da^i, da^r, $d\breve{a}_T$, $d\breve{a}_L$ are related to da by

$$da^i = da^r = da\cos\theta^i, \quad d\breve{a}_T = da\cos\breve{\theta}_T, \quad d\breve{a}_L = da\cos\breve{\theta}_L.$$

Here the first equality is a direct consequence of Snell's law. Then, by (7.18)-(7.21) and use of the relation $k_1 = k^i \sin\theta^i/\sin\theta$ we can express the ratio of the transmitted energy

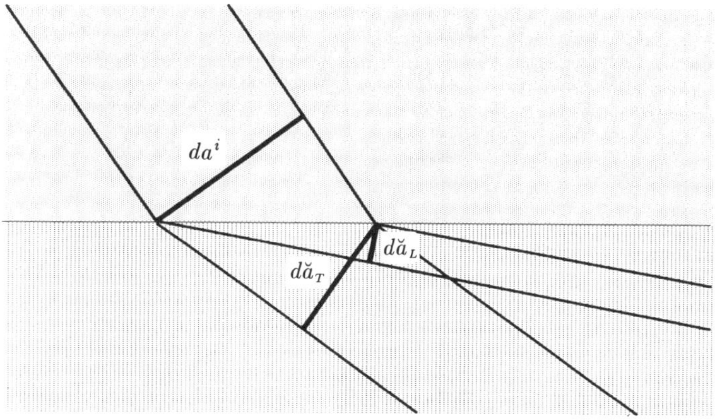

Fig. 4.7 An incident (homogeneous) wave produces two transmitted waves in a dissipative solid.

to the incident one, at the plane $z = 0$, where $\mathbf{x}$ is orthogonal to $\breve{\mathbf{k}}_{L2}$, as

$$\frac{\langle \breve{\mathcal{I}}_L \rangle d\breve{a}_L}{\langle \mathcal{I}^i \rangle da^i} = \frac{\breve{\rho} \tan \theta^i}{\rho \tan \breve{\theta}_L}(1 + \frac{4\breve{\mu}_1 \breve{k}_{L2} \sin^2 \breve{\theta}_L}{\breve{\rho}\omega^2})|\mathcal{T}_L|^2 =: W_L. \tag{7.22}$$

Similarly we find

$$\frac{\langle \breve{\mathcal{I}}_T \rangle d\breve{a}_T}{\langle \mathcal{I}^i \rangle da^i} = \frac{\breve{\rho} \tan \theta^i}{\rho \tan \breve{\theta}_T}(1 + \frac{4\breve{\mu}_1 \breve{k}_{T2} \sin^2 \breve{\theta}_T}{\breve{\rho}\omega^2})|\mathcal{T}_T|^2 =: W_T, \tag{7.23}$$

and

$$\frac{\langle \mathcal{I}^r \rangle da^r}{\langle \mathcal{I}^i \rangle da^i} = |\mathcal{R}_L|^2 =: V. \tag{7.24}$$

So the effective transmission and reflection coefficients are given by (7.22)-(7.24) in terms of the material properties incorporated in a, b, c and may be viewed as parameterized by the frequency ω and the incidence angle θ^i. By the natural idea that energy is conserved in the reflection-refraction process, we expect that the sum of the quantities (7.22) to (7.24) equal unity. The numerical check through Fig. 4.8 shows that such is really the case thus confirming the validity of the arguments about geometry and energy aspects of reflection and refraction.

A more detailed dependence on the incidence angle θ^i and the frequency ω can be given by determining the expressions of $\sin \breve{\theta}$ and $\tan \breve{\theta}$. For definiteness, look at the transmitted, transverse wave and then examine the dependence of $W_T = \langle \breve{\mathcal{I}}_T \rangle d\breve{a}_T / \langle \mathcal{I}^i \rangle da^i$ on θ^i. In view of (7.6) and (7.8), with

$$a = \frac{\breve{\rho}\omega^2 \breve{\mu}_1}{\breve{\mu}_1^2 + \breve{\mu}_2^2}, \qquad b = -\frac{\breve{\rho}\omega^2 \breve{\mu}_2}{\breve{\mu}_1^2 + \breve{\mu}_2^2}, \qquad c = \frac{\omega}{\sqrt{p_\rho}} \sin \theta^i,$$

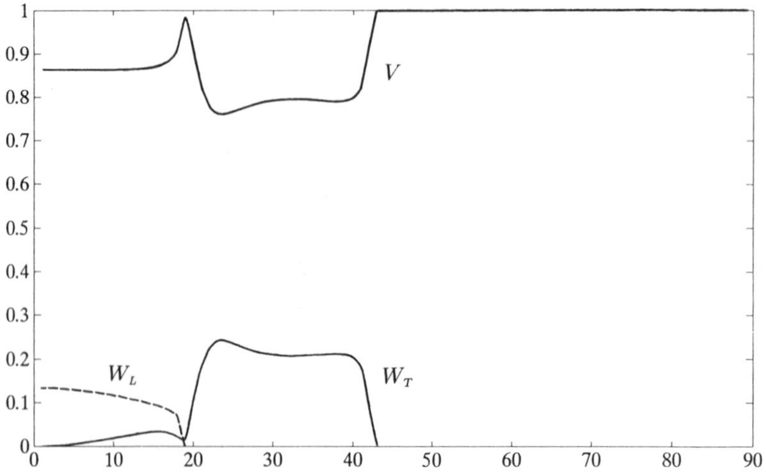

Fig. 4.8 Effective transmission and reflection coefficients W_L, W_T, V versus the incidence angle θ^i.

and (7.16) we obtain

$$W_T(\theta^i) = \frac{\breve{\rho}\sqrt{p_\rho}}{2\omega\rho\cos\theta^i}\sqrt{2(a - c^2(\theta^i)) + 2\sqrt{(a - c^2(\theta^i))^2 + b^2}}\times$$

$$\Big(1 + \frac{8\breve{\mu}_1 \sin^2\theta^i\sqrt{2(-a + c^2(\theta^i)^2 + 2\sqrt{(a - c^2(\theta^i))^2 + b^2}}}{\breve{\rho}p_\rho[2(a + c^2(\theta^i)) + 2\sqrt{(a - c^2(\theta^i))^2 + b^2}]}\Big)|\mathcal{T}_T(\theta^i)|^2$$

where the writing $c(\theta^i)$ is a reminder that c depends on θ^i, in the known way.

5 SURFACE WAVES

A surface wave may be viewed as a wave that propagates in a direction tangential to a surface, while its amplitude decreases (exponentially) in the normal direction. This means that the amplitude of surface waves varies in planes of constant phase. This in turn shows that inhomogeneous waves are the natural framework for the investigation of surface waves. The surface that guides the wave may be the external boundary of a body or a material discontinuity such as an interface between different materials. The existence and the form of surface waves depends on the conditions at the surface.

The simplest mathematical description of surface waves may be given by a function of the form $f(z)\exp[i(kx - \omega t)]$ where x is a direction in the plane boundary surface and z is orthogonal; f expresses the amplitude decay with distance from the surface. Surface waves are easily seen to occur in incompressible viscous fluids. The corresponding propagation condition, or secular equation, determines k in terms of the frequency ω and the viscosity of the fluid. The corresponding surface wave is shown to be the superposition of two inhomogeneous waves. The analysis of surface wave solutions on elastic half-spaces is a preliminary step toward the more involved case of viscoelastic half-spaces. In this regard emphasis is given to the role of the secular equation and the corresponding polynomial analogue, usually called Rayleigh equation. Non-trivial solutions are found to hold when the polarization is in the sagittal plane.

The search for surface wave solutions in viscoelasticity leads formally to a complex analogue of the (elastic) Rayleigh equation. The procedure to select the physically admissible solutions is highly non-trivial and involves both the decay condition and the thermodynamic restrictions. Obviously more complicated is the analysis of Stoneley waves, namely surface waves at the interface between two half-spaces with different material properties.

Surface waves in solids are customarily investigated by neglecting gravity effects. The smallness of these effects is not obvious at all. Indeed, the gravity acceleration induces a prestress in the pertinent body. Such prestress turns out to affect significantly the Rayleigh equation and then the surface wave solutions for any material properties of the half space.

5.1 Surface waves on viscous fluids

The simplest context where surface waves are shown to occur is that of incompressible, viscous fluids. Roughly speaking, surface wave solutions are sought in the form of functions which oscillate along a direction of the free surface of the fluid and decay with depth in the fluid (or distance from the surface).

Following a standard approach, consider an incompressible, viscous fluid occupying the half-space $z \le 0$ and describe the horizontal surface $z = 0$ with the Cartesian coordinates x, y. We take the component v_y of the velocity to vanish and the components v_x, v_z to be independent of y. Then we write the linearized equations of motion as

$$\frac{\partial v_x}{\partial t} = \nu\Big(\frac{\partial^2 v_x}{\partial x^2} + \frac{\partial^2 v_x}{\partial z^2}\Big) - \frac{1}{\rho}\frac{\partial p}{\partial x}, \tag{1.1}$$

$$\frac{\partial v_z}{\partial t} = \nu\Big(\frac{\partial^2 v_z}{\partial x^2} + \frac{\partial^2 v_z}{\partial z^2}\Big) - \frac{1}{\rho}\frac{\partial p}{\partial z} - g, \tag{1.2}$$

where $\nu = \mu/\rho$. The components v_x, v_z have to satisfy the incompressibility constraint

$$\frac{\partial v_x}{\partial x} + \frac{\partial v_z}{\partial z} = 0. \tag{1.3}$$

Let $\mathbf{n}$ be the unit vector of the z-axis. The traction $\mathbf{t} = \mathbf{T}\mathbf{n}$ must vanish at the free surface, i.e.

$$0 = T_{zz} = -p + 2\mu\frac{\partial v_z}{\partial z}, \qquad 0 = T_{xz} = \mu\Big(\frac{\partial v_x}{\partial z} + \frac{\partial v_z}{\partial x}\Big), \tag{1.4}$$

at the free surface.

We look for v_x and v_z in the form of waves propagating in the x-direction with amplitude dependent on z. Then we set

$$v_x = \phi(z)\exp[i(kx - \omega t)], \qquad v_z = \psi(z)\exp[i(kx - \omega t)]$$

and require that ϕ, ψ vanish as $z \to -\infty$. Substitution in (1.1)-(1.3) yields

$$i\omega\phi = \nu\Big(k^2\phi - \frac{d^2\phi}{dz^2}\Big) + \frac{1}{\rho}\frac{\partial p}{\partial x}\exp[-i(kx - \omega t)], \tag{1.5}$$

$$i\omega\psi = \nu\Big(k^2\psi - \frac{d^2\psi}{dz^2}\Big) + \Big(\frac{1}{\rho}\frac{\partial p}{\partial z} + g\Big)\exp[-i(kx - \omega t)], \tag{1.6}$$

$$ik\phi + \frac{d\psi}{dz} = 0. \tag{1.7}$$

Let

$$\phi(z) = A\exp(kz) + B\exp(\beta z)$$

where $\operatorname{Re} k$, $\operatorname{Re}\beta > 0$. Then (1.7) gives

$$\psi(z) = -i[A\exp(kz) + \frac{k}{\beta}B\exp(\beta z)].$$

Equations (1.5)-(1.6) are then expected to determine the function $p(x,z)$. By (1.5) and the choice $\beta = k^2 - i\omega/\nu$ we have

$$\frac{p}{\rho} = \frac{\omega}{k}\exp[i(kx-\omega t)]A\exp(kz) + \gamma(z)$$

where γ is arbitrary. Substitution into (1.6) yields $\gamma(z) = -gz$. We now have to determine the "amplitudes" A, B through the boundary conditions (1.4). Letting $z = \zeta$ in $(1.4)_1$ leads to linear terms in A, B and to non-linear ones. In the linear approximation, time differentiation of $(1.4)_1$ at $z = 0$ yields

$$\left[\rho\left(\frac{\omega^2}{k} - g\right) + i2\mu\omega k\right]A + \left(i2\mu\omega k - \frac{k}{\beta}\rho g\right)B = 0.$$

Evaluation of $(1.4)_2$ at $z = 0$ gives

$$2kA + \left(\beta - \frac{k^2}{\beta}\right)B = 0.$$

So we have a homogeneous linear system in A, B. Non-trivial solutions are allowed if and only if the determinantal equation holds, namely

$$\left(2 - i\frac{\omega}{\nu k^2}\right)^2 + \frac{g}{\nu^2 k^3} = 4\sqrt{1 - i\frac{\omega}{\nu k^2}}.$$

The sought complex-valued function $k = k(\omega)$ is then chosen such that $\operatorname{Re} k > 0$. Meanwhile A and B turn out to be related by

$$B = -\frac{2\sqrt{1 - i\omega/\nu k^2}}{2 - i\omega/\nu k^2}A.$$

The corresponding solution for v_x and v_z shows the characteristic feature of surface waves. Owing to the occurrence of the exponentials $\exp(kz)$ and $\exp(\beta z)$, surface waves (in fluids) are waves whose amplitude decreases exponentially with distance from the surface.

A natural question arises as to whether surface waves are inhomogeneous waves. Represent the complex numbers $k(\omega)$ and $\beta(\omega)$ as

$$k = k_1 + ik_2, \qquad \beta = \beta_1 + i\beta_2;$$

we know that $k_1, \beta_1 > 0$. Then, for example,

$$v_x = [A\exp(kz) + B\exp(\beta z)]\exp[i(kx - \omega t)]$$

can be written as the superposition of two inhomogeneous waves, namely

$$v_x = \exp(-i\omega t)\{A\exp[i(k_1x + k_2z) + (-k_2x + k_1z)] + B\exp[i(k_1x + \beta_2z) + (-k_2x + \beta_1z)]\},$$

which correspond to the complex wave vectors

$$\mathbf{k} = k_1\mathbf{e}_1 + k_2\mathbf{e}_3 + i(-k_2\mathbf{e}_1 + k_1\mathbf{e}_3),$$

$$\mathbf{k} = k_1\mathbf{e}_1 + \beta_2\mathbf{e}_3 + i(-k_2\mathbf{e}_1 + \beta_1\mathbf{e}_3).$$

This in turn indicates that inhomogeneous waves are the natural framework for the analysis of surface waves.

5.2 Rayleigh waves on elastic solids

Following again the view that surface waves propagate along a boundary surface while the amplitude decreases with distance from the surface, in this section we examine the standard description of surface waves on elastic solids with a twofold purpose: to recall basic properties of surface waves on elastic half-spaces and to establish a useful reference for later developments. Consider the half-space $z \geq 0$ and look for harmonic wave solutions of the form

$$u_x = A\exp(bz)\exp[i(kx - \omega t)],$$

$$u_z = B\exp(bz)\exp[i(kx - \omega t)],$$

$$u_y = 0.$$

for real k and ω. The (x, z)-plane determined by the direction of propagation x in the plane boundary surface and the orthogonal direction z is usually called sagittal plane. The decay with distance is guaranteed by the condition $\operatorname{Re} b < 0$. Substitution into the equation of motion $\rho\ddot{\mathbf{u}} - \nabla \cdot \mathbf{T} = 0$, where the body force term is disregarded, yields

$$[\omega^2\rho + \mu(b^2 - k^2) - (\mu + \lambda)k^2]A + i(\mu + \lambda)kbB = 0,$$

$$i(\mu + \lambda)kbA + [\omega^2\rho + \mu(b^2 - k^2) + (\mu + \lambda)b^2]B = 0.$$

Non-trivial solutions for A and B hold if and only if

$$(b^2c_L^2 + \omega^2 - c_L^2k^2)(b^2c_T^2 + \omega^2 - c_T^2k^2) = 0,$$

where $c_L^2 = (2\mu + \lambda)/\rho$, $c_T^2 = \mu/\rho$. Letting $c = \omega/k$ and recalling that $\operatorname{Re} b < 0$, we find that this occurs if b takes one of the two values

$$b_1 = -ka_L, \qquad b_2 = -ka_T$$

where $a_L = \sqrt{1 - c^2/c_L^2}$, $a_T = \sqrt{1 - c^2/c_T^2}$. Of course b_1 and b_2 are real if c is real and $c < c_T < c_L$. In correspondence with b_1 and b_2 we have the two solutions

$$\left(\frac{B}{A}\right)_1 = -ia_L, \qquad \left(\frac{B}{A}\right)_2 = -\frac{i}{a_T}.$$

Then we write the displacement components as

$$u_x = \big[A_1 \exp(-ka_L z) + A_2 \exp(-ka_T z)\big] \exp[i(kx - \omega t)],$$

$$u_z = -i\big[a_L A_1 \exp(-ka_L z) + \frac{1}{a_T} A_2 \exp(-ka_T z)\big] \exp[i(kx - \omega t)],$$

where A_1, A_2, k are still to be determined. This displacement field is admissible provided the real part of a_L and a_T are both negative so that both exponentials vanish at infinity ($z = -\infty$).

At the boundary surface $z = 0$ the traction $\mathbf{t} = \mathbf{T}\mathbf{n}$, $\mathbf{n}$ being the unit normal, must vanish. This condition results in the system

$$2a_L A_1 + \frac{1 + a_T^2}{a_T} A_2 = 0,$$

$$(1 + a_T^2) A_1 + 2A_2 = 0$$

for A_1 and A_2. Non-trivial solutions occur if and only if the determinantal equation

$$4a_L a_T - (1 + a_T^2)^2 = 0 \tag{2.1'}$$

holds, namely

$$4\sqrt{1 - c^2/c_L^2}\,\sqrt{1 - c^2/c_T^2} - (2 - c^2/c_T^2)^2 = 0. \tag{2.1}$$

The determinantal equation (2.1) provides the admissible phase speeds c of the surface wave. On squaring both sides and disregarding the irrelevant value (root) $c = 0$ we have the more familiar form (cf. [23], §4.3.1)

$$s^3 - 8s^2 + 16(\tfrac{3}{2} - q)s - 16(1 - q) = 0 \tag{2.2}$$

where $s = c^2/c_T^2$ and $q = c_T^2/c_L^2 < 1$. Equation (2.2) is called Rayleigh equation (for the elastic half-space).

Surface waves in the present form may exist only if (2.1) is satisfied for at least one value of c. In this regard observe that if $c = c_T$ the left-hand side of (2.1) is equal to -1. If, instead, $c = \epsilon c_T$, ϵ being small, then the left-hand side goes as $2(1 - q)\epsilon^2$ which is strictly positive. This means that (2.1) has a root $c \in (0, c_T)$.

Since the left-hand side of (2.2) is a polynomial, it is easier to look for the roots s of (2.2). However, the squaring process is likely to have introduced spurious roots. To select the roots of (2.2) which are also roots of (2.1) we follow an analysis developed by Hayes and Rivlin [87].

Let $s^{(1)}$ be the real root (between 0 and 1) and let $s^{(2)} = s_1 + is_2$, $s^{(3)} = s_1 - is_2$ be the complex (conjugate) roots. If s, and then c, is complex, the corresponding values of a_L, a_T are complex. Denote by $\pm(a_{L1} + ia_{L2})$ and $\pm(a_{T1} + ia_{T2})$ the values which correspond to $s^{(3)}$ and by $\pm(a_{L1} - ia_{L2})$ and $\pm(a_{T1} - ia_{T2})$ the values which correspond to $s^{(2)}$; e.g., $a_{L1}^2 - a_{L2}^2 + 2ia_{L1}a_{L2} = 1 - qs_1 + iqs_2$. Then

$$a_L a_T = \pm(a_{L1} + ia_{L2})(a_{T1} + ia_{T2}) \qquad \text{if} \qquad s = s^{(2)},$$

$$a_L a_T = \pm(a_{L1} - ia_{L2})(a_{T1} - ia_{T2}) \qquad \text{if} \qquad s = s^{(3)}.$$

Moreover,

$$\frac{a_{L1}a_{L2}}{a_{T1}a_{T2}} = q,$$

which implies that a_{L1}/a_{T1} and a_{L2}/a_{T2} have the same sign. Now, by (2.1') we have

$$[1 + (a_{T1} + ia_{T2})^2]^2 = \pm 4(a_{L1} + ia_{L2})(a_{T1} + ia_{T2}).$$

Equating the imaginary terms yields

$$1 + a_{T1}^2 - a_{T2}^2 = \pm\Big(\frac{a_{L2}}{a_{T2}} + \frac{a_{L1}}{a_{T1}}\Big) \tag{2.3}$$

while

$$1 + a_{T1}^2 - a_{T2}^2 = 1 + \operatorname{Re} a_T^2 = 2 - s_1. \tag{2.4}$$

Since the sum of the roots of (2.2) equals 8 we have

$$s_1 = 4 - \tfrac{1}{2}s^{(1)}.$$

Accordingly, $s^{(1)} \in (0,1)$ implies that $s_1 \in (3.5, 4)$ and hence that $2 - s_1 < 0$. Comparison of (2.3) and (2.4) shows that

$$\frac{a_{L2}}{a_{T2}} + \frac{a_{L1}}{a_{T1}} < 0 \quad (> 0)$$

if the positive (negative) sign is taken. When the inequality $<$ applies, since a_{L1}/a_{T1} and a_{L2}/a_{T2} have the same sign they must both be negative. Hence a_{L1} and a_{T1} cannot both be negative. By the same token we see that also when the inequality $>$ applies the quantities a_{L1} and a_{T1} cannot both be negative. The same conclusion follows in connection with the conjugate root for which

$$[1 + (a_{T1} - ia_{T2})^2]^2 = \pm 4(a_{L1} - ia_{L2})(a_{T1} - ia_{T2}).$$

Accordingly, complex roots of the Rayleigh equation (2.2) do not correspond to admissible displacement fields. Only the real root $s^{(1)} \in (0,1)$ represents a surface wave.

Though mathematically correct, to our mind this statement is open to doubts about the consistency of the model. Complex roots for s correspond to complex values of c. We know that elastic solids allow for complex values of the wave vector only in the limit case of $\mathbf{k}_1, \mathbf{k}_2$ orthogonal to each other, which is reflected in the dependence $\exp(bz)\exp(ikx)$; the corresponding analysis might be performed in terms of evanescent waves [144]. Then it should come as no surprise that complex roots do not represent admissible surface waves. By consistency requirements, complex roots of the Rayleigh equation are likely to be physically admissible in dissipative bodies, such as viscoelastic solids.

All this applies to waves whose polarization lies in the sagittal plane which then may be viewed as a superposition of P and SV waves. The analogous problem for waves whose polarization is orthogonal to the sagittal plane, i.e. SH waves, is trivial. Let

$$u_y = A\exp(bz)\exp[i(kx-\omega t)],$$

$$u_x = 0, \qquad u_z = 0.$$

Then

$$T_{xy} = T_{yx} = i\mu k A\exp(bz)\exp[i(kx-\omega t)], \qquad T_{yz} = T_{zy} = \mu b A\exp(bz)\exp[i(kx-\omega t)],$$

are the non-vanishing components of $\mathbf{T}$. Substitution in the equation of motion yields the condition

$$\rho\omega^2 + \mu(b^2 - k^2) = 0.$$

Hence we assume that $c = \omega/k < c_T = \sqrt{\mu/\rho}$ and take

$$b = -ka_T.$$

The traction-free condition

$$0 = \mathbf{t} = \mathbf{T}\mathbf{n},$$

at $z = 0$, results in two identities and

$$\mu\, b\, A = 0$$

whence $A = 0$, namely SH waves are ruled out. This shows that surface waves of SH type cannot exist in elastic traction-free half-spaces. Meanwhile this suggests that SH waves may exist when the medium is not elastic and/or the boundary is not traction-free. In fact SH waves may exist when a layer of elastic material is superimposed on an elastic half-space (Love waves).

5.3 Rayleigh waves on viscoelastic half-spaces

Sometimes the secular equation for viscoelastic solids is derived by appealing to the so-called correspondence principle (cf. [14], p. 75). In this regard one should perhaps conclude that, in viscoelasticity too, only the real root represents a surface wave. Such need not be the case [54]. An appropriate investigation might be based on the structure of surface waves so that the admissibility of a root, as representative of a surface wave, may clearly be ascertained. Considered here are waves that are a superposition of inhomogeneous waves and, of course, are essentially confined to a neighbourhood of the boundary.

Look at a viscoelastic solid which occupies a half-space. The plane $z = 0$ is identified with the boundary surface $\mathcal{P}$ and the solid half-space is the region $z \geq 0$. Still $\mathbf{n} = -\mathbf{e}_z$ is the outward normal to $\mathcal{P}$.

Consider an incident pair and let k_x, k_y be the common x- and y-components of the wave vectors. Snell's law is taken to hold and then k_x and k_y are also the common values for all wave vectors involved. Correspondingly we have

$$\beta_{L,T}^2 = \kappa_{L,T} - k_x^2 - k_y^2. \tag{3.1}$$

The scalar and vector potentials $\phi, \boldsymbol{\psi}$ may be represented as

$$\phi = \Phi^+ \exp(i\beta_L z) + \Phi^- \exp(-i\beta_L z), \tag{3.2}$$

$$\boldsymbol{\psi} = \boldsymbol{\Psi}^+ \exp(i\beta_T z) + \boldsymbol{\Psi}^- \exp(-i\beta_T z). \tag{3.3}$$

The superscripts $-$ and $+$ label the incident and reflected waves; as usual, the common factor $\exp[i(k_x x + k_y y - \omega t)]$ is understood and not written. Following the standard view of reflection (and refraction) we might choose the root β of (3.1) by requiring that $\mathrm{Re}\,\beta > 0$. Really this restriction proves unnecessary for surface waves. The only significant requirement is - the condition (3.8) - on $\mathrm{Im}\,\beta$ thus ensuring the confinement of the surface wave.

The characteristic features of the reflected waves, and of the secular equation that defines Rayleigh waves, are determined by the vanishing of the traction. This condition has been examined in §4.4 and shown to provide the linear system (4.4.1)-(4.4.3). For convenience we rewrite the system in the form

$$\beta_L \Phi^+ - k_y \Psi_x^+ + (k_x - \tfrac{1}{2}\kappa_T/k_x)\Psi_y^+ = \beta_L \Phi^- + k_y \Psi_x^- - (k_x - \tfrac{1}{2}\kappa_T/k_x)\Psi_y^-, \tag{3.4}$$

$$\beta_L \Phi^+ - (k_y - \tfrac{1}{2}\kappa_T/k_y)\Psi_x^+ + k_x \Psi_y^+ = \beta_L \Phi^- + (k_y - \tfrac{1}{2}\kappa_T/k_y)\Psi_x^- - k_x \Psi_y^+, \tag{3.5}$$

$$(\tfrac{1}{2}\kappa_T - k_x^2 - k_y^2)\Phi^+ - k_y\beta_T \Psi_x^+ + k_x\beta_T \Psi_y^+ = -(\tfrac{1}{2}\kappa_T - k_x^2 - k_y^2)\Phi^- - k_y\beta_T \Psi_x^- + k_x\beta_T \Psi_y^+. \tag{3.6}$$

Once the system (3.4)-(3.6) in the unknowns Φ^+, Ψ_x^+, and Ψ_y^+ is solved, the reflected waves are determined.

We characterize a Rayleigh wave by the following two properties. First, it is the superposition of the admissible inhomogeneous waves, here a longitudinal wave and a transverse one, satisfying the traction-free condition, formally in the absence of any incident wave. Second, the amplitudes of the single components decay with distance from the surface, i.e. when z increases. In the present case this means that

$$\Phi^- = 0, \qquad \Psi_x^- = \Psi_y^- = 0 \tag{3.7}$$

and

$$\operatorname{Im}\beta_L > 0, \qquad \operatorname{Im}\beta_T > 0. \tag{3.8}$$

The conditions (3.7) make the system (3.4)-(3.6) homogeneous. Non-trivial solutions hold only if the determinant vanishes, which leads to the secular equation

$$\beta_L\beta_T(k_x^2 + k_y^2) + (k_x^2 + k_y^2 - \tfrac{1}{2}\kappa_T)^2 = 0. \tag{3.9}$$

Then surface waves may be viewed as solutions (k_x, k_y, β_L, β_T) to the system (3.1),(3.9) subject to the inequalities (3.8).

Let $s = \kappa_T/(k_x^2 + k_y^2)$ and $q = \kappa_L/\kappa_T$. By (3.1), the determinantal equation (3.9) may be written in the form

$$4\sqrt{1-s}\sqrt{1-qs} - (2-s)^2 = 0 \tag{3.10}$$

thus also showing that for dissipative bodies the determinantal equation contains one unknown only. On squaring (3.10) we have the Rayleigh equation for viscoelastic solids,

$$s^3 - 8s^2 + 16(\tfrac{3}{2} - q)s + 16(q-1) = 0, \tag{3.11}$$

which coincides formally with the Rayleigh equation (2.2) for elastic solids but now the parameter q and the unknown s are complex-valued.

The particular case $k_y = 0$ represents the configuration where $\mathbf{k}_2$ belongs to the plane $\mathbf{k}_1, \mathbf{n}$. It follows at once from (3.9), (3.10), and (3.2.1), (3.2.3) that any result which is valid for $k_y = 0$ can be carried over to $k_y \neq 0$ by simply replacing k_x^2 with $k_x^2 + k_y^2$. So, no conceptual generality is lost by considering the simplified configuration $k_y = 0$ (cf. [38]). With this in mind we can say that surface waves are solutions k_x, β_L, β_T to the system

$$k_x^2\beta_L\beta_T + (k_x^2 - \tfrac{1}{2}\kappa_T)^2 = 0, \tag{3.12}$$

$$\beta_L^2 + k_x^2 = \kappa_L, \tag{3.13}$$

$$\beta_T^2 + k_x^2 = \kappa_T, \tag{3.14}$$

subject to the constraints (3.8). Of course (3.10) and (3.11) follow again by letting $s = \kappa_T/k_x^2$. To find such solutions we indicate the following procedure [43].

- Determine the solutions to (3.11) in the complex unknown s. The sought values of s should be determined through (3.10) because (3.11), obtained by a squaring process, might contain spurious roots. However, the observation that it is much simpler to find the roots of a polynomial and that the roots of (3.10) are necessarily also solutions of (3.11) suggests that we first look for the roots of (3.11) and then select the admissible ones.

- Evaluate β_T through (3.14) under the constraint $\operatorname{Im}\beta_T > 0$. Obviously, (3.14) yields two complex values of β_T; we choose that compatible with (3.8).

- Evaluate β_L through (3.12) and check whether $\operatorname{Im}\beta_L > 0$. If (3.12) yields β_L with $\operatorname{Im}\beta_L < 0$ then the root s under consideration does not correspond to an admissible surface wave.

- If $\operatorname{Im}\beta_L > 0$, evaluate k_x through $k_x^2 = \kappa_T/s$, $\operatorname{Re} k_x > 0$.

In the procedure so outlined the relation (3.13) has been disregarded. Indeed, we might follow an alternative, equivalent procedure where the roles of β_L and β_T are interchanged and (3.14), instead of (3.13), is disregarded. It can be shown very easily that disregarding (3.13) is allowed in that (3.13) is a consequence of (3.11), (3.12), and (3.14). Letting

$$Z := \beta_L^2 + k_x^2 - \kappa_L$$

we regard (3.13) as the vanishing of Z. Now, squaring (3.12), using (3.14) and substituting $k_x^2 = \kappa_T/s$, $\kappa_L = q\kappa_T$ yields

$$\beta_L^2 = \kappa_T \frac{(1-\frac{1}{2}s)^4}{s(s-1)}.$$

Then, upon some rearrangement, we obtain

$$Z = \frac{\kappa_T}{16(s-1)}[s^3 - 8s^2 + 16(\tfrac{3}{2} - q)s + 16(q-1)].$$

The validity of (3.11) implies the vanishing of Z.

It is worth emphasizing that in the procedure we do *not* make the fairly customary assumption that $\operatorname{Im} k_x > 0$. It is true that if $\operatorname{Im} k_x < 0$ then the amplitude grows as the wave propagates along the surface. However, this is not at all in contrast with the property that bulk waves in dissipative media decay as they propagate. A surface wave is a superposition of longitudinal and transverse waves. As $\operatorname{Re}\beta_L$, $\operatorname{Re}\beta_T \neq 0$, the two wave vectors $\mathbf{k}_L, \mathbf{k}_T$ are not horizontal and then $\operatorname{Im} k_x < 0$ does not imply the growth of the amplitude along the direction of propagation. Indeed, as we know, amplitude decay along the direction of propagation is guaranteed by the thermodynamic conditions $\operatorname{Im}\mu < 0$, $\operatorname{Im}(2\mu + \lambda) < 0$. Meanwhile, the condition $\operatorname{Im} k_x > 0$ might prove overly restrictive thus dropping out admissible wave solutions. Incidentally, the value of k_x occurs only at the last step of the procedure. Then it can be disregarded if a thorough characterization of the surface waves is not in order (but, e.g., simply their number) thus reducing the pertinent calculations.

Assume that the first step has been accomplished, namely we know the solutions $s^{(1)}, s^{(2)}, s^{(3)}$ to (3.11). Then substitution of $k_x^2 = \kappa_T/s$ in (3.14), the condition $\operatorname{Im}\beta_T > 0$, and some rearrangement lead to

$$\operatorname{Re}\beta_T = S\sqrt{\frac{\rho\omega^2\{\mu_1(s_1^2+s_2^2) - \mu_1 s_1 + \mu_2 s_2 + \sqrt{[(s_1-1)^2+s_2^2](\mu_1^2+\mu_2^2)(s_1^2+s_2^2)}\}}{2(\mu_1^2+\mu_2^2)(s_1^2+s_2^2)}},$$

$$\operatorname{Im}\beta_T = \sqrt{\frac{\rho\omega^2\{-\mu_1(s_1^2+s_2^2)+\mu_1 s_1-\mu_2 s_2+\sqrt{[(s_1-1)^2+s_2^2](\mu_1^2+\mu_2^2)(s_1^2+s_2^2)}\}}{2(\mu_1^2+\mu_2^2)(s_1^2+s_2^2)}},$$

where $s_1 = \operatorname{Re} s$, $s_2 = \operatorname{Im} s$ and

$$S = \operatorname{sgn}[-\mu_2(s_1^2+s_2^2)+\mu_2 s_1+\mu_1 s_2]$$

is the sign of the quantity in square brackets. Hence, by (3.12) we obtain

$$\operatorname{Im}\beta_L = \rho\omega^2\frac{(\zeta_T\mu_1-\sigma_T\mu_2)s_2(s_1^2+s_2^2-4)+(\zeta_T\mu_2+\sigma_T\mu_1)[4(s_1^2+s_2^2)-s_1(s_1^2+s_2^2+4)]}{4(\mu_1^2+\mu_2^2)(\zeta_T^2+\sigma_T^2)(s_1^2+s_2^2)}. \tag{3.15}$$

Accordingly, given a root s, the existence of the corresponding surface wave is ascertained by checking the positivity of the numerator in (3.15).

It is worth remarking that the solutions $s^{(1)}, s^{(2)}, s^{(3)}$ to (3.11), and then the phase speed and damping, depend on the parameter $q = \mu/(2\mu+\lambda)$ which in turn depends on ω through μ and λ. So, given the value of ω for the incident perturbation which excites the surface wave, we determine the value of q and hence the three solutions to (3.11).

Seemingly, the general case of viscoelastic solids does not allow more definite conclusions. In this sense it may be instructive to consider a particular example of viscoelastic solid. By analogy with Rayleigh [147] who had investigated elastic bodies with $\lambda_0 = \mu_0$, Currie, Hayes & O'Leary [54] examined the case when $\lambda_1 = \mu_1$ and λ_2, μ_2 are small compared with μ_1. For this case we apply the procedure indicated above. Let

$$\lambda = \mu_1(1-i\epsilon\alpha), \qquad \mu = \mu_1(1-i\epsilon\beta), \quad \epsilon > 0.$$

The thermodynamic requirement (2.4.7) becomes

$$\beta > 0, \qquad 3\alpha + 2\beta > 0.$$

The smallness of the imaginary parts is made operative by ignoring terms of order ϵ^2 and higher. Then the roots of the Rayleigh equation (3.11) turn out to be

$$\begin{aligned} s^{(1)} &= 2 - \frac{2}{\sqrt{3}} - i\epsilon(\alpha-\beta)\frac{3\sqrt{3}-5}{3\sqrt{3}}, \\ s^{(2)} &= 4 + 2i\epsilon(\alpha-\beta), \\ s^{(3)} &= 2 + \frac{2}{\sqrt{3}} - i\epsilon(\alpha-\beta)\frac{3\sqrt{3}+5}{3\sqrt{3}}. \end{aligned}$$

Look at the root $s^{(2)}$. By (3.14), $\beta_T = \pm\sqrt{\kappa_T(s-1)/s}$. The value of β_T with $\operatorname{Im}\beta_T > 0$ turns out to be

$$\beta_T = \sqrt{\frac{3\rho\omega^2}{4\mu_1}}\left[1 + i\frac{1}{12}\epsilon(\alpha+5\beta)\right]$$

in that $\alpha + 5\beta = (3\alpha + 2\beta)/3 + (5 - 2/3)\beta > 0$. By (3.12) we have

$$\beta_L = -\frac{\kappa_T}{s\beta_T}(1 - \tfrac{1}{2}s)^2$$

and then we obtain

$$\beta_L = \sqrt{\frac{\rho\omega^2}{12\mu_1}}\Big[1 + \frac{i\epsilon}{12}(11\beta - 17\alpha)\Big],$$

which is admissible if $\alpha/\beta < 11/17 = 0.65$. By the same token, for the root $s^{(3)}$ we have an admissible surface wave if

$$1.19 < \frac{\alpha}{\beta} < 4.46.$$

For the root $s^{(1)}$, for which Re $s^{(1)} \in (0,1)$, we find that the surface wave is always admissible. For reasons that become clear in a moment, such a wave is the analogue of the elastic wave; let us call it quasi-elastic wave. This shows that if $0.65 \leq \alpha/\beta \leq 1.19$ or $4.46 \leq \alpha/\beta$ then only the quasi-elastic wave occurs. Otherwise two Rayleigh waves occur, one of them always being the quasi-elastic wave.

More explicit, a priori conditions on the admissibility of roots can be obtained in the particular case of elastic solids where $q = \mu_0/(2\mu_0 + \lambda_0)$ is real-valued and $s = \rho\omega^2/\mu_0 k_x^2$. Let $s = s_1 + is_2$ be a root of (3.11). Set

$$\beta_L = \zeta_L + i\sigma_L, \qquad \beta_T = \zeta_T + i\sigma_T.$$

By (3.13) and (3.14) we have

$$\zeta_L^2 - \sigma_L^2 = \kappa_L - \frac{\kappa_T s_1}{s_1^2 + s_2^2}, \tag{3.16}$$

$$\zeta_L \sigma_L = \frac{\kappa_T s_2}{2(s_1^2 + s_2^2)}, \tag{3.17}$$

$$\zeta_T^2 - \sigma_T^2 = \kappa_T\Big(1 - \frac{s_1}{s_1^2 + s_2^2}\Big), \tag{3.18}$$

$$\zeta_T \sigma_T = \frac{\kappa_T s_2}{2(s_1^2 + s_2^2)}. \tag{3.19}$$

By (3.17) and (3.19), the requirement (3.8) yields

$$\operatorname{sgn} \zeta_L, \zeta_T = \operatorname{sgn} s_2 \quad \text{if} \quad s_2 \neq 0, \tag{3.20}$$

$$\zeta_L, \zeta_T = 0 \quad \text{if} \quad s_2 = 0. \tag{3.21}$$

It follows that, if $s_2 \neq 0$, the solution to (3.16)-(3.20) for $(\zeta_L, \sigma_L, \zeta_T, \sigma_T)$ is given by

$$\begin{Bmatrix} \zeta_L \\ \sigma_L \end{Bmatrix} = \begin{Bmatrix} \operatorname{sgn} s_2 \\ 1 \end{Bmatrix} \frac{1}{\sqrt{2}} \sqrt{\sqrt{\kappa_L^2 + \frac{\kappa_T^2}{s_1^2 + s_2^2} - \frac{2\kappa_L\kappa_T s_1}{s_1^2 + s_2^2}} \begin{Bmatrix} + \\ - \end{Bmatrix} \Big(\kappa_L - \frac{\kappa_T s_1}{s_1^2 + s_2^2}\Big)},$$

$$\begin{Bmatrix} \zeta_T \\ \sigma_T \end{Bmatrix} = \begin{Bmatrix} \operatorname{sgn} s_2 \\ 1 \end{Bmatrix} \frac{1}{\sqrt{2}} \sqrt{\sqrt{\frac{(1+s_1^2+s_2^2)\kappa_T^2}{s_1^2+s_2^2} - \frac{2\kappa_T^2 s_1}{s_1^2+s_2^2}} \begin{Bmatrix} + \\ - \end{Bmatrix} \left(\kappa_T - \frac{\kappa_T s_1}{s_1^2+s_2^2}\right)},$$

To select the roots of (3.12) among those of (3.11) it is convenient to write (3.12) as

$$\zeta_L\zeta_T - \sigma_L\sigma_T + i(\sigma_L\zeta_T + \zeta_L\sigma_T) = -\kappa_T\left(\frac{s_1 - is_2}{s_1^2+s_2^2} + \frac{s_1+is_2}{4} - 1\right).$$

Equating the imaginary parts yields

$$\sigma_L\zeta_T + \sigma_T\zeta_L = \kappa_T s_2 \frac{4-(s_1^2+s_2^2)}{4(s_1^2+s_2^2)}.$$

Then (3.20) and (3.8) provide

$$s_1^2 + s_2^2 < 4. \tag{3.22}$$

Accordingly, for elastic half-spaces we can search for possible complex roots of the cubic (3.11) and say that they are roots of (3.9) too if and only if (3.22) holds. If there were roots - with $s_2 \neq 0$ - then we would have reached the result that our description of surface waves allows solutions which are ruled out in the standard description (cf. [87] and §5.2). Really, a numerical check shows that no root exists in the circle (3.22) as $q \in (0,1)$.

Consider the real root $s^{(1)}$ of (3.11). By (3.21), equation (3.16) yields

$$0 < \sigma_L^2 = -\kappa_L + \frac{\kappa_T}{s^{(1)}}$$

whence it follows that

$$0 < s^{(1)} < \frac{2\mu_0 + \lambda_0}{\mu_0}.$$

Similarly, by (3.18) and (3.21) we have

$$0 < \sigma_T^2 = \kappa_T \frac{1 - s^{(1)}}{s^{(1)}}$$

whence

$$0 < s^{(1)} < 1.$$

In conclusion, the real root $s^{(1)}$ must belong to the interval $(0,1)$ as it happens in the standard context.

5.4 Stoneley waves

Waves confined to the neighbourhood of a surface occur also at the interface between two half-spaces. Such waves are often called Stoneley waves though sometimes Stoneley waves are meant as surface waves between two solid half-spaces. We call Stoneley waves all surface waves at the interface between two half-spaces.

Look at the interface between two elastic half-spaces. We know from the previous section that wave motion is allowed, in the half-space $z > 0$, if u_x and u_z have the form

$$u_x = \left[A_1 \exp(-ka_L z) + A_2 \exp(-ka_T z)\right] \exp[i(kx - \omega t)],$$

$$u_z = -i\left[a_L A_1 \exp(-ka_L z) + \frac{1}{a_T} A_2 \exp(-ka_T z)\right] \exp[i(kx - \omega t)],$$

while $u_y = 0$. Similar expressions hold for the lower half-space $z < 0$ and the pertinent quantities are labelled by the accent $\breve{}$, namely

$$\breve{u}_x = \left[\breve{A}_1 \exp(k\breve{a}_L z) + \breve{A}_2 \exp(k\breve{a}_T z)\right] \exp[i(kx - \omega t)],$$

$$\breve{u}_z = -i\left[\breve{a}_L \breve{A}_1 \exp(k\breve{a}_L z) + \frac{1}{\breve{a}_T} \breve{A}_2 \exp(k\breve{a}_T z)\right] \exp[i(kx - \omega t)].$$

The condition that the displacement and the stress are continuous at the interface $z = 0$ yields four homogeneous equations in the four unknowns $A_1, A_2, \breve{A}_1, \breve{A}_2$, viz.

$$A_1 + A_2 - \breve{A}_1 - \breve{A}_2 = 0,$$

$$a_L A_1 + \frac{1}{a_T} A_2 + \breve{a}_L \breve{A}_1 + \frac{1}{\breve{a}_T} \breve{A}_2,$$

$$2\mu a_L A_1 + \mu \frac{1 + a_T^2}{a_T} A_2 + 2\breve{\mu}\breve{a}_L \breve{A}_1 + \breve{\mu} \frac{1 + \breve{a}_T^2}{\breve{a}_T} \breve{A}_2 = 0,$$

$$[2\mu a_L^2 + \lambda(1 + a_L^2)]A_1 + 2(\mu + \lambda)A_2 - [2\breve{\mu}\breve{a}_L^2 + \breve{\lambda}(1 + \breve{a}_L^2)]\breve{A}_1 - 2(\breve{\mu} + \breve{\lambda})\breve{A}_2 = 0.$$

Requiring that the determinant of the coefficients vanish yields the secular equation in the unknown speed c. Here we remark that the wavenumber k does not appear in the coefficients, and then in the secular equation. Accordingly, the corresponding solutions, namely the Stoneley waves, are not dispersive. The analysis of the secular equation is developed in the book by Cagniard [28]; the generalization to dissipative bodies seems to be a formidable problem.

Definite results are obtained here by restricting attention to the interface between a dissipative solid and a fluid. Our procedure parallels that examined for the Rayleigh waves. Let the half-space $z < 0$ be occupied by a viscoelastic solid and the upper part by an inviscid fluid. For the time being we assume that an incident wave comes from the fluid. A reflected wave in the fluid and two transmitted waves in the solid emanate from the interface. We represent the incident, reflected, and transmitted waves as

$$\phi^i = \Phi^i \exp[i(k_x x - \beta_F z)],$$

$$\phi = \Phi \exp[i(k_x x + \beta_F z)],$$

$$\breve{\phi} = \breve{\Phi} \exp[i(k_x x - \breve{\beta}_L z)],$$

$$\breve{\boldsymbol{\psi}} = \breve{\Psi} \mathbf{e}_y \exp[i(k_x x - \breve{\beta}_T z)],$$

where the subscripts L and T label longitudinal and transverse components in the solid while F is a reminder that the quantity is relative to the fluid; the $+$ and $-$ signs are omitted in that are unnecessary. These representations have common values of k_x, as a consequence of Snell's law, while k_y is taken to be zero; to fix ideas we let $\operatorname{Re} k_x > 0$. No assumption is made about the sign of $\operatorname{Im} k_x$. Positive values for $\operatorname{Re} \beta$ and $\operatorname{Im} \beta$ would merely confirm our picture of the waves involved. The absence of a transverse, horizontal wave is due to the upper medium being a fluid. The representation of the vector potential shows that only the y-component is involved; the proof that the x- and z-components vanish is given in §4.7.

We expect that, for every surface wave solution, the field vanishes at infinity ($z = \pm\infty$), which corresponds to strictly positive values of $\operatorname{Im} \beta_F$, $\operatorname{Im} \breve{\beta}_L$, $\operatorname{Im} \breve{\beta}_T$. This view is strengthened by the dissipative character of the viscoelastic solid. Leaky waves seem to contradict this view or, at least, seem to be outside the realm of surface waves. For leaky waves we should allow negative values of $\operatorname{Im} \beta_F$.

The continuity of normal displacement and traction leads to the system of equations

$$\rho \Phi + 2k_x \frac{\breve{\rho}}{\breve{\kappa}_T} \left[(k_x - \tfrac{1}{2} \breve{\kappa}_T / k_x) \breve{\Phi} + \breve{\beta}_T \breve{\Psi} \right] = -\rho \Phi^i, \tag{4.1}$$

$$\breve{\beta}_L \breve{\Phi} - (k_x \quad \tfrac{1}{2} \breve{\kappa}_T / k_x) \breve{\Psi} = 0, \tag{4.2}$$

$$\beta_F \Phi - \tfrac{1}{2} \breve{\kappa}_T / k_x \breve{\Psi} = \beta_F \Phi^i, \tag{4.3}$$

in the unknown amplitudes Φ, $\breve{\Phi}$, and $\breve{\Psi}$; this system is equivalent to (4.7.12), (4.7.10), (4.7.9).

Now look at the case when no incident wave occurs. The homogeneous system associated with (4.1)-(4.3) allows for non-trivial solutions only if the matrix of the coefficients is singular, namely when the secular equation

$$\rho \breve{\beta}_L \frac{\breve{\kappa}_T}{2k_x} + \beta_F \frac{2\breve{\rho} k_x}{\breve{\kappa}_T} \left[(\breve{\beta}_T \breve{\beta}_L + \left(k_x - \frac{\breve{\kappa}_T}{2k_x}\right)^2 \right] = 0 \tag{4.4}$$

holds. By analogy with [142] we call any solution to (4.4) free mode. A surface wave is defined as a free mode such that the amplitude of the displacement field decreases with distance from the interface. Similarly, a leaky wave is a free mode such that the displacement field decreases with distance from the interface in the solid, and increases with distance in the fluid ($\operatorname{Im} \beta_F > 0$). Of course we do not regard as leaky a wave such that $\operatorname{Im} \beta_F > 0$ but $\operatorname{Re} \beta_F < 0$ because that case represents an incident rather than a reflected wave.

For any wave we have

$$\beta^2 + k_x^2 = \kappa$$

κ denoting any of $\kappa_F, \breve{\kappa}_L, \breve{\kappa}_T$. Since $\kappa_F = \omega^2/p_\rho$ is real then $\mathbf{k}_1$ and $\mathbf{k}_2$ in the fluid are orthogonal. Then we have

$$\beta = \pm\sqrt{\kappa - k_x^2}; \tag{4.5}$$

the choice of the sign is to be determined through the requirement that $\operatorname{Im}\beta$ be positive. Substitution of (4.5) into (4.4) and some rearrangement gives

$$4\sqrt{1-qs}\,\sqrt{1-s} - (2-s)^2 = s^2 \frac{\rho}{\breve{\rho}} \frac{\sqrt{1-qs}}{\sqrt{1-rs}} \tag{4.6}$$

where

$$v = \frac{\rho}{\breve{\rho}}, \qquad q = \frac{\breve{\kappa}_L}{\breve{\kappa}_T}, \qquad r = \frac{\kappa_F}{\breve{\kappa}_T}, \qquad s = \frac{\breve{\kappa}_T}{k_x^2}.$$

So (4.6) is an equation in the complex-valued unknown s, parameterized by q, r, and v. In (4.6) a $\pm$ sign in front of any root is understood.

To determine s from (4.6) we have recourse to a polynomial form. Multiplication by $\sqrt{1-rs}$ yields

$$4\sqrt{1-qs}\,\sqrt{1-s}\,\sqrt{1-rs} = (2-s)^2\sqrt{1-rs} + vs^2\sqrt{1-qs}.$$

Upon squaring, rearranging the terms, and dividing through by s we obtain

$$(1-rs)[s^3 - 8s^2 + (24-16q)s - 16(1-q)] - v^2s^3(1-qs) = 2vs(2-s)^2\sqrt{1-rs}\sqrt{1-qs}$$

whence

$$(r+v^2q)s^4 - (1+8r+v^2)s^3 + (8+24r-16qr)s^2 + 8(2q-3-2r+2qr)s + 16(1-q)$$
$$= 2vs(2-s)^2\sqrt{1-rs}\sqrt{1-qs}.$$

Upon squaring again and rearranging we arrive at the Stoneley equation

$$\sum_{h=0}^{8} c_h s^h = 0 \tag{4.7}$$

where

$$c_0 = E^2, \qquad c_1 = 2DE, \qquad c_2 = D^2 + 2CE - 16F,$$

$$\begin{aligned}
c_3 &= -2BE + 2CD + F[32 + 16(r+q)],\\
c_4 &= C^2 + 2AE - 2BD - F[24 + 32(r+q) + 16rq],\\
c_5 &= 2AD - 2BC + F[8 + 24(r+q) + 32rq],\\
c_6 &= B^2 + 2AC - F[1 + 8(r+q) + 24rq],\\
c_7 &= -2AB + F(r+q+8rq),\\
c_8 &= A^2 - Frq,
\end{aligned}$$

while

$$A = r + v^2 q, \qquad B = 1 + 8r + v^2, \quad C = 8 + 24r - 16qr,$$
$$D = 8(2q - 3 - 2r + 2qr), \qquad E = 16(1 - q), \quad F = 4v^2.$$

In a free mode, namely when the secular equation (4.4) holds, the three amplitudes Φ, $\breve{\Phi}$, and $\breve{\Psi}$ are linearly dependent. For instance, we have

$$\breve{\Phi} = \frac{k_x^2 - \frac{1}{2}\breve{\kappa}_T}{k_x \breve{\beta}_L} \breve{\Psi},$$

$$\Phi = \frac{\breve{\kappa}_T}{2k_x \breve{\beta}_L} \breve{\Psi}$$

and then we regard the free mode as parameterized by $\breve{\Psi}$. Meanwhile we let the (phase) speed V of the wave correspond to the phase propagation along the interface $\mathcal{P}$. Then elementary geometrical considerations show that

$$V = \frac{\omega}{\operatorname{Re} k_x}. \tag{4.8}$$

This definition of phase speed is the same as that adopted in previous investigations (cf. [54]).

5.5 The limit case of a rarefied medium

We now examine the behaviour of surface waves when the upper medium is a comparatively rarefied fluid, namely when $v \ll 1$. On the one hand, this allows us to obtain more detailed results. On the other, this provides the possibility of investigating continuity properties in the behaviour of waves relative to the density of the fluid.

As $s \neq 0, 1/r$, equation (4.6) amounts to the vanishing of

$$f(s) = \sqrt{1 - rs}\, \frac{4\sqrt{1 - qs}\,\sqrt{1 - s} - (2 - s)^2}{s} - vs\sqrt{1 - qs}.$$

Look at the case when q, r, and s are real, which occurs when the lower medium is elastic. Then (4.6) has a real root in the interval $(0, 1/r)$. This is easily seen by observing that

$$f(0) = 2(1 - q) > 0 \qquad , \qquad f(1/r) = -v\frac{1}{r}\sqrt{1 - q/r} < 0.$$

Since

$$\frac{1}{r} = \frac{\breve{\kappa}_T}{\kappa_F} = \left(\frac{c_F}{\breve{b}}\right)^2, \qquad s = \frac{\breve{\kappa}_T}{k_x^2}$$

where c_F and $\breve{b}$ denote the phase speed of the wave in the fluid and of the transverse wave in the solid, then $s < 1/r$, i.e. $k_x^2 > \kappa_F$, means that usually the surface wave is slower than the wave in the fluid.

Now we determine an estimate of the root through successive approximations by regarding v as a small parameter. Physically, it might seem natural to let r be parameterized by ρ, and hence v. Quite easily, though, it follows that we have to assume that the sound speed tends to a non-zero finite value as the density of the fluid tends to zero. Hence, for simplicity, we let r be independent of v. Since $f(0) > 0$ then $f(s)$ vanishes as $v = 0$ if $s = s_0 > 0$ is such that

$$4\sqrt{1 - qs_0}\,\sqrt{1 - s_0} - (2 - s_0)^2 = 0,$$

s_0 being the so-called Rayleigh solution. In the next step we let $s = s_0 + \Delta s$ and, to the linear order in Δs, we have

$$\Delta s = \frac{s_0^2\sqrt{(1 - qs_0)/(1 - rs_0)}}{2\{2 - s_0 + 4[4qs_0 - (q + 1)]/(2 - s_0)^2\}}\, v. \tag{5.1}$$

Through the result (5.1) we derive the first-order correction, to the surface wave speed, induced by the upper half-space. In other words, we may view (5.1) as the first-order correction in passing from the Rayleigh wave to the corresponding Stoneley wave. This perturbation analysis can be straightforwardly extended to viscoelastic solids. The results obtained from (5.1) turn out to be in full agreement with the values of the roots exhibited in the next section.

It is also worth commenting on a well-established result for elastic solids [22], §7.4, whereby

$$V = c_F\sqrt{1 - \frac{v^2}{4r^2(1 - q)^2}}; \tag{5.2}$$

really, in (7.4) of [22], 8 occurs instead of 4, but this seems to be a misprint. The result (5.2) is hardly reliable in that, in the absence of the fluid, the surface wave would propagate with the phase speed in the fluid. The reason for this paradoxical property is due to the procedure adopted which is illustrated as follows. Write the vanishing of $f(s)$ as

$$4\sqrt{1 - qs}\,\sqrt{1 - s} - (2 - s)^2 = vs^2\sqrt{\frac{1 - qs}{1 - rs}}. \tag{5.3}$$

Now approximate the left-hand side to the first order in s, namely

$$4\sqrt{1 - qs}\,\sqrt{1 - s} - (2 - s)^2 \simeq 2(1 - q)s.$$

Then square (5.3) in the approximate form to write

$$1 - rs = \frac{v^2(1 - qs)}{4(1 - q)^2}\, s^2. \tag{5.4}$$

Regard the right-hand side as a perturbative quantity so that $s = 1/r$ is the starting value for the desired root. Substitution of $s = 1/r$ in the right-hand side and neglect of q/r relative to 1 gives

$$rs = 1 - \frac{v^2}{4r^2(1-q)^2}.$$

The observation that

$$s = \frac{\breve{\kappa}_T}{k_x^2} = \frac{\kappa_F}{rk_x^2} = \frac{\omega^2}{c_F^2 r k_x^2}$$

and (4.8) yield (5.2).

In (5.4) the right-hand side vanishes as $v = 0$ and the left-hand side as $s = 1/r = \breve{\kappa}_T p_\rho/\omega^2$, a quantity which is unrelated to $v = 0$. Then $s = 1/r$ is not a good starting value for the desired root and this clarifies the origin of the paradoxical property of (5.2).

5.6 Admissible roots of the Stoneley equation

The Stoneley equation (4.7) has been obtained from the secular equation (4.4) by two successive squaring procedures, and this implies that the roots of (4.7) do not necessarily satisfy equation (4.4). In addition, only those roots are to be selected which are also physically admissible, in that the amplitude decays with distance from the surface, with an exception related to the possible existence of leaky waves within the fluid. Thus, once we have chosen the appropriate sign in (4.5), we have to rule out the spurious roots.

To this end, in correspondence with every root of (4.7) we first evaluate k_x through $k_x^2 = \breve{\kappa}_T/s$, under the non-restrictive condition that $\operatorname{Re} k_x > 0$. Then β is determined through (4.5), where the sign is chosen so as to guarantee that $\operatorname{Im}\beta_F$, $\operatorname{Im}\breve{\beta}_L$, $\operatorname{Im}\breve{\beta}_T > 0$; the choice $\operatorname{Im}\beta_F < 0$ is also allowed, and corresponds to leaky waves. These conditions are then substituted into the secular equation. If it happens that this equation results into an identity we conclude that the given root determines a surface wave. Incidentally, in this section $\beta_F, \beta_L, \beta_T$ are only required to have a positive imaginary part to guarantee the peculiar character of surface waves.

For definiteness we let the solid be copper and the upper half-space be empty or filled with air or water. On adopting cgs units, for air and water we let $c_F = \sqrt{p_\rho}$ be

$$c_F = 0.331\,10^5 \quad \text{and} \quad c_F = 1.5\,10^5,$$

respectively. As regards copper, we take the values of §4.7, viz.

$$\breve{\mu}_0 = 44.44\,10^{10}, \qquad \breve{\lambda}_0 = 103.70\,10^{10},$$

$$\breve{\mu}'(\tau) = -\breve{\mu}_0[1.567\,10^{-6}\exp(-0.1616\,10^{-5}\tau) + 1.565\,10^{-2}\exp(-0.5157\,\tau)],$$

and

$$\breve{\lambda}'(\tau) = -\breve{\lambda}_0[1.567\,10^{-6}\exp(-0.1616\,10^{-5}\tau) + 1.565\,10^{-2}\exp(-0.5157\,\tau)].$$

The density is taken as $\breve{\rho} = 8.9$. Observe that $\breve{\mu}'$ and $\breve{\lambda}'$ in the form of exponentials satisfy the thermodynamic restrictions (2.4.7).

To examine the memory effects we let the frequency ω take the values $0.1, 1, 10, ..., 10^5$. The results, though, are weakly affected by ω.

With these constitutive parameters we have obtained the results in Table 1; s is the root of the secular equation, c_F, c_L, and c_T are the phase speeds ω/k_1 of the bulk wave in the fluid and of the longitudinal and transverse waves in the solid. The notation $F\,L\,T$ means that the solution is the result of a (bulk) wave in the fluid (F) and a longitudinal (L) and a transverse wave (T) in the solid. The label $+$ ($-$) means that the pertinent wave is going upward (downward). The symbol $\star$ means that the root corresponds to a leaky wave. We omit writing the imaginary part of the root whenever it is less than 10^{-2} times the corresponding real part.

We observe that, in the air-solid case, leaky waves occur with a phase speed nearly equal to that of the non-leaky waves; differences occur when double precision is employed. This proves that experiments on surface waves are quite subtle in that sometimes leaky and non-leaky waves are hardly discernible.

It is worth observing that a numerical analysis of (4.7) and (4.4) in the water-solid case would provide solutions representing $F_-L_-T_-$ modes with $\operatorname{Im}\beta_F < 0$. To us this mode is by no means a surface wave but, rather, represents a refraction of the wave F_- which then plays the role of incident wave. To strengthen this view we recall that the analysis has been performed by letting $\Phi^i = 0$.

As we should have expected, there is a continuity between Rayleigh waves and Stoneley waves in that, to each Rayleigh wave for the vacuum-solid interface corresponds a Stoneley wave, with close parameters, for the air-solid and water-solid interfaces. The converse is not true in that there are Stoneley waves without the corresponding Rayleigh wave (cf. the solutions $s = 0.4407 + 0.0025\,i$ and $s = 0.4281$). Indeed, the results show a bifurcation for Stoneley waves which correspond to a Rayleigh wave.

Unlike the case of homogeneous waves in elastic solids, here we may have $c_L \le c_T$. This is a peculiar feature of inhomogeneous waves for which the difference $k_1^2 - k_2^2$ is determined by the material parameters through the real part of (3.2.1), (3.2.3) and (3.3.3), (3.3.4). This implies that the phase speed ω/k_1 is necessarily bounded (from above).

A detailed analysis is in order for the modes $F_-L_-T_-$, which correspond to solutions of (4.4) but are not considered as surface waves. To fix ideas consider the $F_-L_-T_-$ mode relative to $s = 0.8760+0.0274\,i$ with $\omega = 0.1$, where F_- is the wave with $k_x = 0.4851\,10^{-6} - i\,0.0062\,10^{-6}$, $\beta_F = -0.4573\,10^{-6} - i\,0.0065\,10^{-6}$. We can view F_- as the incident wave, in a reflection-refraction problem, and L_-, T_- as the transmitted waves. It is then natural to ask for the occurrence of a reflected wave R_+. By Snell's law the transmitted and reflected

	$\omega = 0.1$	$\omega = 10$
vacuum-solid	$s = 0.8743$	$s = 0.8743$
	$c_L = 2.0586\,10^5$	$c_L = 2.0893\,10^5$
	$c_T = 2.0586\,10^5$	$c_T = 2.0893\,10^5$
	$V = 2.0586\,10^5$	$V = 2.0893\,10^5$
air-solid	$s = 0.8742$	$s = 0.8742$
	$c_F = 0.3310\,10^5$	$c_F = 0.3310\,10^5$
	$c_L = 2.0586\,10^5$	$c_L = 2.0893\,10^5$
	$c_T = 2.0586\,10^5$	$c_T = 2.0893\,10^5$
	$V = 2.0586\,10^5$	$V = 2.0893\,10^5$
	$\theta_F = 80.74°$	$\theta_F = 80.88°$
	$\theta_L = 0.15°$	$\theta_L = 0.04°$
	$\theta_T = 0.06°$	$\theta_T = 0.01°$
	$F_- \; L_+ \; T_+$	$F_- \; L_+ \; T_+$
	$s = 0.8742 \star$	$s = 0.8742 \star$
	$c_F = 0.3310\,10^5$	$c_F = 0.3310\,10^5$
	$c_L = 2.0586\,10^5$	$c_L = 2.0893\,10^5$
	$c_T = 2.0586\,10^5$	$c_T = 2.0893\,10^5$
	$V = 2.0586\,10^5$	$V = 2.0893\,10^5$
	$\theta_F = 80.74°$	$\theta_F = 80.88°$
	$\theta_L = 0.15°$	$\theta_L = 0.04°$
	$\theta_T = 0.06°$	$\theta_T = 0.01°$
	$F_+ \; L_+ \; T_+$	$F_+ \; L_+ \; T_+$
water-solid	$s = 0.8763 - 0.0274\,i \star$	$s = 0.8761 - 0.0267\,i \star$
	$c_F = 1.4997\,10^5$	$c_F = 1.4998\,10^5$
	$c_L = 2.0614\,10^5$	$c_L = 2.0918\,10^5$
	$c_T = 2.0597\,10^5$	$c_T = 2.0902\,10^5$
	$V = 2.0619\,10^5$	$V = 2.0922\,10^5$
	$\theta_F = 43.33°$	$\theta_F = 44.20°$
	$\theta_L = 1.15°$	$\theta_L = 1.01°$
	$\theta_T = 2.59°$	$\theta_T = 2.47°$
	$F_+ \; L_+ \; T_+$	$F_+ \; L_+ \; T_+$
	$s = 0.4627 + 0.0027\,i$	$s = 0.4494$
	$c_F = 1.4977\,10^5$	$c_F = 1.4979\,10^5$
	$c_L = 1.4977\,10^5$	$c_L = 1.4979\,10^5$
	$c_T = 1.4977\,10^5$	$c_T = 1.4979\,10^5$
	$V = 1.4977\,10^5$	$V = 1.4979\,10^5$
	$\theta_F = 0.03°$	$\theta_F = 0.01°$
	$\theta_L = 0.02°$	$\theta_L = 0.01°$
	$\theta_T = 0.10°$	$\theta_T = 0.03°$
	$F_- \; L_- \; T_-$	$F_- \; L_- \; T_-$

Table 1 Properties of surface waves.

waves have the same value of k_x. So we characterize the incident wave F_-, the transmitted waves L_-, T_- and the reflected wave R_+ as

$$\begin{aligned}
F_-: &\quad \phi^i = \Phi^i \exp[(0.0062x + 0.0065z)10^{-6}]\exp[i(0.4851x - 0.4573z)10^{-6}],\\
R_+: &\quad \phi = \Phi \exp[(0.0062x - 0.0065z)10^{-6}]\exp[i(0.4851x + 0.4573z)10^{-6}],\\
L_-: &\quad \breve{\phi} = \breve{\Phi} \exp[(0.0062x + 0.4333z)10^{-6}]\exp[i(0.4851x - 0.0072z)10^{-6}],\\
T_-: &\quad \breve{\psi} = \breve{\Psi} \exp[(0.0062x + 0.1716z)10^{-6}]\exp[i(0.4851x - 0.0209z)10^{-6}].
\end{aligned}$$

The matrix of the coefficients in (4.1)-(4.3) is non-singular and then we can evaluate the amplitude of L_-, T_-, and R_+; evidently the amplitude of R_+ turns out to be zero. So reflection and transmission coefficients are fully established. It is of interest to examine this reflection and transmission from the experimental viewpoint.

Experiments involve homogeneous waves ($\mathbf{k}_2 = 0$) as incident waves or. Accordingly the incident wave is not the F_- wave corresponding to the solution of (4.4). For definiteness, consider the homogeneous wave which follows by letting $\mathbf{k}_2 = 0$ and keeping k_x as in the previous incident wave, i.e.

$$F_-: \quad \phi^i = \Phi^i \exp[i(0.4851x - 0.4573z)10^{-6}].$$

Correspondingly the emergent waves are characterized as

$$\begin{aligned}
R_+: &\quad \phi = \Phi \exp[i(0.4851x + 0.4573z)10^{-6}],\\
L_-: &\quad \breve{\phi} = \breve{\Phi} \exp(0.4332\,10^{-6}\,z)\exp[i(0.4851x - 0.0003z)10^{-6}],\\
T_-: &\quad \breve{\psi} = \breve{\Psi} \exp(0.1705\,10^{-6}\,z)\exp[i(0.4851x - 0.0035z)10^{-6}]
\end{aligned}$$

and their amplitudes are given by

$$\Phi = -(0.6745 + 0.1169\,i)\Phi^i,$$

$$\breve{\Phi} = (-0.1361 + 2.2663\,i)\Phi^i, \qquad \breve{\Psi} = -(3.6027 + 0.2302\,i)\Phi^i.$$

The amplitude of the reflected wave is about 68% of that of the incident one while it should vanish if the incident wave were a component of the $F_-L_-T_-$ mode. This result is closely related to the detection of surface waves. According to Table 1 there is a surface wave solution which occurs at $s = 0.8763 - 0.0274\,i$ while the mode $F_-L_-T_-$ corresponds to $s = 0.8760 + 0.0274\,i$. The fact that the two roots s are so near has been ascertained for a number of values of the material parameters. This indicates as highly plausible that in practice when a surface wave is thought to be revealed the $F_-L_-T_-$ mode is involved in an approximate way which allows R_+ to have a non-zero amplitude.

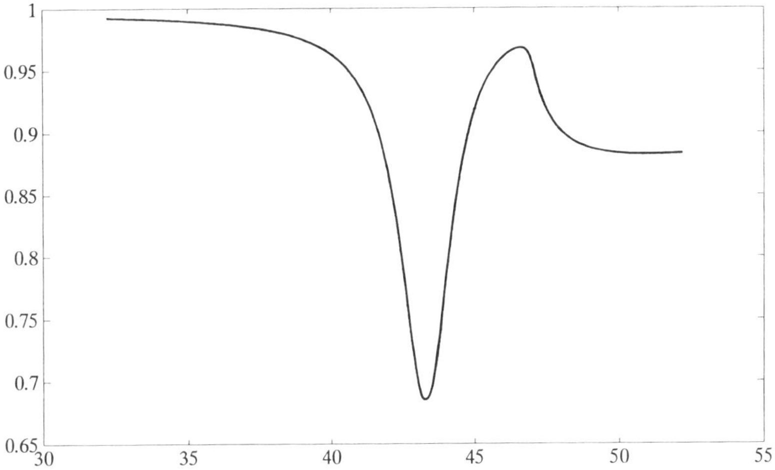

Fig. 5.1 Modulus of the relative amplitude of the reflected wave versus the angle (in degrees) of the homogeneous, incident wave.

To emphasize this view we have examined the modulus of the relative amplitude of the reflected wave, by letting the incident wave be homogeneous, in terms of the incidence angle.

As shown in Fig. 5.1, the amplitude attains a minimum at the critical angle ($\theta \simeq 43°$) in good agreement with similar results which have appeared in the literature. For example, in [134] the Rayleigh angle is considered in connection with the focal shift of beams and is found to be 42°. In [108] the Rayleigh angle is found to be 45.31°.

5.7 Surface waves on prestressed half-spaces

Following §2.6, we review the main steps for the analysis of heavy solids. Consider a solid which is unstressed in a reference placement $\mathcal{B}$. The solid is at equilibrium in an intermediate placement $\mathcal{B}_i$ under a suitable stress field (prestress), induced by the gravity force. As a consequence of the motion, the body occupies the present placement $\mathcal{B}_t$ at the present time t. The position $\mathbf{x}$ in $\mathcal{B}_i$ is given by the deformation $\mathbf{x} = \mathbf{x}(\mathbf{X})$ and the position $\tilde{\mathbf{x}}$ in $\mathcal{B}_t$ by the motion $\tilde{\mathbf{x}} = \tilde{\mathbf{x}}(\mathbf{X}, t)$. By the invertibility of $\mathbf{x} = \mathbf{x}(\mathbf{X})$ we can specify the displacement $\mathbf{u} = \tilde{\mathbf{x}} - \mathbf{x}$ as $\mathbf{u} = \mathbf{u}(\mathbf{x}, t)$. We follow the approximation that the displacement gradient $\mathbf{H} = \nabla \mathbf{u}^\dagger$ remains small, i.e. $|\mathbf{H}| \ll 1$ at every $\mathbf{x} \in \mathcal{B}_i$ and $t \in \mathbb{R}$. Accordingly we neglect quadratic terms in $\mathbf{H}$ and higher.

Consider the equilibrium equation

$$\nabla_{\mathbf{X}} \cdot \mathbf{S}^0 + \rho_0 \mathbf{b} = 0 \tag{7.1}$$

and the equation of motion

$$\rho_0 \ddot{\mathbf{u}} = \nabla_{\mathbf{X}} \cdot \mathbf{S} + \rho_0 \mathbf{b}. \tag{7.2}$$

We let $\mathbf{b}$ be the (constant) gravity acceleration $\mathbf{g}$. Subtraction of (7.1) from (7.2) yields

$$\rho_0 \ddot{\mathbf{u}} = \nabla_{\mathbf{X}} \cdot (\mathbf{S} - \mathbf{S}^0). \tag{7.3}$$

Consider the second Piola-Kirchhoff stresses $\mathbf{Y}^0, \mathbf{Y}^0 + \hat{\mathbf{Y}}$ corresponding to $\mathbf{x}(\mathbf{X})$, $\tilde{\mathbf{x}}(\mathbf{X}, t)$ and regard $\hat{\mathbf{Y}}$ as small so that quadratic terms in $\hat{\mathbf{Y}}$ and higher are neglected. Then (7.3) becomes

$$\rho_0 \ddot{\mathbf{u}} = \nabla_{\mathbf{X}} \cdot (\mathbf{H}\mathbf{Y}^0 + \mathbf{F}\hat{\mathbf{Y}}). \tag{7.4}$$

Divide by J, observe that $\partial \mathbf{u}/\partial \mathbf{X} = \mathbf{H}\mathbf{F}$ and make use of the identity $(F_{iK}/J)_{,i} = 0$. Then (7.4) can be written as

$$\rho \ddot{\mathbf{u}} = \nabla \cdot \big(\frac{1}{J}\mathbf{H}\mathbf{F}\mathbf{Y}^0\mathbf{F}^\dagger + \frac{1}{J}\mathbf{F}\hat{\mathbf{Y}}\mathbf{F}^\dagger\big).$$

Now, $\mathbf{F}\mathbf{Y}^0\mathbf{F}^\dagger/J$ is just the Cauchy stress $\mathbf{T}^0$ in the equilibrium placement $\mathcal{B}_i$ and then

$$\rho \ddot{\mathbf{u}} = \nabla \cdot \big(\mathbf{H}\mathbf{T}^0 + \frac{1}{J}\mathbf{F}\hat{\mathbf{Y}}\mathbf{F}^\dagger\big). \tag{7.5}$$

Application of (7.5) to linear viscoelastic solids with incremental isotropy and regarding λ and μ as constant yield

$$\begin{aligned}\rho \ddot{\mathbf{u}} = \nabla \cdot (\mathbf{T}^0 \nabla \mathbf{u})^\dagger &+ \mu_0 \Delta \mathbf{u}(t) + (\mu_0 + \lambda_0)\nabla(\nabla \cdot \mathbf{u})(t) \\ &+ \int_0^\infty \mu'(s)\Delta \mathbf{u}(t-s)\,ds + \int_0^\infty (\mu'(s) + \lambda'(s))\nabla(\nabla \cdot \mathbf{u})(t-s)\,ds.\end{aligned} \tag{7.6}$$

Remark. In the paper [107], incompressible motions are considered and the Cauchy stress in the placement $\mathcal{B}_t$ is taken as the equilibrium (isotropic) tensor, $\sigma\mathbf{1}$ say, plus the term due to the motion. The corresponding first Piola-Kirchhoff stress $\mathbf{S}$ in $\mathcal{B}_i$ is

$$\mathbf{S} = \sigma \mathbf{1} \frac{\partial}{\partial \tilde{\mathbf{x}}} \mathbf{x} \simeq \sigma \mathbf{1} - \sigma \mathbf{H}^\dagger$$

and the effect on the equation of motion is

$$\nabla \cdot \mathbf{S} = \nabla \sigma - \nabla \sigma \mathbf{H} - \sigma \nabla \mathrm{tr}\, \mathbf{H}.$$

According to (7.6), though, we would have $\sigma \Delta \mathbf{u}$ instead of $-\sigma \nabla(\nabla \cdot \mathbf{u})$. What may seem obvious is the superposition property for the Cauchy stress. This property is not obvious at all. We can use superposition relative to the unstressed placement $\mathcal{B}$.

We now investigate the possibility of surface wave solutions, in prestressed solids, for which the amplitude decays with distance from a given surface (and vanishes at infinite distance). To fix ideas we consider the half-space $z \geq 0$ and let the prestress $\mathbf{T}^0$ be determined by the gravity force $\rho\mathbf{g}$, namely

$$\nabla \cdot \mathbf{T}^0 + \rho\mathbf{g} = 0.$$

In addition, $\mathbf{T}^0\mathbf{n} = 0$ on the boundary of the half-space (the plane $z = 0$ and the semi-spherical surface). By symmetry $\mathbf{T}^0$ depends on z only and then

$$T^0_{13}, T^0_{23} = 0, \quad T^0_{33,3} = -\rho g,$$

$$T^0_{11}, T^0_{12}, T^0_{22} = 0.$$

Since T^0_{33} vanishes at $z = 0$, if ρ is allowed to depend on z then we have

$$T^0_{33} = f(z) := -g\int_0^z \rho(\xi)d\xi, \qquad T^0_{ik} = 0 \quad \text{if} \quad i,k \neq 3. \tag{7.7}$$

Alternatively, we might take $\mathbf{T}^0$ as isotropic and, specifically,

$$\mathbf{T}^0 = f(z)\mathbf{1}. \tag{7.8}$$

Then, letting $\mathbf{u} = \mathbf{U}(\mathbf{x})\exp(-i\omega t)$ we can write the equation of motion as

$$\rho\omega^2 U_i = f' U_{i,3} + f U_{i,33} + \mu\Delta U_i + (\mu + \lambda)U_{j,ji}, \tag{7.9}$$

or

$$\rho\omega^2 U_i = f' U_{i,3} + (\mu + f)\Delta U_i + (\mu + \lambda)U_{j,ji}, \tag{7.10}$$

according as (7.7) or (7.8) is considered, a prime denoting differentiation with respect to z.

To fix ideas consider (7.10). By analogy with the theory of waves in unstressed bodies, we look for surface wave solutions in the form

$$U_1 = \Phi \exp\Big[-\int_0^z \alpha(\xi)d\xi\Big]\exp(ikx),$$

$$U_3 = \Psi \exp\Big[-\int_0^z \alpha(\xi)d\xi\Big]\exp(ikx),$$

and $U_2 = 0$. Substitution in (7.10) yields

$$\begin{aligned}&[\rho\omega^2 + (f+\mu)(\alpha^2 - \alpha' - k^2) - k^2(\lambda+\mu) - \alpha f']\Phi - ik\alpha(\lambda+\mu)\Psi = 0,\\ &-ik\alpha(\lambda+\mu)\Phi + [\rho\omega^2 - \alpha f' + (\mu+f)(\alpha^2 - \alpha' - k^2) + (\lambda+\mu)(\alpha^2 - \alpha')]\Psi = 0.\end{aligned} \tag{7.11}$$

Non-trivial solutions for Φ, Ψ are allowed only if the determinantal equation

$$[\rho\omega^2+(f+\mu)(\alpha^2-\alpha'-k^2)-\alpha f'][\rho\omega^2+(f+2\mu+\lambda)(\alpha^2-\alpha'-k^2)-\alpha f']+k^2(\lambda+\mu)^2\alpha'=0 \tag{7.12}$$

holds.

In (7.12), ρ and f are given functions of z while k may be viewed as a parameter. Then (7.12) is a first-order differential equation in the unknown function $\alpha(z)$. To determine the solution α to (7.12) analytically is out of the question. Instead we consider the approximate form of (7.12) when $|f| \ll |\mathrm{Re}\,\mu|$ and $\rho \simeq$ constant, $f' = -\rho g \simeq$ constant. Accordingly we look for a solution $\alpha' \simeq 0$ and then α constant. In such a case (7.12) splits into the two equations

$$\rho\omega^2 + \mu(\alpha^2 - k^2) + \rho g\alpha = 0, \tag{7.13}$$

$$\rho\omega^2 + (2\mu + \lambda)(\alpha^2 - k^2) + \rho g\alpha = 0. \tag{7.14}$$

The surface wave character indicates that only the values of α are admissible that satisfy $\mathrm{Re}\,\alpha > 0$. Letting ν stand for μ or $2\mu + \lambda$, we obtain from (7.13) and (7.14) that

$$\alpha = k\sqrt{1 - \frac{c^2}{c_\nu^2} + \left(\frac{\rho g}{2\nu k}\right)^2} - \frac{\rho g}{2\nu}$$

where $c_\nu = \sqrt{\nu/\rho}$, $c = \omega/k$ and the square root is understood as that with positive real part; of course only those with $\mathrm{Re}\,\alpha > 0$ are admissible values. The two possible values for ν, and then c_ν, yield the values α_L, α_T for α.

If we take formally $g = 0$ we have the well-known values $\alpha = k\sqrt{1 - c^2/c_\nu^2}$ for the attenuation coefficient of surface waves in unstressed bodies. If, instead, we regard the gravity as a small correction to the unstressed case, $\rho^2 g^2 \ll 4\nu^2 k^2(1 - c^2/c_\nu^2)$, we can write

$$\alpha = k\sqrt{1 - \frac{c^2}{c_\nu^2}} + \frac{\rho g}{2\nu}\left(\frac{1}{2k}\sqrt{1 - \frac{c^2}{c_\nu^2}} - 1\right).$$

In this framework the system (7.11) reduces to

$$[\rho\omega^2 + \mu\alpha^2 + \rho g\alpha - (2\mu + \lambda)k^2]\Phi - ik(\mu + \lambda)\alpha\Psi = 0,$$

$$-ik(\mu + \lambda)\alpha\Phi + [\rho\omega^2 + (2\mu + \lambda)\alpha^2 + \rho g\alpha - \mu k^2]\Psi = 0.$$

The two solutions for Ψ/Φ are then given by

$$\left(\frac{\Psi}{\Phi}\right)_L = \frac{i\alpha_L}{k}, \qquad \left(\frac{\Psi}{\Phi}\right)_T = \frac{ik}{\alpha_T}.$$

Accordingly,

$$U_1 = [\Phi_L \exp(-\alpha_L z) + \Phi_T \exp(-\alpha_T z)]\exp(ikx), \tag{7.15}$$

$$U_3 = i[\frac{\alpha_L}{k}\Phi_L \exp(-\alpha_L z) + \frac{k}{\alpha_T}\Phi_T \exp(-\alpha_T z)] \exp(ikx). \tag{7.16}$$

The vanishing of the traction $\mathbf{t} = \mathbf{Tn}$ at $z = 0$ amounts to

$$T_{13}, T_{33} = 0 \quad \text{at} \quad z = 0,$$

$T_{23} = 0$ being trivially true. Since $\mathbf{T}^0\mathbf{n} = 0$, at $z = 0$, by (7.15) and (7.16) we have

$$0 = T_{13}|_{z=0} = -\mu\big[2\alpha_L\Phi_L + \big(\alpha_T + \frac{k^2}{\alpha_T}\big)\Phi_T\big] \exp(ikx),$$

$$0 = T_{33}|_{z=0} = -i\big\{\big[2\mu\frac{\alpha_L^2}{k} + \lambda\big(\frac{\alpha_L^2}{k} - k\big)\big]\Phi_L + 2\mu k\Phi_T\big\} \exp(ikx).$$

Non-trivial solutions for Φ_L and Φ_T are possible if the corresponding determinantal equation holds, namely

$$4\frac{\alpha_L}{k}\frac{\alpha_T}{k} - \Big(1 + \frac{\alpha_T^2}{k^2}\Big)\Big[\frac{2\mu+\lambda}{\mu}\frac{\alpha_L^2}{k^2} - \frac{\lambda}{\mu}\Big] = 0.$$

Now observe that, by (7.14) and (7.13),

$$\frac{2\mu+\lambda}{\mu}\frac{\alpha_L^2}{k^2} - \frac{\lambda}{\mu} = 2 - \frac{\rho\omega^2}{\mu k^2} - \frac{\rho g \alpha_L}{\mu k^2},$$

$$\frac{\rho\omega^2}{\mu k^2} = -\frac{\alpha_T^2}{k^2} + 1 - \frac{\rho g \alpha_T}{\mu k^2}.$$

Then we can write the determinantal equation as

$$4\frac{\alpha_L}{k}\frac{\alpha_T}{k} - \Big(1 + \frac{\alpha_T^2}{k^2}\Big)\Big[1 + \frac{\alpha_T^2}{k^2} + \frac{\rho g}{\mu k}\Big(\frac{\alpha_T}{k} - \frac{\alpha_L}{k}\Big)\Big] = 0. \tag{7.17}$$

Equation (7.17) provides the admissible phase speeds $c = \omega/k$ of the surface wave. As it must be, the standard form of the determinantal equation (cf. [2], §5.11, for elastic solids) is recovered by dropping out the gravity effect $(\rho g)(\mu k)[(\alpha_T/k) - (\alpha_L/k)]$.

Remark. The results so derived are based on the approximation that $|f| \ll |\mathrm{Re}\,\mu|$. The extent of validity of this approximation may be estimated as follows. Letting ρ be constant we have $|f| = \rho g z$ and then $|f| \ll |\mathrm{Re}\,\mu|$ means $z \ll z_0 := \mathrm{Re}\,\mu/\rho g$. This in turn means that the approximation is good for high values of z_0 and this occurs for high values of the phase speed $\sqrt{\mathrm{Re}\,\mu/\rho}$. For soil $z_0 \simeq 6\,10^3$ m, for metals $z_0 \simeq 10^6$ m.

It is worth considering how (7.17) can be written in terms of a single unknown quantity, which is essential for determining the phase speed c. To save writing let $\eta_L = \alpha_L/k, \eta_T = \alpha_T/k$ and $\gamma = \rho g/\mu k$. Then the equation (7.17) takes the form

$$\frac{4\eta_L\eta_T}{(1+\eta_T^2)^2} - 1 = \frac{\gamma}{1+\eta_T^2}(\eta_T - \eta_L). \tag{7.18}$$

Meanwhile η_L and η_T satisfy the analogue of (7.13) and (7.14), namely

$$\eta_T^2 + \gamma\eta_T - \left(1 - \frac{c^2}{c_T^2}\right) = 0, \tag{7.19}$$

$$\eta_L^2 + q\gamma\eta_L - \left(1 - \frac{c^2}{c_L^2}\right) = 0, \tag{7.20}$$

where, as usual, $q = (c_T/c_L)^2$. Now we take the view that η_T is the unknown quantity and then we need to write η_L and c in terms of η_T.

By (7.19)

$$1 + \eta_T^2 = 2 - \frac{c^2}{c_T^2} - \gamma\eta_T.$$

Substitution in (7.18) yields

$$4\eta_L\eta_T = \left(2 - \frac{c^2}{c_T^2} - \gamma\eta_T\right)\left(2 - \frac{c^2}{c_T^2} - \gamma\eta_L\right)$$

whence

$$(4 - \gamma^2)\eta_L\eta_T = \left(2 - \frac{c^2}{c_T^2}\right)^2 - \gamma\left(2 - \frac{c^2}{c_T^2}\right)(\eta_L + \eta_T). \tag{7.21}$$

Solving (7.21) with respect to η_L gives

$$\eta_L = \frac{\left(2 - c^2/c_T^2\right)^2 - \gamma\left(2 - c^2/c_T^2\right)\eta_T}{\gamma\left(2 - c^2/c_T^2\right) + (4 - \gamma^2)\eta_T}. \tag{7.22}$$

By (7.19) we have

$$\frac{c^2}{c_T^2} = 1 - \eta_T^2 - \gamma\eta_T \tag{7.23}$$

and then, by substitution in (7.22), we determine η_L as a function of η_T. Precisely, since

$$2 - \frac{c^2}{c_T^2} = \eta_T^2 + \gamma\eta_T + 1,$$

we obtain

$$\eta_L = \frac{(\eta_T^2 + \gamma\eta_T + 1)(\eta_T^2 + 1)}{\gamma\eta_T^2 + 4\eta_T + \gamma}. \tag{7.24}$$

To express c/c_L in terms of η_T we observe that

$$\frac{c^2}{c_L^2} = q\frac{c^2}{c_T^2}$$

and hence, by (7.19),

$$\frac{c^2}{c_L^2} = q(1 - \eta_T^2 - \gamma\eta_T). \tag{7.25}$$

Substitution of (7.24) and (7.25) in (7.20) and some rearrangement yields the desired equation

$$\eta_T^8 + 2\gamma\eta_T^7 + (4+\gamma^2)\eta_T^6 + 2\gamma(3-2q)\eta_T^5 + (\gamma^2 - 2\gamma^2 q - 16q + 6)\eta_T^4 - \gamma(8q+2)\eta_T^3 + (16q - \gamma^2 - 12)\eta_T^2 + \gamma(12q-6)\eta_T + 1 + 2\gamma^2 q - \gamma^2 = 0. \quad (7.26)$$

So, once we have solved (7.26) in η_T, by (7.24) we determine also the attenuation factor $k\eta_L$ of the longitudinal component and by (7.23) we find the phase speed c.

Of course if gravity is neglected then (7.19) must reduce, or be equivalent, to the standard Rayleigh equation

$$s^3 - 8s^2 + (24 - 16q)s - 16(1-q) = 0,$$

in the unknown $s = c^2/c_T^2$. To ascertain that this is so let $\gamma = 0$ in (7.26) and observe that, in such a case, $\eta_T^2 = 1 - s$. Direct substitution in (7.26) shows that the usual Rayleigh equation follows once the common factor s is, as usual, canceled in that we are looking for non-zero values of c.

6 WAVE PROPAGATION IN MULTILAYERED MEDIA

Time-harmonic wave propagation in layered media is of interest in many fields of research, such as in the analysis of laminated composite structures, geophysics, and submarine acoustics. Layered media may be modelled in various ways and the choice of the pertinent model is a matter of closeness to the physical reality. Discretely layered media are a finite sequence of homogeneous layers sandwiched between two homogeneous half-spaces. Continuously layered media consist in fact of a layer between two homogeneous half-spaces; the material properties of the layer vary continuously and depend only on the coordinate orthogonal to the boundaries. The discrete model, besides being of interest in itself, may be regarded as a discretization of the continuous one, and is likely to be of use in the elaboration of numerical procedures. For definiteness, the material in both half-spaces and layers is taken to be an isotropic, viscoelastic solid. Moreover, all interfaces or boundaries are parallel planes.

Essentially, the problem consists in evaluating the response of the material system to an inhomogeneous wave (really a conjugate pair of waves) impinging on the layers (or the layer) from a half-space. The response is given by the reflected waves and the transmitted waves, in the two half-spaces, and the field in the layers (or the layer). The mathematical approach to wave propagation is strictly related to the model of layered medium. In discretely layered media the field within each layer is decomposed into up and downgoing waves. Iteration of the behaviour at a single interface is the core of the procedure which leads to the determination of the field in each layer and of the reflection and transmission matrices of the sequence of layers. In continuously layered media the main difficulty lies in the determination of the field in the layer. A simpler procedure, based on fundamental systems of solutions, is developed subject to the crucial feature that the dynamics of the layer is eventually described by a Helmholtz equation. This is the case, for example, when the incident wave vectors are orthogonal to the interfaces. A general procedure is accomplished through the use of the propagator matrix along with fundamental systems of solutions. Definite results are obtained for the reflection and transmission matrices of the layer.

6.1 Discretely layered media

Consider $n-1$ plane parallel layers between two half-spaces. Each layer and each half-space consist of a linear viscoelastic homogenoeus and isotropic solid. The material parameters are constant in any layer and any half-space. Adjacent layers are in welded contact and then displacement and traction are required to be continuous across the interface. Consider Cartesian coordinates with the z-axis perpendicular to the layers, and denote by the apex $n+1$ the upper half-space ($z \to \infty$), by n,..., 2 the interior layers, and by 1 the lower half-space ($z \to -\infty$). We denote by $\mathcal{P}_s$ the interface between the layers s and $s+1$; $1 \leq s \leq n$. A conjugate pair of inhomogeneous waves, incident from the upper half-space, impinges on the interface $\mathcal{P}_n$ while no wave is incident from the lower half-space. The natural problem is to determine the reflected pair, in the upper half-space, and the transmitted pair, in the lower half-space, in terms of the material properties of the layers.

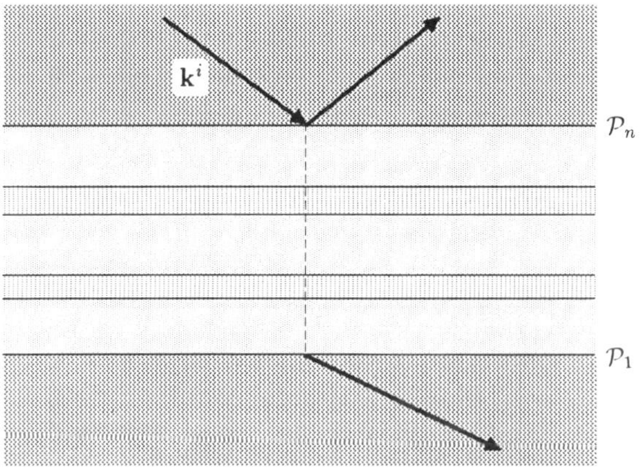

Fig. 6.1 A multilayered medium modelled as a sequence of layers sandwiched between two homogeneous half-spaces.

Since the incident field is in fact a conjugate pair of waves and the media involved are homogenous and isotropic, we look for the field in any layer as the superposition of up and downgoing longitudinal and transverse inhomogeneous waves, generated by multiple reflections and transmissions at the interfaces. The pertinent waves are described through their vector and scalar potentials. By the general expressions (4.1.7) and (4.1.8), it follows that the continuity of $\mathbf{U}$ and $\mathbf{t}$ at every interface implies the validity of Snell's law for the wave vectors. Consequently, the up and downgoing waves within each layer constitute two conjugate pairs. In addition it follows from §4.2 that, at each layer, the wave vectors are completely determined in terms of the incident ones and of the material parameters of the corresponding continua. More precisely, the x- and y-components of the pertinent wave vectors take the same values, say k_x and k_y, that are determined by the waves of the incident pair. Of course, the z-components are equal, to within the sign, to

$$\beta_L^s = \sqrt{\kappa_L^s - k_x^2 - k_y^2}, \qquad \beta_T^s = \sqrt{\kappa_T^s - k_x^2 - k_y^2},$$

where the root is meant with positive real part (or positive imaginary part if the real part vanishes). Then it follows that

$$\mathbf{k}_L^s = k_x \mathbf{e}_x + k_y \mathbf{e}_y \pm \beta_L^s \mathbf{e}_z, \qquad \mathbf{k}_T^s = k_x \mathbf{e}_x + k_y \mathbf{e}_y \pm \beta_T^s \mathbf{e}_z,$$

where the sign is + or − according as we are considering an up or downgoing wave. The corresponding scalar and vector potentials contain the common factor $\exp[i(k_x x + k_y y)]$, which is usually omitted. The superscript s, $2 \leq s \leq n$ labels quantities pertaining to the s-layer while no label or the accent ˘ indicate quantities of the upper or lower half-space.

By analogy with the procedure of Ch. 4 for conjugate pairs, we now establish recurrence formulae relating the amplitudes of waves in adjacent layers. Then, by generalizing the Thomson-Haskell technique [162, 82, 5, 22, 102, 45] to viscoelastic materials we determine the up and downgoing waves in each layer along with the reflection and transmission matrices. Such matrices in turn allow the evaluation of the net effect of the stalk of layers on the incident pair. As a comment we observe that a similar approach is applied in [109] to determine the acoustic material signature of a layered plate, while analogous arguments are developed for wave propagation in periodically stratified solid and fluid layers [152, 149] or elastic but anisotropic layers [136]. Our procedure works under *generic conditions*, which means that neither numerical nor analytic singularities occur in the expressions involved in the development of our computations. Anomalous cases will be considered later; a typical one is that of a layer so thick that a transverse wave originated at an interface is essentially extinguished, because of dissipation, before the next interface is reached. This makes some entries of the matrices, entering the general procedure, become numerically singular.

Letting $z = 0$ be the interface $\mathcal{P}_n$ we describe the fields of the incident and reflected pairs as (cf. (4.3.1) and (4.3.2))

$$\phi = \left[\Phi^+ \exp(i\beta_L z) + \Phi^- \exp(-i\beta_L z)\right] \exp[i(k_x x + k_y y)], \tag{1.1}$$

$$\boldsymbol{\psi} = \left[\boldsymbol{\Psi}^+ \exp(i\beta_T z) + \boldsymbol{\Psi}^- \exp(-i\beta_T z)\right] \exp[i(k_x x + k_y y)]. \tag{1.2}$$

To describe the transmitted field in the lower half-space it is convenient to identify the plane $z = 0$ with the interface $\mathcal{P}_1$; then the scalar and vector potentials are written as

$$\breve{\phi} = \breve{\Phi}^- \exp(-i\breve{\beta}_L z) \exp[i(k_x x + k_y y)], \qquad \breve{\boldsymbol{\psi}} = \breve{\boldsymbol{\Psi}}^- \exp(-i\breve{\beta}_T z) \exp[i(k_x x + k_y y)]. \tag{1.3}$$

Trivially, when considering the stalk of layers as a whole we have to fix the origin of the z-axis and then to account properly for the common origin.

The effect of the stalk on the incident pair is evaluated by determining the amplitudes Φ^+, Ψ^+, $\breve{\Phi}^-$, $\breve{\Psi}^-$ of the reflected and transmitted pairs in terms of Φ^- and Ψ^-. Really, by the transversality of Ψ we can write Ψ_z as a linear combination of Ψ_x and Ψ_y and then we have eventually to determine the amplitudes Φ^+, Ψ_x^+, Ψ_y^+ $\breve{\Phi}^-$, $\breve{\Psi}_x^-$, $\breve{\Psi}_y^-$ in terms of Φ^-, Ψ_x^-, Ψ_y^- (cf. (4.3.11)).

Consider the s-th layer and choose the boundary $\mathcal{P}_{s-1}$ between the layers s and $s-1$ as the plane $z = 0$. We write

$$\phi^s = [\Phi^{s-} \exp(-i\beta_L^s z) + \Phi^{s+} \exp(i\beta_L^s z)] \exp[i(k_x x + k_y y)], \tag{1.4}$$

$$\boldsymbol{\psi}^s = [\Psi^{s-} \exp(-i\beta_T^s z) + \Psi^{s+} \exp(i\beta_T^s z)] \exp[i(k_x x + k_y y)], \tag{1.5}$$

where the unknown constants Φ^{s-} and Ψ^{s-} refer to the downgoing waves, while Φ^{s+} and Ψ^{s+} refer to the upgoing ones. If $\operatorname{Re}\beta_{L,T}^s = 0$ then $\operatorname{Im}\beta_{L,T}^s > 0$ and, by (1.4), the amplitude of the + wave decreases as z increases; the opposite happens to the amplitude of the − wave. This is consistent with the view that the two waves are originated through interactions at $\mathcal{P}_{s-1}$ and $\mathcal{P}_s$, respectively. In this connection we also make a slight abuse of language by calling the two waves up and downgoing, although the real part of the wave vector is horizontal.

Now we look for the expressions of displacement and traction in the layer s in that $\mathbf{U}$ and $\mathbf{t}$ are continuous at each plane of discontinuity in the material parameters. Explicit expressions for $\mathbf{U}^s$ and $\mathbf{t}^s$ in terms of the amplitudes entering (1.4) and (1.5) follow straightforwardly as an application of (4.1.7) and (4.1.8). Then the statement of continuity at the each interface is changed into a $1-1$ correspondence between the complex parameters yielding the amplitudes of the waves affecting any two adjacent layers.

We consider first $\mathbf{U}^s$ and $\mathbf{t}^s$ at $z = 0$. The components of these vectors are found from (4.3.4)-(4.3.9), through obvious changes in the notation. In the analysis of the behaviour of inhomogeneous waves at an interface these components have been grouped in a very special way, that brings into evidence a common dependence on the same combinations of the amplitudes and simplifies the development of the calculations involved. This strongly suggests that we adopt a similar formulation in the present framework. Therefore we define the *traction-displacement vector* $\mathbf{Z}^s$ as

$$\mathbf{Z}^s = \left(\frac{U_x^s}{i}, \frac{U_y^s}{i}, \frac{t_z^s}{2}, \frac{U_z^s}{i}, \frac{t_x^s}{2k_x}, \frac{t_y^s}{2k_y}\right)^\dagger. \tag{1.6}$$

Of course $\mathbf{Z}$ is continuous within each layer and, because of Snell's law, is also continuous across any interface. It follows from (4.1.7) and (4.1.8) that

$$\mathbf{Z}^s(z) = C^s\, E^s(z)\mathbf{A}^s, \tag{1.7}$$

where the 6×6 complex matrix C^s has the block structure

$$C^s = \begin{pmatrix} C_1^s & C_1^s \\ C_2^s & -C_2^s \end{pmatrix};$$

C_1^s and C_2^s denote the 3×3 constant matrices (cf. (4.5.2) and (4.5.5))

$$C_1^s = \begin{pmatrix} k_x & -k_x k_y/\beta_T^s & -k_y^2/\beta_T^s - \beta_T^s \\ k_y & k_x^2/\beta_T^s + \beta_T^s & k_x k_y/\beta_T^s \\ \mu^s(\frac{1}{2}\kappa_T^s - k_x^2 - k_y^2) & -\mu^s k_y \beta_T^s & \mu^s k_x \beta_T^s \end{pmatrix}$$

$$C_2^s = \begin{pmatrix} \beta_L^s & -k_y & k_x \\ \mu^s \beta_L^s & -\mu^s k_y & \mu^s(k_x - \frac{1}{2}\kappa_T^s/k_x) \\ \mu^s \beta_L^s & -\mu^s(k_y - \frac{1}{2}\kappa_T^s/k_y) & \mu^s k_x \end{pmatrix}.$$

The symbol E^s denotes the diagonal matrix

$$E^s = \mathrm{diag}\,(\exp(i\beta_L^s z), \exp(i\beta_T^s z), \exp(i\beta_T^s z), \exp(-i\beta_L^s h^s), \exp(-i\beta_T^s z), \exp(-i\beta_T^s z)), \tag{1.8}$$

The matrix E^s is non-singular; indeed, $\det E^s = 1$. Further, $\mathbf{A}^s$ is a column matrix, which is called vector amplitude, and is given by

$$\mathbf{A}^s = (\mathbf{A}^{s+}, \mathbf{A}^{s-})^\dagger,$$

where

$$\mathbf{A}^{s+} = (\Phi^{s+}, \Psi_x^{s+}, \Psi_y^{s+})^\dagger, \qquad \mathbf{A}^{s-} = (\Phi^{s-}, -\Psi_x^{s-}, -\Psi_y^{s-})^\dagger.$$

As $z \to 0_+$, equation (1.7) gives

$$\mathbf{Z}^s(0_+) = C^s \mathbf{A}^s$$

thus showing that the matrix C^s relates the value of traction and displacement at $z = 0$ to the vector amplitude.

The continuity condition at the plane $\mathcal{P}_{s-1}$, i.e. at $z = 0$, can be written in the form

$$\mathbf{Z}^s(0_+) = \mathbf{Z}^{s-1}(0_-).$$

Comparison with the expression for $\mathbf{Z}^s(0_+)$ shows that

$$C^s \mathbf{A}^s = \mathbf{Z}^{s-1}(0_-),$$

whence the vector amplitude $\mathbf{A}^s$ follows in terms of traction and displacement in the layer $s-1$ as

$$\mathbf{A}^s = C^{-s} \mathbf{Z}^{s-1}(0_-),$$

where

$$C^{-s} := (C^s)^{-1} = \frac{1}{2} \begin{pmatrix} C_1^{-s} & C_2^{-s} \\ C_1^{-s} & -C_2^{-s} \end{pmatrix}, \tag{1.9}$$

and (cf. (4.5.3) and (4.5.6))

$$C_1^{-s} := (C_1^s)^{-1} = \frac{1}{\kappa_T^s \beta_T^s} \begin{pmatrix} 2k_x\beta_T^s & 2k_y\beta_T^s & 2\beta_T^s/\mu^s \\ -k_x k_y & (\beta_T^s)^2 - k_y^2 & -2k_y/\mu^s \\ k_x^2 - (\beta_T^s)^2 & k_x k_y & 2k_x/\mu^s \end{pmatrix}, \tag{1.10}$$

$$C_2^{-s} := (C_2^s)^{-1} = \frac{2}{\mu^s \kappa_T^s \beta_L^s} \begin{pmatrix} \mu^s(\frac{1}{2}\kappa_T^s - k_x^2 - k_y^2) & k_x^2 & k_y^2 \\ -\mu^s k_y \beta_L^s & 0 & k_y\beta_L^s \\ \mu^s k_x \beta_L^s & -k_x\beta_L^s & 0 \end{pmatrix}. \tag{1.11}$$

Substitution of the expression for $\mathbf{A}^s$ into (1.7) yields the traction and displacement in the layer s in terms of the corresponding values in the layer $s-1$, that is

$$\mathbf{Z}^s(z) = C^s\, E^s(z)\, C^{-s}\, \mathbf{Z}^{s-1}(0_-).$$

As usual, the common factor $\exp[i(k_x x + k_y y)]$ is understood and not written. It is natural to regard $C^s E^s(z) C^{-s}$ as the *propagator* matrix in the s-layer. Also we have

$$\det\,(C^s\, E^s(z)\, C^{-s}) = 1,$$

which shows that the transformation is non-singular. On setting $z = h^s$, where h^s is the width of the layer s, we find the limit from below of $\mathbf{Z}^s$ at the plane $\mathcal{P}_s$. Iteration of the procedure allows us to show that $\mathbf{Z}$ at the upper interface, say simply $\mathbf{Z}^{n+1}$, and $\mathbf{Z}$ at the lower interface, say $\mathbf{Z}^1$, are related by

$$\mathbf{Z}^{n+1} = C^n\, E^n\, C^{-n} ... C^2\, E^2\, C^{-2} \mathbf{Z}^1, \tag{1.12}$$

where each E^s is evaluated at the corresponding height h^s.

Now we establish the connection between $\breve{\mathbf{A}}^-$ and $\mathbf{A}^+, \mathbf{A}^-$. First, set the plane $z = 0$ at the boundary of the layers 1 and 2, and restrict attention to the lower half-space. Then observe that the limit of $\mathbf{Z}^1$ in terms of the parameters $\breve{\Phi}^-$, $\breve{\Psi}_x^-$, and $\breve{\Psi}_y^-$ is easily recovered by comparison with the right-hand sides of (4.5.1) and (4.5.4). We have

$$\mathbf{Z}^1(0_-) = \begin{pmatrix} \breve{C}_1 \\ -\breve{C}_2 \end{pmatrix} \breve{\mathbf{A}}^-.$$

Substitution into (1.12) yields

$$\mathbf{Z}^{n+1} = \begin{pmatrix} B_1\breve{C}_1 - B_2\breve{C}_2 \\ B_3\breve{C}_1 - B_4\breve{C}_2 \end{pmatrix} \breve{\mathbf{A}}^-, \tag{1.13}$$

at $\mathcal{P}_n$, where the B's are 3×3 matrices defined by

$$\begin{pmatrix} B_1 & B_2 \\ B_3 & B_4 \end{pmatrix} := C^n E^n C^{-n} ... C^2 E^2 C^{-2}.$$

Notice that the matrices B depend on the material parameters and the width of the layers. Second, to establish a connection between $\mathbf{Z}^{n+1}$ and $\mathbf{A}^+, \mathbf{A}^-$, we consider the upper half-space and regard the boundary $\mathcal{P}_n$ as the plane $z = 0$. Comparison with the definition (1.6) of the traction-displacement vector $\mathbf{Z}$ and the left-hand sides of (4.5.1) and (4.5.4) shows that the continuity at the interface results in

$$\mathbf{Z}^{n+1} = C\mathbf{A} = C\begin{pmatrix}\mathbf{A}^+\\ \mathbf{A}^-\end{pmatrix}. \tag{1.14}$$

Notice that the column vector $\mathbf{A}$ has been decomposed into its up and downgoing constituents, namely $\mathbf{A}^+$ and $\mathbf{A}^-$, since $\mathbf{A}^+$ is unknown whereas $\mathbf{A}^-$ is given. Comparison with (1.13) gives

$$C\begin{pmatrix}\mathbf{A}^+\\ \mathbf{A}^-\end{pmatrix} = \begin{pmatrix}B_1\check{C}_1 - B_2\check{C}_2\\ B_3\check{C}_1 - B_4\check{C}_2\end{pmatrix}\check{\mathbf{A}}^-.$$

By solving (1.14) for $\mathbf{A}^+$ and $\mathbf{A}^-$ and substituting the expression of C^{-1} from (1.9) we have

$$\mathbf{A}^+ = \tfrac{1}{2}\left[C_1^{-1}(B_1\check{C}_1 - B_2\check{C}_2) + C_2^{-1}(B_3\check{C}_1 - B_4\check{C}_2)\right]\check{\mathbf{A}}^-$$
$$\mathbf{A}^- = \tfrac{1}{2}\left[C_1^{-1}(B_1\check{C}_1 - B_2\check{C}_2) - C_2^{-1}(B_3\check{C}_1 - B_4\check{C}_2)\right]\check{\mathbf{A}}^-.$$

These equations provide the desired transmission and reflection matrices. More precisely, going back to the more familiar notation in terms of scalar amplitudes, we find that

$$\begin{pmatrix}\check{\Phi}^-\\ \check{\Psi}_x^-\\ \check{\Psi}_y^-\end{pmatrix} = T\begin{pmatrix}\Phi^-\\ \Psi_x^-\\ \Psi_y^-\end{pmatrix} \tag{1.15}$$

where

$$T = 2\,\mathbb{I}\left[C_1^{-1}(B_1\check{C}_1 - B_2\check{C}_2) - C_2^{-1}(B_3\check{C}_1 - B_4\check{C}_2)\right]^{-1}\mathbb{I},$$

and $\mathbb{I} = \mathrm{diag}(1, -1, -1)$. By the same token, we obtain

$$\begin{pmatrix}\Phi^+\\ \Psi_x^+\\ \Psi_y^+\end{pmatrix} = R\begin{pmatrix}\Phi^-\\ \Psi_x^-\\ \Psi_y^-\end{pmatrix}, \tag{1.16}$$

where

$$R = \tfrac{1}{2}\left[C_1^{-1}(B_1\check{C}_1 - B_2\check{C}_2) + C_2^{-1}(B_3\check{C}_1 - B_4\check{C}_2)\right]\mathbb{I}\,T,$$

or, in a simpler form,

$$R = \mathbb{I} + C_2^{-1}(B_3\check{C}_1 - B_4\check{C}_2)\,\mathbb{I}\,T.$$

The reflection and transmission matrices depend on the material properties and the width of the layers $2, \dots n$, through the matrices B, and on the material properties of layers $n+1$ and 1 through the matrices C and $\check{C}$. If we set $B_2 = B_3 = 0$ and $B_1 = B_4 = \mathbb{1}$ then the transmission and reflection matrices reduce to those characterizing reflection and transmission for a plane interface between two half-spaces. As regards the transmission matrix we find that

$$T = 2\,\mathbb{I}\{C_1^{-1}[\breve{C}_1 + C_1(C_2)^{-1}\breve{C}_2)]\}^{-1}\,\mathbb{I},$$

whence (cf. §4.5)

$$T = 2\,\mathbb{I}[\breve{C}_1 + C_1(C_2)^{-1}\breve{C}_2)]^{-1}C_1\,\mathbb{I}.$$

Analogously, upon substitution of the expressions for the B's we find for R the same result as in §4.5. To sum up, the transmitted pair is described by (1.3), with $\breve{\Phi}^-$, $\breve{\Psi}_x^-$, $\breve{\Psi}_y^-$ given by (1.15), while $\breve{\Psi}_z^-$ follows from (4.3.11). The reflected pair consists of the + waves, in (1.1) and (1.3). The amplitudes are provided by (1.16) and (4.3.11).

Suppose now that $k_y = 0$. To establish the continuity condition it is convenient to modify (1.6) by setting $Z_6^s = t_y^s/2$. Then the last rows of C_2^s and $\breve{C}_2$ read $(0, \mu^s\kappa_T^s/2, 0)$ and $(0, \breve{\mu}\breve{\kappa}_T/2, 0)$, respectively. The other entries of the matrices C_1^s, C_2^s, $\breve{C}_1$ and $\breve{C}_2$ are found through the substitution $k_y = 0$. The same holds for C_1^{-s} and C_2^{-s}, provided the second row of the matrix in the expression C_2^{-s} is changed to $(0, 0, \beta_L^s)$. Then everything goes as in the case $k_y \neq 0$.

Due to the structure of the layer matrices E^s, the numerical implementation of these results may present some problems when the real exponential contributions $\exp(-\mathrm{Im}\,\beta_L^s\,h^s)$ or $\exp(-\mathrm{Im}\,\beta_T^s\,h^s)$ are too far from unity. The physical meaning of this condition is that the attenuation of the amplitude coefficients due to absorption effects is too strong and the wave cannot propagate across sufficiently thick layers. Similarly, if $\beta_L^s = 0$ for some s then the matrix C_2^s becomes singular and hence no inversion is allowed. On the contrary, elements of C_1^s are unbounded if $\beta_T^s = 0$. These drawbacks are examined in §§6.2, 6.3.

6.2 Thick layers

Wave propagation in a stack of layers has been described in the previous section by using a matrix method which parallels the Thomson-Haskell technique [162, 82]. The theoretical model used is exact, in the sense that no approximation is introduced. The method allows for any kind of incident wave, through the use of inhomogeneous waves, and embodies linear elastic layers as a particular case. Although this matrix formulation is simple and attractive, the implementation is affected by some limitations. In this section we examine a class of difficulties that occur because of the structure (1.8) of the matrix E^s entering (1.7) or (1.12).

Consider the entries of the matrix E^s, that is $\exp(\pm i\beta_{L,T}^s h^s)$, and observe that because

$$\beta_{L,T}^s = \zeta_{L,T}^s + i\sigma_{L,T}^s,$$

we can write the entries as

$$\exp\left(\pm i\beta^s_{L,T}h^s\right) = \exp\left(\mp\ \sigma^s_{L,T}h^s\right)\exp\left(\pm\ i\zeta^s_{L,T}h^s\right).$$

The occurrence of real-valued exponents gives rise to numerical singularities if $\sigma^s_{L,T}$, or h^s or both become very large. In physical terms this reflects the fact that an attenuating wave cannot propagate across a layer, from side to side, if the layer is thick enough. According to the current literature, similar features are ascribed to the case when the incident wave vector is beyond the critical angle for one or more waves in the layer. A number of procedures have been elaborated to eliminate drawbacks connected with the Thomson-Haskell technique; these include recursive algorithms [102], methods employing minors of the matrices [5, 65, 75, 1], the global matrix method [151], decoupling algorithms [68], and recourse to pseudo-materials [45].

The method set up in this section is based on a development of the matrix technique that allows for the possibility that up and downgoing waves extinguish within a layer. The continuity conditions at each interface are still required and the determination of reflected and transmitted amplitudes is transformed into an algebraic problem which is no longer affected by the overly large or small numbers entering the expression of E^s. The procedure applies also to elastic layers, at high frequencies, when evanescent modes occur. We only need preliminarily to know when the exponential decay factor is regarded as small; let say

$$\exp\left(-|\sigma^s_L|h^s\right) < \varepsilon \quad \text{or} \quad \exp\left(-|\sigma^s_T|h^s\right) < \varepsilon, \tag{2.1}$$

where ε is a suitably small, given positive number. More generally, we regard as negligible those quantities whose moduli are smaller than the given value ε.

Here we study reflection and transmission matrices for the layered medium under the assumption that at least one of the two inequalities (2.1) holds for layer s, while the same inequalities are false for the other layers. A straightforward modification of the discussion leads to the algorithm to be adopted when (2.1) holds for two or more layers. The inequalities (2.1) can be tested directly, without involving the amplitudes of the pertinent waves, provided only that the material parameters of the layers and the common components k_x and k_y of the incident pair are known.

Restrict attention to the case when the product σh is large in a layer s. Then it is convenient to introduce two smaller stacks, namely, stack *I*, formed by the layers from n to $s+1$, and stack *II*, formed by the layer $s-1$ down to 2 (the lower half-space is labelled by 1). We study in detail wave propagation inside s and then use the results to establish a connection between waves transmitted by stack *I* and those incident on stack *II*, so as to set up a well-posed algebraic problem for reflection and transmission matrices of the whole stratified medium.

To make the procedure operative, we observe that, according to (4.1.4) and (4.1.5), k_L and k_T are generally near to each other, say $k_L \simeq k_T$. For any complex-valued vector

$\mathbf{w}$, we evaluate the modulus $|\mathbf{w}|$ through

$$|\mathbf{w}| = \sqrt{\mathbf{w} \cdot \mathbf{w}^*}.$$

Denote by $\mathbf{U}[\phi]$ and $\mathbf{U}[\boldsymbol{\psi}]$ the displacement vectors generated by the potentials ϕ and $\boldsymbol{\psi}$, respectively. Then (4.1.4) and (4.1.5) show that

$$|\mathbf{U}[\phi]| \simeq |k_L|\,|\phi|, \qquad |\mathbf{U}[\boldsymbol{\psi}]| \simeq |k_T|\,|\boldsymbol{\psi}|,$$

whence it follows that

$$\frac{|\mathbf{U}[\phi]|}{|\mathbf{U}[\boldsymbol{\psi}]|} \simeq \frac{|\phi|}{|\boldsymbol{\psi}|}.$$

Similarly, an inspection of the representation (4.1.8) for $\mathbf{t}$ shows that the tractions $\mathbf{t}[\phi]$, $\mathbf{t}[\boldsymbol{\psi}]$ satisfy

$$|\mathbf{t}[\phi]| \simeq \rho\omega^2|\phi|, \qquad |\mathbf{t}[\boldsymbol{\psi}]| \simeq \rho\omega^2|\boldsymbol{\psi}|,$$

whence

$$\frac{|\mathbf{t}[\phi]|}{|\mathbf{t}[\boldsymbol{\psi}]|} \simeq \frac{|\phi|}{|\boldsymbol{\psi}|}.$$

Accordingly, the amplitudes of the scalar and vector potentials may be used to estimate the values of displacement and traction.

Confine the attention to the z-dependence of potentials and, owing to (1.4) and (1.5), let

$$\phi^{s+} = \Phi^{s+}\exp(i\beta_L^s z), \qquad \phi^{s-} = \Phi^{s-}\exp(-i\beta_L^s z)$$

$$\boldsymbol{\psi}^{s+} = \Psi^{s+}\exp(i\beta_L^s z), \qquad \boldsymbol{\psi}^{s-} = \Psi^{s-}\exp(-i\beta_L^s z).$$

Of course, these expressions refer to up and downgoing longitudinal and transverse waves. In fact, the addition of the two scalar quantities and multiplication by the common exponential factor $\exp[i(k_x x + k_y y)]$ gives back the scalar potential (1.4). A similar remark applies to the vector quantities. It is immediately seen that

$$|\phi^{s+}(h^s)| = |\phi^{s+}(0)|\exp(-\sigma_L^s h^s), \tag{2.2}$$

$$|\phi^{s-}(h^s)| = |\phi^{s-}(0)|\exp(\sigma_L^s h^s), \tag{2.3}$$

$$|\boldsymbol{\psi}^{s+}(h^s)| = |\boldsymbol{\psi}^{s+}(0)|\exp(-\sigma_T^s h^s), \tag{2.4}$$

$$|\boldsymbol{\psi}^{s-}(h^s)| = |\boldsymbol{\psi}^{s-}(0)|\exp(\sigma_T^s h^s), \tag{2.5}$$

the plane $z = 0$ being positioned at $\mathcal{P}_{s-1}$. Since we are dealing with an incident pair such that at least one of the inequalities (2.1) holds, the first problem is to find which, if any, of the four waves travelling in the s-th layer may be considered as extinguished. To this end we refer to (2.2)-(2.5) but also need some estimate of the quantities involved. This is achieved by considering the stack I and regarding the layer s as a homogeneous lower half-space; then the algorithms of the previous section apply and the transmission matrix is determined, since no singular behaviour occurs. Accordingly, the related values of the transmitted fields $|\phi^{s-}(h^s)|$ and $|\boldsymbol{\psi}^{s-}(h^s)|$ are easily found. We consider these quantities as giving the required estimate. Insertion of the values into (2.3) and (2.5) gives

$$|\phi^{s-}(0)| = |\phi^{s-}(h^s)| \exp(-\sigma_L^s h^s), \tag{2.3'}$$

$$|\boldsymbol{\psi}^{s-}(0)| = |\boldsymbol{\psi}^{s-}(h^s)| \exp(-\sigma_T^s h^s). \tag{2.5'}$$

By means of these relations we can perform a detailed analysis of wave propagation inside the layer s. We say that the downgoing wave ϕ^{s-} or $\boldsymbol{\psi}^{s-}$ is effective according as $|\phi^{s-}(0)| > \varepsilon$ or $|\boldsymbol{\psi}^{s-}(0)| > \varepsilon$. In other words, waves generated at $\mathcal{P}_s$ are called effective if they reach $\mathcal{P}_{s-1}$ with appreciable amplitude. Similarly, the upgoing waves ϕ^{s+} and $\boldsymbol{\psi}^{s+}$ are said to be effective if $|\phi^{s+}(h^s)| > \varepsilon$ or $|\boldsymbol{\psi}^{s+}(h^s)| > \varepsilon$. Four cases are now considered.

Case 1: $|\phi^{s-}(0)| < \varepsilon$ and $|\boldsymbol{\psi}^{s-}(0)| < \varepsilon$. The waves transmitted by the stack *I*, that is ϕ^{s-} and $\boldsymbol{\psi}^{s-}$, are not effective. Accordingly, the waves propagating inside the stack *II* are disregarded, and hence there is no transmitted wave in the half-space corresponding to $s = 1$. The waves reflected into the upper half-space come from the analysis of the previous section applied to the stack *I*.

Case 2: $|\phi^{s-}(0)| < \varepsilon$ and $|\boldsymbol{\psi}^{s-}(0)| > \varepsilon$. The downgoing transverse wave $\boldsymbol{\psi}^{s-}$ is effective; hence it is regarded as a wave incident on the upper boundary $\mathcal{P}_{s-1}$ of the stack *II*. The longitudinal wave ϕ^{s-} is not effective and is considered to extinguish within the layer s. The analysis of the propagation through the stack *II* yields the moduli of the reflected fields, that is, $|\phi^{s+}(0)|$ and $|\boldsymbol{\psi}^{s+}(0)|$. Then (2.2) and (2.4) give the estimates of the corresponding values at $z = h^s$, to be compared with ε. If both $\phi^{s+}(h^s)$ and $\boldsymbol{\psi}^{s+}(h^s)$ are negligible then the reflection matrix of the original layer is determined entirely from the analysis of the behaviour of the stack *I*, as in Case 1. Here, though, also a transmission occurs through the stack *II* connected with the incident wave $\boldsymbol{\psi}^{s-}$. If only ϕ^{s+} is negligible then the layer s is regarded as affected by up and downgoing transverse waves, $\boldsymbol{\psi}^{s-}$ and $\boldsymbol{\psi}^{s+}$. If instead $\boldsymbol{\psi}^{s+}$ is negligible, while ϕ^{s+} is effective, we have to consider the waves $\boldsymbol{\psi}^{s-}$ and ϕ^{s+}. Finally, if both ϕ^{s+} and $\boldsymbol{\psi}^{s+}$ are effective then the layer s is affected by the waves $\boldsymbol{\psi}^{s-}$, ϕ^{s+}, and $\boldsymbol{\psi}^{s+}$.

Case 3: $|\phi^{s-}(0)| > \varepsilon$ and $|\boldsymbol{\psi}^{s-}(0)| < \varepsilon$. The downgoing transverse wave $\boldsymbol{\psi}^{s-}$ is to be neglected at $\mathcal{P}_{s-1}$ and the longitudinal wave ϕ^{s-} is incident on the stack *II*. Then everything goes as in Case *2*, except that the roles of ϕ^{s-} and $\boldsymbol{\psi}^{s-}$ are interchanged.

Case 4: $|\phi^{s-}(0)| > \varepsilon$ and $|\boldsymbol{\psi}^{s-}(0)| > \varepsilon$. Both downgoing waves ϕ^{s-} and $\boldsymbol{\psi}^{s-}$ are effective; they constitute a pair incident on the stack *II*. An estimate for the amplitudes of the upgoing longitudinal and tranverse waves - namely the waves reflected from the stack

II - is then obtained straightforwardly. Then, inside the layer s, we have to deal with the downgoing waves ϕ^{s-} and $\boldsymbol{\psi}^{s-}$, and the possible upgoing waves that are effective.

We now discuss the details of the calculations involved. The underlying idea is that, by definition, only effective waves are taken to reach the interface opposite to that where they emanate while the non-effective waves are only considered at the interface where they emanate. To be specific, we consider Case *3* and assume that the two longitudinal waves are the only effective waves in the layer s. Under these conditions, the upgoing transverse wave leaving the boundary $\mathcal{P}_{s-1}$ does not reach $\mathcal{P}_s$; similarly, the downgoing transverse wave starting at $\mathcal{P}_s$ does not arrive at $\mathcal{P}_{s-1}$. Accordingly, we have to analyze wave propagation in *I* by assuming that, at the interface $\mathcal{P}_s$, the lower limit of the displacement-traction vector results from superposition of a transmitted (downgoing) transverse wave, with an up and a downgoing longitudinal wave. Once this is done, we turn to wave propagation in *II* and evaluate the upper limit of $\mathbf{Z}$ at $\mathcal{P}_{s-1}$ as a result of an incident (downgoing) and a reflected (upgoing) longitudinal wave along with a reflected (upgoing) transverse wave. Combination of the results for the stacks *I* and *II*, and account of propagation of the two longitudinal waves inside the layer s yields the required result.

Examine formally the procedure. Concerning stack *I*, by analogy with (1.12) and (1.14) we find that

$$C\mathbf{A} = C^n\, E^n\, C^{-n} \ldots C^{s+1} E^{s+1} C^{-(s+1)} \mathbf{Z}^{s+1}(h^s_+), \tag{2.6}$$

the plane $z = 0$ being set at $\mathcal{P}_{s-1}$. Because of the continuity for $\mathbf{Z}$ we have $\mathbf{Z}^{s+1}(h^s_+) = \mathbf{Z}^s(h^s_-)$. To find the expression of $\mathbf{Z}^s(h^s_-)$ it is convenient to represent the scalar potential as in (1.4). According to our definitions we have $\Phi^{s+} = \phi^{s+}(0)$ and $\Phi^{s-} = \phi^{s-}(0)$. The vector potential is expressed in the form

$$\boldsymbol{\psi}^s = (\bar{\Psi}^{s-}_x \mathbf{e}_x + \bar{\Psi}^{s-}_y \mathbf{e}_y + \bar{\Psi}^{s-}_z \mathbf{e}_z) \exp[-i\beta^s_T(z - h^s)] \exp[i(k_x x + k_y y)].$$

Actually, the representation of the vector potential takes into account the fact that only the transmitted field is considered; for convenience, the constant amplitude vector has been related to the value of the potential at $z = h^s$. Comparison with (4.1.7), (4.1.8), (4.3.11) and the definition (1.6) of $\mathbf{Z}$ shows that

$$\mathbf{Z}^s(h^s_-) = C^s \begin{pmatrix} \bar{\mathbf{A}}^{s+} \\ \bar{\mathbf{A}}^{s-} \end{pmatrix}$$

where

$$\bar{\mathbf{A}}^{s+} = \left(\Phi^{s+} \exp(i\beta^s_L h^s),\, 0,\, 0,\right)^\dagger,$$

$$\bar{\mathbf{A}}^{s-} = \left(\Phi^{s-} \exp(-i\beta^s_L h^s),\, -\bar{\Psi}^{s-}_x,\, -\bar{\Psi}^{s-}_y\right)^\dagger.$$

Upon substitution of the expression of $\mathbf{Z}^s(h^s_-)$ and multiplication by C^{-1}, (2.6) reduces to the equivalent form

$$\begin{pmatrix} \mathbf{A}^+ \\ \mathbf{A}^- \end{pmatrix} = D \begin{pmatrix} \bar{\mathbf{A}}^{s+} \\ \bar{\mathbf{A}}^{s-} \end{pmatrix},$$

where

$$D = \begin{pmatrix} D_1 & D_2 \\ D_3 & D_4 \end{pmatrix} = C^{-1}\, C^n\, E^n\, C^{-n} ... C^{s+1}\, E^{s+1}\, C^{-(s+1)}\, C^s.$$

Equivalently we can write

$$\mathbf{A}^+ = D_1 \bar{\mathbf{A}}^{s+} + D_2 \bar{\mathbf{A}}^{s-}, \tag{2.7}$$

$$\mathbf{A}^- = D_3 \bar{\mathbf{A}}^{s+} + D_4 \bar{\mathbf{A}}^{s-}. \tag{2.8}$$

Consider now the stack *II*, which is subject to the action of the incident wave ϕ^{s-} and originates two waves ϕ^{s+} and $\boldsymbol{\psi}^{s+}$, such that the transverse wave does not reach the next interface (is non-effective). The scalar field in the (upper) layer s is given by (1.4), as before. The vector potential follows from (1.5) once the downgoing (minus) contribution is removed. Following the same notation we have

$$C^s \begin{pmatrix} \mathbf{A}^{s+} \\ \mathbf{A}^{s-} \end{pmatrix} = C^{s-1}\, E^{s-1}\, C^{-(s-1)} ... C^2\, E^2\, C^{-2} \begin{pmatrix} \breve{C}_1 \\ -\breve{C}_2 \end{pmatrix} \breve{\mathbf{A}}^-$$

where

$$\mathbf{A}^{s+} = \left(\Phi^{s+},\, \Psi_x^{s+},\, \Psi_y^{s+},\right)^\dagger,$$

$$\mathbf{A}^{s-} = \left(\Phi^{s-},\, 0,\, 0\right)^\dagger.$$

Multiplication of both sides by C^{-s} yields

$$\mathbf{A}^{s+} = F_1 \breve{\mathbf{A}}^-, \tag{2.9}$$

$$\mathbf{A}^{s-} = F_2 \breve{\mathbf{A}}^-; \tag{2.10}$$

the definition of the 3×3 matrices F_1 and F_2 is given by the matrix relation

$$\begin{pmatrix} F_1 \\ F_2 \end{pmatrix} = C^{-s}\, C^{s-1}\, E^{s-1}\, C^{-(s-1)} ... C^2\, E^2\, C^{-2} \begin{pmatrix} \breve{C}_1 \\ -\breve{C}_2 \end{pmatrix}.$$

The last step consists in solving the system of 12 complex equations (2.7)-(2.10) in the 12 complex unknowns Φ^+, Ψ_x^+, Ψ_y^+, Φ^{s-}, $\bar{\Psi}_x^{s-}$, $\bar{\Psi}_y^{s-}$, Φ^{s+}, Ψ_x^{s+}, Ψ_y^{s+}, $\breve{\Phi}^-$, $\breve{\Psi}_x^-$, $\breve{\Psi}_y^-$ whereas Φ^-, Ψ_x^-, Ψ_y^- are given data. Actually, what we are really interested in is the determination of the reflected pair in the upper half-space and the transmitted pair in the lower half-space, namely the first three and the last three unknowns. In this regard it is convenient to multiply (2.8) by D_4^{-1} and (2.10) by F_2^{-1}. The pertinent system of equations turn out to be

$$\begin{pmatrix} \Phi^+ \\ \Psi_x^+ \\ \Psi_y^+ \end{pmatrix} = D_1 \begin{pmatrix} \Phi^{s+} \exp(i\beta_L^s h^s) \\ 0 \\ 0 \end{pmatrix} + D_2 \begin{pmatrix} \Phi^{s-} \exp(-i\beta_L^s h^s) \\ -\bar{\Psi}_x^{s-} \\ -\bar{\Psi}_y^{s-} \end{pmatrix}, \tag{2.7'}$$

$$\begin{pmatrix} \Phi^{s-}\exp(-i\beta_L^s h^s) \\ -\bar{\Psi}_x^{s-} \\ -\bar{\Psi}_y^{s-} \end{pmatrix} + D_4^{-1}D_3 \begin{pmatrix} \Phi^{s+}\exp(i\beta_L^s h^s) \\ 0 \\ 0 \end{pmatrix} = D_4^{-1}\begin{pmatrix} \Phi^- \\ -\Psi_x^- \\ -\Psi_y^- \end{pmatrix}, \tag{2.8'}$$

$$\begin{pmatrix} \check{\Phi}^- \\ -\check{\Psi}_x^- \\ -\check{\Psi}_y^- \end{pmatrix} = F_2^{-1}\begin{pmatrix} \Phi^{s-} \\ 0 \\ 0 \end{pmatrix}. \tag{2.10'}$$

Comparison with (2.10') allows (2.9) to be rewritten in the form

$$\begin{pmatrix} \Phi^{s+} \\ \Psi_x^{s+} \\ \Psi_y^{s+} \end{pmatrix} = F_1F_2^{-1}\begin{pmatrix} \Phi^{s-} \\ 0 \\ 0 \end{pmatrix}. \tag{2.9'}$$

Consider now the first rows of (2.8') and (2.9') which constitute the following linear system for Φ^{s-} and Φ^{s+}:

$$\{F_1F_2^{-1}\}_{11}\Phi^{s-} - \Phi^{s+} = 0,$$

$$\Phi^{s-}\exp(-i\beta_L^s h^s) + \{D_4^{-1}D_3\}_{11}\Phi^{s+}\exp(i\beta_L^s h^s)$$
$$= \{D_4^{-1}\}_{11}\Phi^- - \{D_4^{-1}\}_{12}\Psi_x^- - \{D_4^{-1}\}_{13}\Psi_y^-.$$

The solution for Φ^{s-} is

$$\Phi^{s-} = \frac{\{D_4^{-1}\}_{11}\Phi^- - \{D_4^{-1}\}_{12}\Psi_x^- - \{D_4^{-1}\}_{13}\Psi_y^-}{\exp(-i\beta_L^s h^s) + \{D_4^{-1}D_3\}_{11}\{F_1F_2^{-1}\}_{11}\exp(i\beta_L^s h^s)}. \tag{2.11}$$

The denominator in (2.11) results from the combination of two exponentials; evaluation of their numerical values allows us to say which one, if any, may be neglected in numerical computations. Substitution of (2.11) into (2.10') yields the field transmitted into the lower half-space. Comparison of (2.7') and (2.8') provides the reflected field in the upper half-space as

$$\begin{pmatrix} \Phi^+ \\ \Psi_x^+ \\ \Psi_y^+ \end{pmatrix} = (D_1 - D_2D_4^{-1}D_3)\begin{pmatrix} \Phi^{s+}\exp(i\beta_L^s h^s) \\ 0 \\ 0 \end{pmatrix} + D_2D_4^{-1}\begin{pmatrix} \Phi^- \\ -\Psi_x^- \\ -\Psi_y^- \end{pmatrix}.$$

Substitution of $\Phi^{s+} = \{F_1F_2^{-1}\}_{11}\Phi^{s-}$ and use of (2.11) yield the reflected field, namely $\Phi^+, \Psi_x^+, \Psi_y^+$, in terms of the incident one, namely $\Phi^-, \Psi_x^-, \Psi_y^-$. The other cases can be dealt with by following the same lines.

Here we have not examined the case of a high degree of attenuation resulting from the combined effects of a number of thin layers instead of a single thick layer. An estimate of the amplitude of the field within the s-th layer could be obtained, as before, by regarding this medium as a half-space; then the amplitudes of the field emanating from the plane interface $\mathcal{P}_s$ could be found through the transmission matrix. In so doing it is easy to verify whether the longitudinal or the tranverse wave, or both, should be neglected. If the answer is positive, the solution of the problem follows straightaway.

6.3 Layers with singular transfer matrices

The general theory of §6.1 applies when the z-components of longitudinal and transverse wave vectors are non-zero. Actually, if

$$\beta_L^s = 0, \qquad \beta_T^s = 0 \tag{3.1}$$

the matrix C^s becomes singular and cannot be inverted to give the transfer matrix for the layer s. Of course, the difficulty is not of computational character. Rather we observe that, because of (3.1), there is no physical meaning in considering up and downgoing inhomogeneous pairs; intuitively, we can say that wave propagation along the z-direction does not occur. Mathematically, merely setting $\beta_{L,T}^s = 0$ in the standard expression of the inhomogeneous waves is much too restrictive. This requires a specific analysis of wave propagation when at least one of the conditions (3.1) hold.

Suppose that the x- and y-components of the incident pair are such that $\beta_L^s = 0$ or, equivalently,

$$k_x^2 + k_y^2 = \kappa_L^s. \tag{3.2}$$

Consider the obvious solution

$$\phi^s = \Phi^s \exp[i(k_x x + k_y y)],$$

with the constraint (3.2), to the Helmholtz equation

$$\Delta\phi^s + \kappa_L^s \phi^s = 0.$$

When inserted in the continuity conditions for displacement and traction at the interfaces it does not provide a consistent algebraic linear system in the pertinent amplitudes, namely the number of unknowns is smaller than the number of equations. Really, if (3.2) holds, the Helmholtz equation allows for the more general solution

$$\phi^s = (m + pz)\exp[i(k_x x + k_y y)], \tag{3.3}$$

where m and p are complex constants. The general solution (3.3) is required to make $\mathbf{U}$ and $\mathbf{t}$ satisfy the continuity conditions, at the boundaries, with appropriate values for m and p. Independently of the value assigned to the constants, the potential (3.3) may be regarded as a wave propagating horizontally, that is in a direction parallel to the interface, with depth-dependent amplitude.

Now we show how the solution (3.3) enters the continuity conditions. First we determine the displacement $\mathbf{U}[\phi^s]$ and the traction $\mathbf{t}[\phi^s]$ associated to the potential ϕ^s. Use of (4.1.2) and (4.1.3) shows that, up to the scalar factor $\exp[i(k_x x + k_y y)]$,

$$\mathbf{U}^s[\phi^s] = i(m + pz)(k_x \mathbf{e}_x + k_y \mathbf{e}_y) + p\mathbf{e}_z,$$

$$\mathbf{t}^s[\phi^s] = -2\mu^s \big[ip(k_x \mathbf{e}_x + k_y \mathbf{e}_y) - \chi^s (m + pz)\mathbf{e}_z\big],$$

where $\chi^s = \frac{1}{2}\kappa_T^s - k_x^2 - k_y^2$. Besides the longitudinal wave (3.3), the s-th layer is affected by up and downgoing tranverse waves with potential $\boldsymbol{\psi}^s$ which, because $\beta_T^s \neq 0$, has the form (1.5). Then comparison with (4.3.4) and (4.3.9) leads to

$$\begin{aligned}\mathbf{U}^s(0_+) =& i\big[k_x m - \frac{k_x k_y}{\beta_T^s}(\Psi_x^{s+} - \Psi_x^{s-}) - \frac{k_y^2 + (\beta_T^s)^2}{\beta_T^s}(\Psi_y^{s+} - \Psi_y^{s-})\big]\mathbf{e}_x \\ &+ i\big[k_y m + \frac{k_x^2 + (\beta_T^s)^2}{\beta_T^s}(\Psi_x^{s+} - \Psi_x^{s-}) + \frac{k_x k_y}{\beta_T^s}(\Psi_y^{s+} - \Psi_y^{s-})\big]\mathbf{e}_y \\ &+ \big[p - ik_y(\Psi_x^{s+} + \Psi_x^{s-}) + ik_x(\Psi_y^{s+} + \Psi_y^{s-})\big]\mathbf{e}_z,\end{aligned}$$

$$\begin{aligned}\mathbf{t}^s(0_+) =& -2\mu^s k_x\big[ip + k_y(\Psi_x^{s+} + \Psi_x^{s-}) - \big(k_x - \tfrac{1}{2}\frac{\kappa_T^s}{k_x}\big)(\Psi_y^{s+} + \Psi_y^{s-})\big]\mathbf{e}_x \\ &- 2\mu^s k_y\big[ip + \big(k_y - \tfrac{1}{2}\frac{\kappa_T^s}{k_y}\big)(\Psi_x^{s+} + \Psi_x^{s-}) - k_x(\Psi_y^{s+} + \Psi_y^{s-})\big]\mathbf{e}_y \\ &+ 2\mu^s\big[\chi^s m - k_y\beta_T^s(\Psi_x^{s+} - \Psi_x^{s-}) + k_x\beta_T^s(\Psi_y^{s+} - \Psi_y^{s-})\big]\mathbf{e}_z.\end{aligned}$$

These expressions indicate that we can take advantage of the results of §6.1. In this regard, merely for technical convenience, we set

$$m = \Phi^{s+} + \Phi^{s-}, \qquad p = (\Phi^{s+} - \Phi^{s-})/h^s,$$

h^s being the width of the s-th layer. Comparison with (4.1.7) and (4.1.8), the expressions of $\mathbf{U}^s[\phi^s]$ and $\mathbf{t}^s[\phi^s]$ in terms of ϕ^{s+} and ϕ^{s-}, and the definition (1.6) of $\mathbf{Z}$ allow us to write

$$\mathbf{Z}^s(z) = \bar{C}^s \bar{E}(z)\begin{pmatrix}\mathbf{A}^{s+}\\ \mathbf{A}^{s-}\end{pmatrix}$$

where, as before,

$$\mathbf{A}^{s+} = \big(\Phi^{s+},\, \Psi_x^{s+},\, \Psi_y^{s+}\big)^\dagger, \qquad \mathbf{A}^{s-} = \big(\Phi^{s-},\, -\Psi_x^{s-},\, -\Psi_y^{s-}\big)^\dagger.$$

The operator $\bar{C}^s$ is written in block form as

$$\bar{C}^s = \begin{pmatrix} C_1^s & C_1^s \\ \bar{C}_2^s & -\bar{C}_2^s\end{pmatrix};$$

C_1^s is defined in §5.1 while

$$\bar{C}_2^s = \begin{pmatrix} -i/h^s & -k_y & k_x \\ -i\mu^s/h^s & -\mu^s k_y & \mu^s(k_x - \kappa_T^s/2k_x) \\ -i\mu^s/h^s & -\mu^s(k_y - \kappa_T^s/2k_y) & \mu^s k_x \end{pmatrix}$$

and the inverse is

$$\bar{C}_2^{-s} = \frac{2i}{\mu^s \kappa_T^s} \begin{pmatrix} \mu^s \chi^s h^s & k_x^2 h^s & k_y^2 h^s \\ i\mu^s k_y & 0 & -ik_y \\ -i\mu^s k_x & ik_x & 0 \end{pmatrix}.$$

The matrix $\bar{E}^s(z)$, represented in the block form

$$\bar{E}^s = \begin{pmatrix} \bar{E}_1^s & \bar{E}_2^s \\ \bar{E}_3^s & \bar{E}_4^s \end{pmatrix},$$

is given by

$$\bar{E}_1^s = \operatorname{diag}\left(1 + z/2h^s,\ \exp(i\beta_T^s z),\ \exp(i\beta_T^s z)\right),$$

$$\bar{E}_2^s = -\bar{E}_3^s = \operatorname{diag}\left(-z/2h^s,\ 0,\ 0\right),$$

$$\bar{E}_4^s = \operatorname{diag}\left(1 - z/2h^s,\ \exp(-i\beta_T^s z),\ \exp(-i\beta_T^s z)\right).$$

The continuity of displacement and traction at $z = 0$ is summarized by

$$\mathbf{Z}^{s-1}(0_-) = \mathbf{Z}^s(0_+) = \bar{C}^s \begin{pmatrix} \mathbf{A}^{s+} \\ \mathbf{A}^{s-} \end{pmatrix}.$$

Hence we have

$$\begin{pmatrix} \mathbf{A}^{s+} \\ \mathbf{A}^{s-} \end{pmatrix} = \bar{C}^{-s}\mathbf{Z}^{s-1},$$

where

$$\bar{C}^{-s} = \tfrac{1}{2} \begin{pmatrix} C_1^{-s} & \bar{C}_2^{-s} \\ C_1^{-s} & -\bar{C}_2^{-s} \end{pmatrix}.$$

Also we obtain

$$\mathbf{Z}^s(h_-^s) = \bar{C}^s \bar{E}^s(h^s) \bar{C}^{-s} \mathbf{Z}^{s-1}(0_-). \tag{3.4}$$

Accordingly the relation (1.12) is changed to

$$\mathbf{Z}^{n+1} = C^n\, E^n\, C^{-n} ... \bar{C}^s\, \bar{E}^s\, \bar{C}^{-s} ... C^2\, E^2\, C^{-2} \mathbf{Z}^1$$

and the reflection and transmission matrices follow.

We now consider the other degenerate case when a transverse wave in the layer s has a vanishing vertical component of the wave vector, viz

$$\beta_T^s = 0, \qquad k_x^2 + k_y^2 = \kappa_T^s. \tag{3.5}$$

Still we keep the plane $z = 0$ at the interface $\mathcal{P}_{s-1}$ between the layers s and $s-1$. By analogy with the previous treatment for the longitudinal wave, we start from the observation that, because of (3.5), the general solution to

$$\Delta \boldsymbol{\psi}^s + \kappa_T^s \boldsymbol{\psi}^s = 0$$

is

$$\boldsymbol{\psi}^s = (\mathbf{m} + \mathbf{p}z)\exp[i(k_x x + k_y y)], \tag{3.6}$$

with $\mathbf{m}$ and $\mathbf{p}$ any complex-valued vectors. The divergence-free condition

$$\nabla \cdot \boldsymbol{\psi}^s = 0,$$

on the viscoelastic vector potential (3.6) yields the linear relations

$$k_x m_x + k_y m_y - ip_z = 0, \tag{3.7}$$

$$k_x p_x + k_y p_y = 0. \tag{3.8}$$

By (3.8), if $k_x \neq 0$,

$$p_x = -\frac{k_y}{k_x} p_y. \tag{3.8'}$$

Then $\boldsymbol{\psi}^s$ is determined as soon as the four complex constants m_x, m_y, m_z, p_y are evaluated. This scheme formally resembles the standard situation when we consider the potential (1.5) and regard Ψ_x^{s-}, Ψ_x^{s+}, Ψ_y^{s-}, Ψ_y^{s+} as unknowns.

We now evaluate the displacement and the traction. Substitution of the vector potential (3.6) into the general expressions (4.1.2) and (4.1.3) and use of (3.5) and (3.8') yield

$$\begin{aligned}\mathbf{U}[\boldsymbol{\psi}^s] = {} & \big[ik_y(m_z + zp_z) - p_y\big]\mathbf{e}_x - \big[k_y p_y/k_x + ik_x(m_z + zp_z)\big]\mathbf{e}_y \\ & + \big[ik_x(m_y + zp_z) - ik_y(m_x - zk_y p_y/k_x\big]\mathbf{e}_z,\end{aligned}$$

$$\begin{aligned}\mathbf{t}[\boldsymbol{\psi}^s] = & -2\mu^s k_x\big[k_y(m_x - zk_y p_y/k_x) - (k_x - \tfrac{1}{2}\kappa_T^s/k_x)(m_y + zp_y)\big]\mathbf{e}_x \\ & - 2\mu^s k_y\big[(k_y - \tfrac{1}{2}\kappa_T^s/k_y)(m_x - zk_y p_y/k_x) - k_x(m_y + zp_y)\big]\mathbf{e}_y - 2i\mu^s\kappa_T^s p_y/k_x\mathbf{e}_z.\end{aligned}$$

Hence we have trivially the values $\mathbf{U}[\boldsymbol{\psi}^s](0_+)$ and $\mathbf{t}[\boldsymbol{\psi}^s](0_+)$. The scalar potential for longitudinal waves is given by (1.4) and the corresponding expressions for displacement and traction, at $z = 0_+$, can be derived from (4.3.4)-(4.3.9). Altogether we have

$$\begin{aligned}\mathbf{U}(0_+) = & [ik_x(\Phi^{s+} + \Phi^{s-}) - p_y + ik_y m_z]\mathbf{e}_x \\ & + [ik_y(\Phi^{s+} + \Phi^{s-}) - k_y p_y/k_x - ik_x m_z]\mathbf{e}_y \\ & + [i\beta_L^s(\Phi^{s+} - \Phi^{s-}) - ik_y m_x + ik_x m_y]\mathbf{e}_z,\end{aligned}$$

$$\begin{aligned}\mathbf{t}(0_+) = &-2\mu^s k_x[-\beta_L^s(\Phi^{s+} - \Phi^{s-}) + k_y m_x - (k_x - \tfrac{1}{2}\kappa_T^s/k_x)m_y]\mathbf{e}_x \\ &-2\mu^s k_y[-\beta_L^s(\Phi^{s+} - \Phi^{s-}) + (k_y - \tfrac{1}{2}\kappa_T^s/k_y)m_x - k_x m_y]\mathbf{e}_y \\ &+2\mu^s[\chi^s(\Phi^{s+} + \Phi^{s-}) - i\kappa_T^s/k_x p_y]\mathbf{e}_z.\end{aligned}$$

In view of the definition (1.6) of $\mathbf{Z}$, the continuity condition, at $z = 0$, yields

$$\mathbf{Z}^{s-1}(0_-) = \mathbf{Z}^s(0_+) = \tilde{C}\boldsymbol{A}$$

where the column vector $\boldsymbol{A}$ contains the unknown amplitudes, of scalar and vector potentials, in the form

$$\boldsymbol{A} = \left(\Phi^{s+} + \Phi^{s-},\ p_y,\ m_z,\ \Phi^{s+} - \Phi^{s-},\ m_x,\ m_y\right)^{\dagger};$$

the 6×6 complex-valued matrix $\tilde{C}$ may be represented in block form as

$$\tilde{C} = \begin{pmatrix} \tilde{C}_1^s & 0 \\ 0 & C_2^s \end{pmatrix},$$

where

$$\tilde{C}_1^s = \begin{pmatrix} k_x & i & k_y \\ k_y & ik_y/k_x & -k_x \\ -\mu^s\kappa_T^s/2 & -i\mu^s\kappa_T^s/k_x & 0 \end{pmatrix}.$$

The inverse $(\tilde{C}_1^s)^{-1} = \tilde{C}^{-s}$ of $\tilde{C}_1^s$ is

$$(\tilde{C}_1^s)^{-1} = \frac{1}{k_x^2 + k_y^2}\begin{pmatrix} 2k_x & 2k_y & 2(k_x^2 + k_y^2)/(\mu^s\kappa_T^s) \\ ik_x^2 & ik_xk_y & 2ik_x(k_x^2 + k_y^2)/(\mu^s\kappa_T^s) \\ k_y & -k_x & 0 \end{pmatrix}$$

while C_2^s and its inverse are given in §6.1. The result is

$$\boldsymbol{A} = \tilde{C}^{-s}\mathbf{Z}^{s-1}(0_-),$$

with

$$\tilde{C}^{-s} = (\tilde{C}^s)^{-1} = \begin{pmatrix} \tilde{C}_1^{-s} & 0 \\ 0 & C_2^{-s} \end{pmatrix}.$$

Once the amplitudes $\boldsymbol{A}$ are obtained, a straightforward calculation based on the evaluation of the components of $\mathbf{U}$ and $\mathbf{t}$ at an arbitrary value of z yields

$$\mathbf{Z}^s(h_-^s) = \tilde{C}^s\tilde{E}^s\tilde{C}^{-s}\mathbf{Z}^{s-1}(0_-), \tag{3.9}$$

where $\tilde{E}^s$ has the block structure

$$\tilde{E}^s = \begin{pmatrix} \tilde{E}_1^s & \tilde{E}_2^s \\ \tilde{E}_3^s & \tilde{E}_1^s \end{pmatrix},$$

where

$$\tilde{E}_1^s = \begin{pmatrix} [\exp(i\beta_L^s h^s) + \exp(-i\beta_L^s h^s)]/2 & 0 & 0 \\ 0 & 1 & 0 \\ 0 & 0 & 1 \end{pmatrix},$$

$$\tilde{E}_2^s = \begin{pmatrix} [\exp(i\beta_L^s h^s) - \exp(-i\beta_L^s h^s)]/2 & 0 & 0 \\ 0 & 0 & 0 \\ 0 & -ik_x h^s & -ik_y h^s \end{pmatrix},$$

$$\tilde{E}_3^s = \begin{pmatrix} [\exp(i\beta_L^s h^s) - \exp(-i\beta_L^s h^s)]/2 & 0 & 0 \\ 0 & -h^s k_y/k_x & 0 \\ 0 & h^s & 0 \end{pmatrix}.$$

The matrix product $\tilde{C}^s \tilde{E}^s \tilde{C}^{-s}$ should replace $C^s E^s C^{-s}$ in (1.12), when it happens that $\beta_T^s = 0$.

The further possibility that the z-components of the wave vectors in the upper or lower half-spaces vanish is not examined in detail. This circumstance may be investigated along the lines described at the end of §4.5 in connection with the study of singular behaviours at the plane interface between two half-spaces.

6.4 Scalar fields in continuously layered media

In this section we investigate layered media whose material properties depend on a single Cartesian coordinate, say, z and the phenomenon under consideration is governed by the scalar Helmholtz equation, namely

$$\Delta u + K\, u = 0 \tag{4.1}$$

where $K = K(z)$ is a known function. Actually, a number of problems in heterogeneous media lead to equations of the form

$$\Delta u - N^2 \frac{\partial^2 u}{\partial t^2} = 0$$

for the unknown function u of space and time variables, N^2 being (the square of) a given function of the space variables (cf., e.g., [15]). More generally, we can find equations where $N^2 \partial^2/\partial t^2$ is replaced by a combination of second- and first-order time derivatives of u and u itself, with space-dependent coefficients. In the case of time-harmonic dependence, namely $u \simeq \exp(-i\omega t)$, the unknown function u satisfies the Helmholtz equation (4.1) where K is a space-dependent function, usually complex-valued and parameterized by the angular frequency ω.

Two examples are given as follows. First, consider a dielectric where the permittivity and the conductivity are given functions of the position and the electromagnetic field is time-harmonic. Then the dielectric is governed by

$$\Delta \mathbf{A} - \kappa \mathbf{A} - \frac{1}{\kappa}(\nabla \cdot \mathbf{A})\nabla\kappa = 0$$

where $\mathbf{A}$ is the vector potential and $\kappa = \epsilon\omega^2/c^2$, ϵ being the complex-valued effective permittivity and c the light speed *in vacuo* (cf. [22], §19). If ϵ, and then κ, depends on a single Cartesian coordinate, say z, then we arrive at (4.1) for the z-component of $\mathbf{A}$ with $K = \kappa + (\partial^2\kappa/\partial z^2)/2\kappa - (\partial\kappa/\partial z)^2/2\kappa^2$. For the x- and y-component of $\mathbf{A}$ (4.1) holds with $K = \kappa$. Second, consider the linearized equations of hydrodynamics and disregard the body force term. By (2.6.10), letting $\mathbf{b} = 0$ we have

$$\ddot{\mathbf{u}} - c^2\nabla(\nabla \cdot \mathbf{u}) = 0.$$

For a time-harmonic dependence we have

$$\Delta\phi + k^2\phi = 0$$

where ϕ is the scalar potential of $\mathbf{u}$ and $k = \omega/c$. Often, the linearized equations of hydrodynamics are claimed to provide

$$\Delta p + k^2 p - \frac{1}{\rho}\nabla\rho \cdot p = 0$$

which becomes the Helmholtz equation upon the change of unknown $u = p/\sqrt{\rho}$ and the identification $K = k^2 + \Delta\rho/2\rho - 3(\nabla\rho)^2/4\rho^2$. These examples show how (4.1) can be regarded as a model for wave propagation in continuously layered media. A further example is given in the next section.

So far a great many investigations of the equation (4.1) have been performed thus providing a large amount of literature on the subject. In particular, methods have been devised or improved for the determination of approximate solutions; some of them (WKB, rays) are examined in the next chapters. In this section we develop an approach which is based on a systematic use of complex-valued operators and fundamental systems of solutions. In particular some problems are solved for an Epstein layer, that is the main model of continuously layered medium, and a general expression for the reflection and the transmission coefficients is determined. Then the thin-layer limit is shown to provide the expected Fresnel's formula. An extension of these results to the description of wave propagation in dissipative continuously layered solids is then developed in the following sections.

Consider now (4.1), where K is a given function of the coordinate z only. Indeed, to fix ideas we consider an Epstein layer in $z \in [0, h]$ and let

$$K = \begin{cases} K_0, & z < 0, \\ K(z), & z \in [0, h], \\ K_1, & z > h. \end{cases}$$

Discontinuities for K are allowed to occur at $z = 0, h$ by letting $K(0), K(h)$ be possibly different than K_0, K_1. Having in mind that we are generalizing the picture of plane wave propagation, we look for solutions to the equation (4.1), by separation of variables in the form

$$u = X(x)Y(y)Z(z), \tag{4.2}$$

that are continuous at $z = 0, h$. Substitution of (4.2) in (4.1) yields

$$\frac{X''}{X} + \frac{Y''}{Y} + \frac{Z''}{Z} + K = 0$$

where a prime denotes differentiation with respect to the pertinent variable. Hence there are constants β, γ (possibly complex-valued) such that

$$Z'' + (K - \beta - \gamma)Z = 0, \tag{4.3}$$

$$X'' + \beta X = 0, \tag{4.4}$$

$$Y'' + \gamma Y = 0. \tag{4.5}$$

By (4.4) and (4.5) we have

$$X \simeq \exp(\pm i\sqrt{\beta}\, x), \qquad Y \simeq \exp(\pm i\sqrt{\gamma}\, y).$$

As regards (4.3), we assume that $K(z)$, possibly complex-valued, is continuous so that the existence of a fundamental system of solutions is allowed as $z \in [0, h]$.

Let $f(z), g(z)$ be a fundamental system of solutions. Then, aside from the $\exp(-i\omega t)$ factor we write

$$u = [af(z) + bg(z)] \exp(\pm i\sqrt{\beta}\, x) \exp(\pm i\sqrt{\gamma}\, y), \qquad 0 < z < h,$$

a, b being complex-valued coefficients. Of course we have

$$Z \simeq \exp(\pm i\sqrt{K_0 - \beta - \gamma}\, z) \qquad \text{as} \quad z < 0$$

and

$$Z \simeq \exp(\pm i\sqrt{K_1 - \beta - \gamma}\, z) \qquad \text{as} \quad z > h.$$

Indeterminacies in signs are solved by looking at particular wave solutions. For example, suppose that an incident wave is coming from $z = -\infty$. We can always choose the y-axis such that $\gamma = 0$. Then the incident wave is taken as

$$u = c_0 \exp(i\sqrt{K_0 - \beta}\, z) \exp(i\sqrt{\beta}\, x),$$

where c_0 and β are given constants.

We investigate the reflection-refraction problem for the Epstein layer, which consists of finding the reflected wave at $z = 0$ and the transmitted wave at $z = h$. Specifically, we have

$$u = \begin{cases} [c_0 \exp(i\sqrt{K_0 - \beta}\, z) + d_0 \exp(-i\sqrt{K_0 - \beta}\, z] \exp(i\sqrt{\beta}\, x), & z < 0, \\ [af(z) + bg(z)] \exp(i\sqrt{\beta}\, x), & z \in [0, h], \\ [c_1 \exp(i\sqrt{K_1 - \beta}\, z)] \exp(i\sqrt{\beta}\, x), & z > h, \end{cases}$$

where c_0 is the known amplitude of the incident wave, while d_0 and c_1 are the unknown amplitudes of the reflected and transmitted waves. It is worth observing that in any region $z < 0$, $z \in [0, h]$, $z > h$, the dependence on x is through the common exponential $\exp(i\sqrt{\beta}\, x)$. Accordingly, looking for wave solutions with separable variables implies that Snell's law is satisfied. The solution u and the derivative u' are now required to be continuous at $z = 0, h$.

Within the factor $\exp(i\sqrt{\beta}\, x)$, the solution u to (4.1) is represented by the solution ψ to

$$\psi'' + \zeta^2 \psi = 0 \tag{4.6}$$

where $\psi = \psi(z)$ and

$$\zeta^2 = \begin{cases} K_0 - \beta, & z < 0, \\ K(z) - \beta, & z \in [0, h], \\ K_1 - \beta, & z > h. \end{cases}$$

Of course it may happen that $K(0) - \beta \neq K_0 - \beta$, $K(h) - \beta \neq K_1 - \beta$. Let $\zeta_0^2 = K_0 - \beta$, $\zeta_1^2 = K_1 - \beta$. To study the reflection and the refraction from the Epstein layer it is natural to look for solutions of the form

$$\psi = \begin{cases} \exp(i\zeta_0 z) + \mathcal{R} \exp(-i\zeta_0 z), & z < 0, \\ af(z) + bg(z), & z \in [0, h], \\ \mathcal{T} \exp(i\zeta_1 z), & z > h. \end{cases}$$

This corresponds to a wavefield in the half-space $z < 0$ given by the superposition of an incident wave of unit amplitude coming from $z = -\infty$ and a reflected wave of amplitude $\mathcal{R}$; the half-space $z > 0$ is only affected by the transmitted wave of amplitude $\mathcal{T}$. The symbols $\mathcal{R}, \mathcal{T}$ are used to emphasize that we regard them as the complex-valued coefficients of reflection and refraction. To determine $\mathcal{R}$ and $\mathcal{T}$ we require the continuity of ψ and ψ' at $z = 0$ and $z = h$, namely

$$\psi(0_-) = \psi(0_+), \qquad \psi'(0_-) = \psi'(0_+),$$

$$\psi(h_-) = \psi(h_+), \qquad \psi'(h_-) = \psi'(h_+),$$

which follow as the counterpart of continuity of displacement and traction. Letting the subscripts $0, 1$ denote the values at $z = 0, h$, we have

$$1 + \mathcal{R} = af_0 + bg_0,$$

$$i\zeta_0 - i\zeta_0\mathcal{R} = af_0' + bg_0',$$

$$af_1 + bg_1 = \mathcal{T}\exp(i\zeta_1 h),$$

$$af_1' + bg_1' = i\zeta_1\mathcal{T}\exp(i\zeta_1 h).$$

In terms of the operators

$$B_0(f) = f_0' - i\zeta_0 f_0, \qquad B_0^*(f) = f_0' + i\zeta_0 f_0 \tag{4.7}$$

$$B_1(f) = f_1' - i\zeta_1 f_1, \qquad B_1^*(f) = f_1' + i\zeta_1 f_1 \tag{4.8}$$

we can write

$$\mathcal{R} - af_0 - bg_0 = -1,$$

$$\mathcal{T}\exp(i\zeta_1 h) - af_1 - bg_1 = 0,$$

$$aB_0^*(f) + bB_0^*(g) = 2i\zeta_0,$$

$$b[B_0^*(f), B_1(g)] = -2i\zeta_0 B_1(f)$$

where $[\cdot,\cdot]$ denotes the commutator, namely

$$[A(f), B(g)] = A(f)B(g) - A(g)B(f)$$

for any functions f, g and operators A, B. Here we have assumed that $B_0^*(f) \neq 0$, which is in fact no significant restriction for our developments.

Letting

$$[B_0^*(f), B_1(g)] \neq 0,$$

we can write

$$\mathcal{R} = \frac{[B_0(g), B_1(f)]}{[B_0^*(f), B_1(g)]}, \tag{4.9}$$

$$\mathcal{T} = \frac{2i\zeta_0\exp(-i\zeta_1 h)J}{[B_0^*(f), B_1(g)]}, \tag{4.10}$$

where J is the (constant value of the) Wronskian of f, g. Accordingly, once we know a fundamental system of solutions f, g, the coefficients of reflection and refraction $\mathcal{R}, \mathcal{T}$ are provided by (4.9) and (4.10).

Observe that, as expected, the values of $\mathcal{R}, \mathcal{T}$ are invariant under the change of the fundamental system of solutions. To show that this is so consider another fundamental system of solutions, say r, s. Of course f, g and r, s are related by a non-singular transformation

$$r(z) = mf(z) + ng(z),$$

$$s(z) = pf(z) + qg(z),$$

where the complex numbers m, n, p, q satisfy $mq - np \neq 0$. In view of the linearity of the differential operators B we have

$$[B_0(r), B_1(s)] = (mq - np)[B_0(f), B_1(g)],$$

$$[B_0^*(r), B_1(s)] = (mq - np)[B_0^*(f), B_1(g)].$$

Then we obtain the invariance result for $\mathcal{R}$, namely

$$\frac{[B_0(s), B_1(r)]}{[B_0^*(r), B_1(s)]} = \frac{[B_0(g), B_1(f)]}{[B_0^*(f), B_1(g)]}.$$

Similarly, we obtain the invariance for the refraction coefficient $\mathcal{T}$.

On the basis of this invariance we can choose the fundamental system of solutions f, g such as

$$f(0) = 1, \quad f'(0) = 0, \qquad g(0) = 0, \quad g'(0) = 1.$$

Then $J = 1$ and

$$\mathcal{R} = -\frac{B_1(f) + i\zeta_0 B_1(g)}{B_1(f) - i\zeta_0 B_1(g)}, \tag{4.9'}$$

$$\mathcal{T} = -\frac{2i\zeta_0 \exp(-i\zeta_1 h)}{B_1(f) - i\zeta_0 B_1(g)}. \tag{4.10'}$$

It is of interest to examine the behaviour of $\mathcal{R}, \mathcal{T}$ in the limit case when the incident wavelength is much longer than the thickness h (thin-layer limit). To emphasize the behaviour of the heterogeneous Epstein layer, henceforth we disregard discontinuities at $z = 0, h$, viz. $K(0) = K_0, K(h) = K_1$. Of course we expect that the limit yields the results associated with a jump discontinuity of the properties at the plane interface.

Define $N(z)$ such that

$$\zeta^2 = \zeta_0^2 N^2(z)$$

and then $N(0) = 1$. It is convenient to introduce the variable $\tau = z/h$ and denote by a superposed dot the derivative with respect to τ. Then consider the pertinent equation in the form

$$\ddot{\psi} + h^2 \zeta_0^2 n^2 \psi = 0$$

where n depends on τ in the form

$$n^2(\tau) = \begin{cases} 1, & \tau < 0, \\ \zeta^2(\tau h)/\zeta_0^2, & \tau \in [0, 1], \\ \zeta_1^2/\zeta_0^2, & \tau > 1. \end{cases}$$

Correspondingly we define the operators $\mathcal{B}_\tau$ and $\mathcal{B}_\tau^*$, parameterized by τ, as

$$\mathcal{B}_\tau(\psi) = \dot{\psi} - i\eta n\psi, \qquad \mathcal{B}_\tau^*(\psi) = \dot{\psi} + i\eta n\psi, \qquad \tau \in [0, 1] \tag{4.11}$$

where $\eta = \zeta_0 h$. Meanwhile we write the differential equation as

$$\ddot{\psi} + \eta^2 n^2 \psi = 0. \tag{4.12}$$

The thin-layer limit corresponds to $\eta \to 0$. To investigate this limit observe that ψ is a function of τ parameterized by η. Assume that the representation

$$\psi(\tau; \eta) = \psi_0(\tau) + \eta\psi_1(\tau) + \eta^2\psi_2(\tau) + o(\eta^2)$$

holds. Substitution in (4.12) provides

$$\ddot{\psi}_0 = 0,$$

$$\ddot{\psi}_1 = 0,$$

$$\ddot{\psi}_2 + n^2\psi_0 = 0.$$

To within η^2-terms, we look for a fundamental system of solutions $v(\tau)$, $w(\tau)$, $\tau \in [0, 1]$, to (4.12) such that

$$v(0) = 1, \; \dot{v}(0) = 0, \qquad w(0) = 0, \; \dot{w}(0) = 1.$$

For convenience two independent solutions for ψ_0 are taken as

$$\psi_0(\tau) = 1, \qquad \psi_0(\tau) = \tau.$$

Then $v(\tau)$ and $w(\tau)$ are represented as

$$v(\tau) = 1 + \eta^2 f(\tau) + o(\eta^2),$$

$$w(\tau) = \tau + \eta^2 g(\tau) + o(\eta^2).$$

This corresponds to setting $\psi_1 = 0$, while ψ_2 is given by f or g accordingly as $\psi_0 = 1$ or $\psi_0 = \tau$. Thus f and g satisfy

$$\ddot{f} + n^2 = 0, \qquad \ddot{g} + n^2\tau = 0,$$

with vanishing initial data. In terms of the geometrical variable z we can write the fundamental system of solutions v, w as

$$v = 1 + \eta^2 f\left(\frac{z}{\eta}\right) + o(\eta^2),$$

$$w = \frac{z}{\eta} + \eta^2 g\left(\frac{z}{\eta}\right) + o(\eta^2),$$

whence

$$v' = \eta \dot{f}(\frac{z}{\eta}) + o(\eta),$$

$$w' = \frac{1}{\eta} + \eta \dot{g}(\frac{z}{\eta}) + o(\eta).$$

Owing to the definitions (4.7) and (4.8) for B_0 and B_1 we can write

$$B_0(v) = \eta \dot{f}_0 - i\zeta_0 + o(\eta), \qquad B_0(w) = \frac{1}{\eta} - i + \eta \dot{g}_0 + o(\eta),$$

$$B_1(v) = \eta \dot{f}_1 - i\zeta_1 + o(\eta), \qquad B_1(w) = \frac{1}{\eta} - i\frac{\zeta_1}{\zeta_0} + \eta \dot{g}_1 + o(\eta).$$

Substitution in (4.9) (here (4.9') does not work because $w'(0) \neq 1$) yields

$$\mathcal{R} = \frac{\zeta_0 - \zeta_1 - i\eta(\dot{f}_1 - \dot{f}_0 + \zeta_1) + o(\eta)}{\zeta_0 + \zeta_1 - i\eta(\dot{f}_0 - \dot{f}_1 + \zeta_1) + o(\eta)}.$$

Then we have

$$\lim_{\eta \to 0} \mathcal{R} = \frac{\zeta_0 - \zeta_1}{\zeta_0 + \zeta_1},$$

that is Fresnel's formula (cf. 4.5.22). Similarly we obtain

$$\lim_{\eta \to 0} \mathcal{T} = \frac{2\zeta_0}{\zeta_0 + \zeta_1},$$

which is the usual form for the transmission coefficient in the scalar theory (cf. (4.5.22)).

Further results can be obtained by proceeding along the same lines. In particular, by starting from (2.9) we can show that $\mathcal{R}$ satisfies a Riccati equation. We refer the interested reader to [32] for the detailed proof.

Another approach to wave propagation in a layer is developed in §8.5 by following the method of successive approximations.

6.5 Traction-displacement vector in continuously layered media

Still we consider a layer, between two parallel surfaces, whose material parameters are allowed to vary smoothly in the direction orthogonal to the surfaces. Above and below the layer are homogeneous half-spaces. The material in the layer and the half-spaces is an isotropic viscoelastic solid. An inhomogeneous wave, incident from a half-space, is reflected and transmitted by the layer. Our purpose is to determine the amplitudes of the reflected and transmitted waves. While the geometry is the same as in §6.4, now we study the process in terms of the traction-displacement vector. We proceed by starting from the pertinent dynamical equations without necessarily restricting attention to Helmholtz

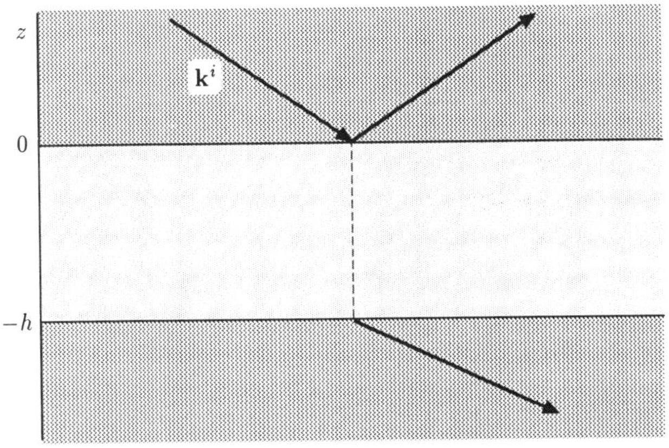

Fig. 6.2 Wave reflection and refraction through a heterogeneous layer between homogeneous half-spaces.

equations. Incidentally, the Helmholtz equation is the exact model of wave propagation in the particular case when the wave vector of the incident pair is perpendicular to the layer.

Let the z-axis be perpendicular to the layer and let $z > 0$ be the half-space where the incident and reflected waves propagate. Denote by h the width of the layer which then occupies the strip $-h < z < 0$ while the other half-space occupies the region $z < -h$. We denote by $\mathcal{P}_{-h}$ and $\mathcal{P}_0$ the plane boundaries of the layer. The mass density ρ and the parameters μ, λ are continuous functions of z as $-h < z < 0$, are constant as $z > 0$ and $z < -h$, and may suffer jump discontinuities at $z = 0$ and $z = -h$.

An incident, downgoing, conjugate pair of waves in the half-space $z > 0$ produces a reflected, upgoing, pair (in $z > 0$) and a transmitted, downgoing, pair in the half-space $z < -h$. We describe the waves in the homogeneous half-spaces through the scalar and vector potentials of longitudinal and transverse waves. Also, consistent with Snell's law, we let k_x and k_y be continuous across $\mathcal{P}_0$ and $\mathcal{P}_{-h}$. In fact this condition can be derived by looking for solutions to the wave equations through separation of variables and requiring the appropriate boundary conditions at the interface; this may be done by paralleling closely the procedure of §6.4. Accordingly we write the solution in the upper half-space as

$$\phi = \left[\Phi^+ \exp(i\beta_L z) + \Phi^- \exp(-i\beta_L z)\right] \exp[i(k_x x + k_y y)], \qquad 0 < z, \tag{5.1}$$

$$\boldsymbol{\psi} = \left[\boldsymbol{\Psi}^+ \exp(i\beta_T z) + \boldsymbol{\Psi}^- \exp(-i\beta_T z)\right] \exp[i(k_x x + k_y y)], \qquad 0 < z, \tag{5.2}$$

where

$$\beta_{L,T} = \sqrt{\kappa_{L,T} - k_x^2 - k_y^2}.$$

The pertinent fields in the lower half-space are

$$\breve{\phi} = \breve{\Phi}^- \exp(-i\breve{\beta}_L z) \exp[i(k_x x + k_y y)], \qquad z < -h, \tag{5.1'}$$

$$\breve{\boldsymbol{\psi}} = \breve{\Psi}^- \exp(-i\breve{\beta}_T z) \exp[i(k_x x + k_y y)], \qquad z < -h. \tag{5.2'}$$

Again Ψ_z is regarded as a linear combination of Ψ_x and Ψ_y through the orthogonality condition (cf. 4.3.11)

$$\mathbf{k}_T \cdot \boldsymbol{\Psi} = 0.$$

The quantities Φ^-, Ψ_x^-, and Ψ_y^- are the given data while Φ^+, Ψ_x^+, Ψ_y^+ and $\breve{\Phi}^-$, $\breve{\Psi}_x^-$, $\breve{\Psi}_y^-$ are the unknown amplitudes of the reflected and transmitted waves. To determine them we need the solution in the (intermediate) layer and the consequent application of the continuity requirements for displacement and traction at $\mathcal{P}_0$ and $\mathcal{P}_{-h}$.

Consider the layer $-h < z < 0$. Although a description of the displacement field in terms of potentials is still allowed, the dependence of the material parameters and density on z rules out the possibility of solutions in the form of complex exponentials. Accordingly, by paralleling analogous procedures for elastic materials [102], it is convenient to examine $\mathbf{u}$ in the usual form $\mathbf{u} = \mathbf{U}\exp(-i\omega t)$ and let

$$\mathbf{U} = \boldsymbol{\mathcal{U}}(z) \exp[i(k_x x + k_y y)], \qquad -h < z < 0. \tag{5.3}$$

The unknown function $\boldsymbol{\mathcal{U}}(z)$ is determined through the equation of motion, in the form

$$\rho\omega^2 \mathbf{U} + \nabla \cdot \mathbf{T} = 0, \tag{5.4}$$

where

$$\mathbf{T} = \lambda(\nabla \cdot \mathbf{U})\mathbf{1} + \mu(\nabla\mathbf{U} + \nabla\mathbf{U}^\dagger).$$

By analogy with the procedure for discretely stratified media (§6.1) we determine the pertinent unknowns through a traction-displacement vector. By maintaining the notation $\mathbf{n} = -\mathbf{e}_z$ we have

$$\mathbf{t} = \mathbf{T}\,\mathbf{n} = -(T_{xz}\mathbf{e}_x + T_{yz}\mathbf{e}_y + T_{zz}\mathbf{e}_z).$$

Thus, substitution into (1.6) yields

$$\mathbf{Z} = -\left(iU_x,\, iU_y,\, \frac{T_{zz}}{2},\, iU_z,\, \frac{T_{xz}}{2k_x},\, \frac{T_{yz}}{2k_y}\right)^\dagger. \tag{5.5}$$

The continuity of displacement and traction at $\mathcal{P}_0$ and $\mathcal{P}_{-h}$ is summarized by

$$\mathbf{Z}(0_+) = \mathbf{Z}(0_-), \qquad \mathbf{Z}(-h_+) = \mathbf{Z}(-h_-)$$

where $-h_+$ and $-h_-$ means $-h$ from above and below, respectively. We now evaluate the expressions of these quantities. Obviously, due to the factor $\exp[i(k_x x + k_y y)]$ in $\mathbf{U}$, derivatives with respect to x and y result in multiplications by ik_x and ik_y. As usual, this common factor is omitted if no ambiguity arises.

The vector $\mathbf{Z}(0_+)$ follows from comparison with (4.5.1) and (4.5.4); we have

$$\mathbf{Z}(0_+) = C \begin{pmatrix} \mathbf{A}^+ \\ \mathbf{A}^- \end{pmatrix}, \tag{5.6}$$

where

$$\mathbf{A}^+ = \left(\Phi^+,\ \Psi_x^+,\ \Psi_y^+\right)^\dagger, \qquad \mathbf{A}^- = \left(\Phi^-,\ -\Psi_x^-,\ -\Psi_y^-\right)^\dagger,$$

and

$$C = \begin{pmatrix} C_1 & C_1 \\ C_2 & -C_2 \end{pmatrix},$$

the matrices C_1 and C_2 being defined in (4.5.2) and (4.5.5). Similarly, substitution of (5.1') and (5.2') into the general expressions (4.1.7) and (4.1.8) for $\mathbf{U}$ and $\mathbf{t}$ yields

$$\mathbf{Z}(-h_-) = \begin{pmatrix} \breve{C}_1 \\ -\breve{C}_2 \end{pmatrix} \breve{\mathbf{A}}^-, \tag{5.7}$$

where $\breve{\mathbf{A}}^-$ is defined as

$$\breve{\mathbf{A}}^- = \left(\breve{\Phi}^- \exp(i\breve{\beta}_L h),\ -\breve{\Psi}_x^- \exp(i\breve{\beta}_T h),\ -\breve{\Psi}_y^- \exp(i\breve{\beta}_T h)\right)^\dagger.$$

Now we have to determine the traction-displacement vector in the layer. Observe that, upon substitution of (5.3) and the constitutive equation for $\mathbf{T}$, the differential equation of motion (5.4) results in a system of three second-order differential equations for the components of $\boldsymbol{\mathcal{U}}$. More profitably, we may enlarge the set of unknowns by regarding the z-dependent components of $\mathbf{t} = -\mathbf{T}\mathbf{e}_z$ as additional unknowns. Denote by a prime the derivative with respect to z. By the constitutive equation for $\mathbf{T}$ we can write

$$\mathbf{T}\mathbf{e}_z = \lambda(\nabla \cdot \mathbf{U})\mathbf{e}_z + \mu(\nabla U_z + \mathbf{U}')$$

whence we obtain the Cartesian components in the form

$$\mathcal{U}_x' = -ik_x\mathcal{U}_z + \frac{1}{\mu}T_{xz}, \tag{5.8}$$

$$\mathcal{U}_y' = -ik_y\mathcal{U}_z + \frac{1}{\mu}T_{yz}, \tag{5.9}$$

$$\mathcal{U}_z' = -i\frac{\lambda}{\lambda + 2\mu}(k_x\mathcal{U}_x + k_y\mathcal{U}_y) + \frac{1}{\lambda + 2\mu}T_{zz}. \tag{5.10}$$

Meanwhile the Cartesian components of the equation of motion (5.4) can be given the form

$$T_{xz}' = \left(4\mu\frac{\lambda + \mu}{\lambda + 2\mu}k_x^2 + \mu k_y^2 - \rho\omega^2\right)\mathcal{U}_x + \mu\frac{3\lambda + 2\mu}{\lambda + 2\mu}k_xk_y\mathcal{U}_y - i\frac{\lambda}{\lambda + 2\mu}k_xT_{zz}, \tag{5.11}$$

$$T'_{yz} = \mu\frac{3\lambda+2\mu}{\lambda+2\mu}k_xk_y\mathcal{U}_x + \Big(4\mu\frac{\lambda+\mu}{\lambda+2\mu}k_y^2 + \mu k_x^2 - \rho\omega^2\Big)\mathcal{U}_y - i\frac{\lambda}{\lambda+2\mu}k_yT_{zz}, \tag{5.12}$$

$$T'_{zz} = -\rho\omega^2\mathcal{U}_z - ik_xT_{xz} - ik_yT_{yz}. \tag{5.13}$$

Here **T** represents the stress to within the factor $\exp[i(k_xx+k_yy)]\exp(-i\omega t)$. The six first-order ordinary differential equations (5.8)-(5.13) constitute the desired system in the unknowns $\mathcal{U}_x, \mathcal{U}_y, \mathcal{U}_z, T_{xz}, T_{yz}, T_{zz}$.

As a particular case and also for a direct, physical motivation of §8.5 we consider the circumstance when the wave vectors of the incident pair are vertical, namely $k_x, k_y = 0$. By (5.3) we have

$$\mathbf{U} = \boldsymbol{\mathcal{U}}(z), \qquad -h < z < 0$$

and the system (5.8)-(5.13) reduces to

$$\mathcal{U}'_x = \frac{1}{\mu}T_{xz}, \qquad \mathcal{U}'_y = \frac{1}{\mu}T_{yz}, \qquad \mathcal{U}'_z = \frac{1}{\lambda+2\mu}T_{zz},$$

$$T'_{xz} = -\rho\omega^2\mathcal{U}_x, \qquad T'_{yz} = -\rho\omega^2\mathcal{U}_y, \qquad T'_{zz} = -\rho\omega^2\mathcal{U}_z.$$

It is apparent that this system decouples in three subsystems of two first-order equations. Upon comparison, each subsystem yields a single second-order equation in a single unknown; one for each of the quantities $\mathcal{U}_x, \mathcal{U}_y, \mathcal{U}_z$. Letting $\mathcal{U}_\parallel$ stand for any of $\mathcal{U}_x, \mathcal{U}_y$ we have

$$(\mu\mathcal{U}'_\parallel)' + \rho\omega^2\mathcal{U}_\parallel = 0,$$

$$[(\lambda+2\mu)\mathcal{U}'_z]' + \rho\omega^2\mathcal{U}_z = 0.$$

The change of unknowns,

$$\tilde{U}_\parallel(z) = \mathcal{U}_\parallel(z)\exp\Big(\tfrac{1}{2}\int_0^z \frac{\mu'}{\mu}(s)\,ds\Big), \qquad \tilde{U}_z(z) = \mathcal{U}_z(z)\exp\Big(\tfrac{1}{2}\int_0^z \frac{\lambda'+2\mu'}{\lambda+2\mu}(s)\,ds\Big)$$

makes $\tilde{U}_\parallel$ and $\tilde{U}_z$ be solutions to the Helmholtz equations

$$\tilde{U}''_\parallel + \Big[\frac{\rho\omega^2}{\mu} - \frac{1}{4}\Big(\frac{\mu'}{\mu}\Big)^2\Big]\tilde{U}_\parallel = 0, \tag{5.14}$$

$$\tilde{U}''_z + \Big[\frac{\rho\omega^2}{\lambda+2\mu} - \frac{1}{4}\Big(\frac{\lambda'+2\mu'}{\lambda+2\mu}\Big)^2\Big]\tilde{U}_z = 0. \tag{5.15}$$

The solution to these equations can then be determined by paralleling the procedure of the previous section, in terms of fundamental sets of solutions. The only formal difference is that the continuity of the derivative, at the interface, has to be replaced by the continuity of the traction, namely of $\mu U'_\parallel$ and $(\lambda+2\mu)U'_z$.

Turn the attention to the general system (5.8)-(5.13). Owing to the definition of **Z** and the fact that k_x and k_y are constant, we can write the system (5.8)-(5.13) in the compact form

$$\mathbf{Z}' = B\mathbf{Z} \tag{5.16}$$

where

$$B = \begin{pmatrix} 0 & B_1 \\ B_2 & 0 \end{pmatrix}$$

and

$$B_1 = i \begin{pmatrix} -k_x & 2k_x/\mu & 0 \\ -k_y & 0 & 2k_y/\mu \\ \rho\omega^2/2 & -k_x^2 & -k_y^2 \end{pmatrix},$$

$$B_2 = -\frac{i}{\lambda + 2\mu} \begin{pmatrix} \lambda k_x & \lambda k_y & -2 \\ \alpha_1 & \mu(3\lambda + 2\mu)k_y/2 & \lambda \\ \mu(3\lambda + 2\mu)k_x/2 & \alpha_2 & \lambda \end{pmatrix},$$

$$\alpha_1 = 2\mu(\lambda + \mu)k_x + (\mu k_y^2 - \rho\omega^2)(\lambda + 2\mu)/(2k_x),$$

$$\alpha_2 = 2\mu(\lambda + \mu)k_y + (\mu k_x^2 - \rho\omega^2)(\lambda + 2\mu)/(2k_y).$$

It is of interest to establish a *conservation law* for the system (5.16) which also will be useful in the determination of the reflection and transmission matrices. Consider two solutions **Z** and $\tilde{\mathbf{Z}}$ of (5.16) and decompose them into 3-dimensional column vectors as

$$\mathbf{Z}^\dagger = (\mathbf{Z}_1^\dagger, \mathbf{Z}_2^\dagger), \qquad \tilde{\mathbf{Z}}^\dagger = (\tilde{\mathbf{Z}}_1^\dagger, \tilde{\mathbf{Z}}_2^\dagger).$$

Owing to (5.16) we have

$$\mathbf{Z}_1' = B_1\mathbf{Z}_2,$$

$$\mathbf{Z}_2' = B_2\mathbf{Z}_1,$$

and the same for $\tilde{\mathbf{Z}}$. For technical reasons it is convenient to consider the matrix

$$\Lambda = 2i \begin{pmatrix} 0 & k_x & 0 \\ 0 & 0 & k_y \\ 1 & 0 & 0 \end{pmatrix} \tag{5.17}$$

and the scalar quantity

$$f = \mathbf{Z}^\dagger \begin{pmatrix} 0 & \Lambda \\ -\Lambda^\dagger & 0 \end{pmatrix} \tilde{\mathbf{Z}}.$$

Now we are in a position to prove that f is independent of z. To this end we observe that, by definition,

$$f = (\mathbf{Z}_1^\dagger, \mathbf{Z}_2^\dagger) \begin{pmatrix} 0 & \Lambda \\ -\Lambda^\dagger & 0 \end{pmatrix} \begin{pmatrix} \tilde{\mathbf{Z}}_1 \\ \tilde{\mathbf{Z}}_2 \end{pmatrix} = \mathbf{Z}_1^\dagger \Lambda \tilde{\mathbf{Z}}_2 - \mathbf{Z}_2^\dagger \Lambda^\dagger \tilde{\mathbf{Z}}_1. \tag{5.18}$$

Then we evaluate the derivative with respect to z to obtain

$$\begin{aligned}(\mathbf{Z}_1^\dagger \Lambda \tilde{\mathbf{Z}}_2 - \mathbf{Z}_2^\dagger \Lambda^\dagger \tilde{\mathbf{Z}}_1)' &= \mathbf{Z}_2^\dagger (B_1^\dagger \Lambda - \Lambda^\dagger B_1)\tilde{\mathbf{Z}}_2 + \mathbf{Z}_1^\dagger (\Lambda B_2 - B_2^\dagger \Lambda^\dagger)\tilde{\mathbf{Z}}_1 \\ &= \mathbf{Z}_2^\dagger \left[B_1^\dagger \Lambda - (B_1^\dagger \Lambda)^\dagger\right]\tilde{\mathbf{Z}}_2 + \mathbf{Z}_1^\dagger \left[\Lambda B_2 - (\Lambda B_2)^\dagger\right]\tilde{\mathbf{Z}}_1. \end{aligned} \tag{5.19}$$

Now, direct evaluation shows that

$$B_1^\dagger \Lambda = 2 \begin{pmatrix} -\rho\omega^2/2 & k_x^2 & k_y^2 \\ k_x^2 & -2k_x^2/\mu & 0 \\ k_y^2 & 0 & -2k_y^2/\mu \end{pmatrix},$$

$$\Lambda B_2 = \frac{2}{\lambda + 2\mu} \begin{pmatrix} \alpha_1 k_x & \mu(3\lambda + 2\mu)k_x k_y/2 & \lambda k_x \\ \mu(3\lambda + 2\mu)k_x k_y/2 & \alpha_2 k_y & \lambda k_y \\ \lambda k_x & \lambda k_y & -2 \end{pmatrix}.$$

The symmetry of $B_1^\dagger \Lambda$ and ΛB_2 implies the vanishing of the right-hand side of (5.19). Hence (5.19) yields the desired conservation law $f =$ constant.

Direct application of the pertinent definitions yields

$$f = \mathcal{U}_x \tilde{t}_x + \mathcal{U}_y \tilde{t}_y - \mathcal{U}_z \tilde{t}_z - t_x \tilde{\mathcal{U}}_x - t_y \tilde{\mathcal{U}}_y + t_z \tilde{\mathcal{U}}_z.$$

Trivially, this expression takes the more compact form $f = \boldsymbol{\mathcal{U}} \cdot \tilde{\mathbf{t}} - \mathbf{t} \cdot \tilde{\boldsymbol{\mathcal{U}}}$ if $\mathcal{U}_z \to i\mathcal{U}_z$, $t_z \to it_z$. The constancy of f carries over into the set of inhomogeneous waves in viscoelastic solids, without any restrictions on the wave vectors, previous results derived for standard elastic plane waves (cf. [102], Ch. 2).

6.6 Reflection and transmission matrices for a layer

To determine the reflection and transmission matrices for a layer $-h < z < 0$ we need to relate $\mathbf{Z}(0_+)$ to $\mathbf{Z}(-h_-)$ through the solution to the system of differential equations (5.16) for $\mathbf{Z}$. To this end we make use of a fundamental system of solutions.

Let $\mathbf{Z}_{(i)}(z)$, $i = 1, .., 6$, be a fundamental system of solutions to (5.16) in the interval $-h < z < 0$. This means that the $\mathbf{Z}_{(i)}$'s are six independent solutions of (5.16) and hence every solution to (5.16) may be represented as a linear combination of the $\mathbf{Z}_{(i)}$'s. We show that the continuity conditions on $\mathbf{Z}$ at $z = 0$ and $z = -h$ determine completely the amplitudes of the waves belonging to the reflected and transmitted conjugate pairs in terms of $\mathbf{A}^-$ and the values of the $\mathbf{Z}_{(i)}$'s at $z = 0$ and $z = -h$. We also prove that, as expected, the amplitudes so obtained are invariant under changes of the fundamental system of solutions.

Represent any solution $\mathbf{Z}(z)$ to (5.16) in terms of a fundamental system as

$$\mathbf{Z}(z) = \sum_{i=1}^{6} a_i \mathbf{Z}_{(i)}(z), \qquad a_i = \text{constant}. \tag{6.1}$$

For formal convenience, we regard the six constants a_i as the components of a vector $\mathbf{a}$,

$$\mathbf{a} = (a_1, ..., a_6)^\dagger,$$

and define the 6×6 matrix W as

$$W_{ji} = Z_{j(i)}.$$

This allows (6.1) to be written as

$$\mathbf{Z}(z) = W(z)\mathbf{a}. \tag{6.2}$$

Usually W is called the *fundamental matrix* or the resolvent operator ([102], §2.2; [5], Ch. 7) of the system (5.16). As an immediate consequence of the definition, W turns out to satisfy the matrix differential equation

$$W' = BW. \tag{6.3}$$

More specifically, represent the fundamental matrix W in block form as

$$W = \begin{pmatrix} W_1 & W_2 \\ W_3 & W_4 \end{pmatrix};$$

in terms of the fundamental systems of solutions $\mathbf{Z} = (\mathbf{Z}_1, \mathbf{Z}_2)^\dagger$, this means

$$(W_1)_{ji} = (Z_1)_{j(i)}, \qquad (W_2)_{ji} = (Z_1)_{j(i+3)},$$

$$(W_3)_{ji} = (Z_2)_{j(i)}, \qquad (W_4)_{ji} = (Z_2)_{j(i+3)},$$

the indices i and j taking values from 1 to 3. Then, owing to the form of B, the system (6.3) decouples in two subsystems as

$$\begin{cases} W_1' = B_1 W_3, \\ W_3' = B_2 W_1, \end{cases} \qquad \begin{cases} W_2' = B_1 W_4, \\ W_4' = B_2 W_2, \end{cases} \tag{6.4}$$

in the unknown matrices W_1, W_3, and W_2, W_4.

To determine the reflection and transmission matrices, we need also to establish a general form for the inverse of the fundamental matrix. In this connection we derive preliminary results on the propagation invariants for the system (6.3). This is accomplished by means of the conserved quantity f associated with each pair of solutions to equation

(5.16). In so doing we are in fact carrying over into the framework of dissipative bodies recent results on propagation of seismic waves in stratified elastic media (cf. [102], Ch. 2).

By analogy with the representation (5.18) of the scalar invariant f, let us define the 3×3 matrix

$$V = W_1^\dagger \Lambda W_4 - W_3^\dagger \Lambda^\dagger W_2 \tag{6.5}$$

and observe that V is independent of z. The result follows by regarding each entry of V as resulting from the same combination of two solutions of (5.16) that was considered in the proof of the invariance of f. Otherwise, a direct calculation and use of (6.4) show that

$$V' = W_3^\dagger (B_1^\dagger \Lambda - \Lambda^\dagger B_1) W_4 - W_1^\dagger (\Lambda B_2 - B_2^\dagger \Lambda^\dagger) W_2 = 0,$$

the last equality being a consequence of the symmetry of the matrices $B_1^\dagger \Lambda$ and ΛB_2. By the same token it is easily shown that

$$(W_2^\dagger \Lambda W_4 - W_4^\dagger \Lambda^\dagger W_2)' = 0,$$

$$(W_1^\dagger \Lambda W_3 - W_3^\dagger \Lambda^\dagger W_1)' = 0.$$

By a proper choice of the initial values of the traction-displacement vectors $\mathbf{Z}_{(i)}$ we can make the constant matrices $W_2^\dagger \Lambda W_4 - W_4^\dagger \Lambda^\dagger W_2$ and $W_1^\dagger \Lambda W_3 - W_3^\dagger \Lambda^\dagger W_1$ vanish. Then by the definition of W we have

$$\begin{pmatrix} W_4^\dagger & W_2^\dagger \\ W_3^\dagger & W_1^\dagger \end{pmatrix} \begin{pmatrix} -\Lambda^\dagger & 0 \\ 0 & \Lambda \end{pmatrix} \begin{pmatrix} W_1 & W_2 \\ W_3 & W_4 \end{pmatrix}$$
$$= \begin{pmatrix} -V^\dagger & W_2^\dagger \Lambda W_4 - W_4^\dagger \Lambda^\dagger W_2 \\ W_1^\dagger \Lambda W_3 - W_3^\dagger \Lambda^\dagger W_1 & V \end{pmatrix} = \begin{pmatrix} -V^\dagger & 0 \\ 0 & V \end{pmatrix}.$$

Left multiplication by the inverse matrix

$$\begin{pmatrix} -V^{-\dagger} & 0 \\ 0 & V^{-1} \end{pmatrix},$$

where $V^{-\dagger} = (V^{-1})^\dagger$, yields

$$\begin{pmatrix} -V^{-\dagger} & 0 \\ 0 & V^{-1} \end{pmatrix} \begin{pmatrix} W_4^\dagger & W_2^\dagger \\ W_3^\dagger & W_1^\dagger \end{pmatrix} \begin{pmatrix} -\Lambda^\dagger & 0 \\ 0 & \Lambda \end{pmatrix} \begin{pmatrix} W_1 & W_2 \\ W_3 & W_4 \end{pmatrix} = \begin{pmatrix} \mathbb{1} & 0 \\ 0 & \mathbb{1} \end{pmatrix}.$$

This in turn shows that the inverse of the matrix W is given by

$$W^{-1} = \begin{pmatrix} -V^{-\dagger} & 0 \\ 0 & V^{-1} \end{pmatrix} \begin{pmatrix} W_4^\dagger & W_2^\dagger \\ W_3^\dagger & W_1^\dagger \end{pmatrix} \begin{pmatrix} -\Lambda^\dagger & 0 \\ 0 & \Lambda \end{pmatrix}. \tag{6.6}$$

We are now in a position to determine the reflection and transmission matrices. Letting $W_0 := W(0_-)$ we can write

$$\mathbf{Z}(0_-) = W_0 \, \mathbf{a}.$$

The continuity condition $\mathbf{Z}(0_-) = \mathbf{Z}(0_+)$ and (5.6) yield

$$C \begin{pmatrix} \mathbf{A}^+ \\ \mathbf{A}^- \end{pmatrix} = W_0 \, \mathbf{a}. \tag{6.7}$$

Similarly, letting $W_{-h} = W(-h_+)$ we find that

$$\mathbf{Z}(-h_+) = W_{-h} \, \mathbf{a}.$$

Therefore, the continuity condition at $z = -h$ and (5.7) yield

$$\begin{pmatrix} \breve{C}_1 \\ -\breve{C}_2 \end{pmatrix} \breve{\mathbf{A}}^- = W_{-h} \, \mathbf{a}. \tag{6.8}$$

Equations (6.7) and (6.8) constitute a linear system of twelve equations in twelve unknowns, viz the components of $\mathbf{a}$, $\mathbf{A}^+$, and $\breve{\mathbf{A}}^-$. By solving the system we determine the amplitudes $\mathbf{A}^+$ and $\breve{\mathbf{A}}^-$ of the reflected and transmitted pairs in terms of $\mathbf{A}^-$, the amplitude of the incident pair.

Incidentally, it is a satisfactory check of consistency to ascertain that the solution for $\mathbf{A}^+$ and $\breve{\mathbf{A}}^-$ is independent of the choice of the fundamental system of solutions. Because the matrix W is nonsingular at each value of z (cf. [102], Ch. 2), we solve (6.7) for $\mathbf{a}$ and insert the result into (6.8) to obtain

$$W_{-h} \, (W_0)^{-1} \, C \begin{pmatrix} \mathbf{A}^+ \\ \mathbf{A}^- \end{pmatrix} = \begin{pmatrix} \breve{C}_1 \breve{\mathbf{A}}^- \\ -\breve{C}_2 \breve{\mathbf{A}}^- \end{pmatrix}. \tag{6.9}$$

Now we show that if (6.9) can be solved for $\mathbf{A}^+$ and $\breve{\mathbf{A}}^-$ then the result is independent of the choice of the $\mathbf{Z}_{(i)}$'s. Specifically, given the same incident amplitude $\mathbf{A}^-$, consider another fundamental system of solutions $\boldsymbol{\mathcal{Z}}_{(i)}(z)$ and denote by $\mathcal{W}$ the corresponding matrix; $\mathcal{W}_0$ and $\mathcal{W}_{-h}$ are the limit values of $\mathcal{W}$ as $z \to 0$ and $z \to -h$ within the layer. Since each $\boldsymbol{\mathcal{Z}}_{(i)}$ is a solution to the system (5.16) there exists a non-singular constant matrix, say K, such that $\mathcal{W} = WK$; then

$$\mathcal{W}_0 = W_0 \, K, \qquad \mathcal{W}_{-h} = W_{-h} \, K.$$

Accordingly we find that

$$\mathcal{W}_{-h}(\mathcal{W}_0)^{-1} = W_{-h}(W_0)^{-1}.$$

Correspondingly the amplitudes, say $\boldsymbol{\mathcal{A}}^+$ and $\breve{\boldsymbol{\mathcal{A}}}^-$, are the solution to the system

$$\begin{pmatrix} \breve{C}_1 \breve{\boldsymbol{\mathcal{A}}}^- \\ -\breve{C}_2 \breve{\boldsymbol{\mathcal{A}}}^- \end{pmatrix} = W_{-h} \, (W_0)^{-1} \, C \begin{pmatrix} \boldsymbol{\mathcal{A}}^+ \\ \mathbf{A}^- \end{pmatrix}.$$

Comparison with the system (6.9) shows that the solution for $\boldsymbol{\mathcal{A}}^+$ and $\breve{\boldsymbol{\mathcal{A}}}^-$ is the same as that for $\mathbf{A}^+$ and $\breve{\mathbf{A}}^-$. Hence the solution is independent of the choice of the fundamental system.

This freedom in the choice of the system of fundamental solutions is exploited to simplify the determination of solutions to the system (6.7), (6.8). Define the *propagator matrix* $P(z)$ as the solution to the

Cauchy problem

$$P' = BP, \qquad P(0) = \mathbb{1}. \tag{6.10}$$

In terms of the block form

$$P = \begin{pmatrix} P_1 & P_2 \\ P_3 & P_4 \end{pmatrix},$$

where P_1, P_2, P_3, P_4 are 3×3 matrices, we have

$$P_1(0) = P_4(0) = \mathbb{1}, \qquad P_2(0) = P_3(0) = 0.$$

The matrices

$$P_2^\dagger \Lambda P_4 - P_4^\dagger \Lambda^\dagger P_2, \qquad P_1^\dagger \Lambda P_3 - P_3^\dagger \Lambda^\dagger P_1$$

are constant and then vanish. Further, by the constancy of

$$V = P_1^\dagger \Lambda P_4 - P_3^\dagger \Lambda P_2$$

we have $V = \Lambda$. Then, owing to (6.6),

$$P^{-1} = \begin{pmatrix} -\Lambda^{-\dagger} & 0 \\ 0 & \Lambda^{-1} \end{pmatrix} \begin{pmatrix} P_4^\dagger & P_2^\dagger \\ P_3^\dagger & P_1^\dagger \end{pmatrix} \begin{pmatrix} -\Lambda^\dagger & 0 \\ 0 & \Lambda \end{pmatrix}, \tag{6.11}$$

and

$$\Lambda^{-1} = \frac{1}{2ik_x k_y} \begin{pmatrix} 0 & 0 & k_x k_y \\ k_y & 0 & 0 \\ 0 & k_x & 0 \end{pmatrix}.$$

In terms of P, the relation (6.9) yields

$$\begin{pmatrix} \mathbf{A}^+ \\ \mathbf{A}^- \end{pmatrix} = C^{-1} (P_{-h})^{-1} \begin{pmatrix} \check{C}_1 \\ -\check{C}_2 \end{pmatrix} \check{\mathbf{A}}^-. \tag{6.12}$$

Comparison with (1.9), for $s = 1$, shows that

$$\check{\mathbf{A}}^- = T_d \mathbf{A}^-,$$

$$\mathbf{A}^+ = R_d \mathbf{A}^-,$$

where the subscript d denotes evaluation in correspondence with an incident downgoing wave. By a slight abuse of language T_d is called the transmission matrix; it is defined through

$$T_d^{-1} = \frac{1}{2} (\, C_1^{-1}, \; -C_2^{-1}\,) \, (P_{-h})^{-1} \begin{pmatrix} \check{C}_1 \\ -\check{C}_2 \end{pmatrix}. \tag{6.13}$$

The reflection matrix R_d is then given by

$$R_d = \frac{1}{2} (C_1^{-1}, \quad C_2^{-1}) (P_{-h})^{-1} \begin{pmatrix} \breve{C}_1 \\ -\breve{C}_2 \end{pmatrix} T_d. \tag{6.14}$$

The matrices C_1, C_2 and their inverses are defined in §4.5, where the explicit expressions are also given. Insertion of the expression (6.11) for P^{-1} leads to the equivalent representations

$$T_d^{-1} = \frac{1}{2} (C_1^{-1} \Lambda^{-\dagger} \quad C_2^{-1} \Lambda^{-1}) \begin{pmatrix} P_4^\dagger & P_2^\dagger \\ P_3^\dagger & P_1^\dagger \end{pmatrix} \begin{pmatrix} \Lambda^\dagger \breve{C}_1 \\ \Lambda \breve{C}_2 \end{pmatrix},$$

$$R_d = \frac{1}{2} (C_1^{-1} \Lambda^{-\dagger} \quad -C_2^{-1} \Lambda^{-1}) \begin{pmatrix} P_4^\dagger & P_2^\dagger \\ P_3^\dagger & P_1^\dagger \end{pmatrix} \begin{pmatrix} \Lambda^\dagger \breve{C}_1 \\ \Lambda \breve{C}_2 \end{pmatrix} T_d.$$

It is understood that the matrices P_i are evaluated at $z = -h$.

Operatively, the effect of a layer on reflection and transmission may be determined as follows. Assume that a propagator matrix P has been evaluated, possibly numerically, by integration of (6.10). Substitution of the pertinent results into (6.13) and (6.14) yields the reflection and transmission matrices R_d, T_d. The effective reflection and transmission matrices, relative to the amplitudes Φ, Ψ_x, Ψ_y are obtained by replacing $\breve{\mathbf{A}}^-$, $\mathbf{A}^+$, $\mathbf{A}^-$ with their expressions to obtain

$$\begin{pmatrix} \breve{\Phi}^- \exp(i\breve{\beta}_L h) \\ -\breve{\Psi}_x^- \exp(i\breve{\beta}_T h) \\ -\breve{\Psi}_y^- \exp(i\breve{\beta}_T h) \end{pmatrix} = T_d \begin{pmatrix} \Phi^- \\ -\Psi_x^- \\ -\Psi_y^- \end{pmatrix},$$

whence $\breve{\Phi}^-$, $\breve{\Psi}_x$, and $\breve{\Psi}_y^-$ and then the effective transmission matrix follow. Finally, replacing $\mathbf{A}^+$ and $\mathbf{A}^-$ with their expressions we have

$$\begin{pmatrix} \Phi^+ \\ \Psi_x^+ \\ \Psi_y^+ \end{pmatrix} = R_d \begin{pmatrix} \Phi^- \\ -\Psi_x^- \\ -\Psi_y^- \end{pmatrix},$$

with R_d given by (6.14), whence we can read the effective reflection matrix. Observe that both T_d and R_d require the determination of the propagator matrix P. Finally, if we are interested in the traction-displacement vector within the layer, we observe that the result follows by replacing into (6.2) the value of $\mathbf{a}$ obtained from (6.7); we have

$$\mathbf{Z}(z) = P(z) C \begin{pmatrix} \mathbf{A}^+ \\ \mathbf{A}^- \end{pmatrix}, \qquad -h < z < 0.$$

Though conceptually nothing new occurs, it is worth having a look at the symmetric process of reflection and refraction when the incident pair is coming from the lower half-space. The field within the lower half-space is described by

$$\breve{\phi} = \breve{\Phi}^+ \exp(i\breve{\beta}_L z) + \breve{\Phi}^- \exp(-i\breve{\beta}_L z), \qquad z < -h,$$

$$\breve{\boldsymbol{\psi}} = \breve{\boldsymbol{\Psi}}^+ \exp(i\breve{\beta}_T z) + \breve{\boldsymbol{\Psi}}^- \exp(-i\breve{\beta}_T z), \qquad z < -h$$

where the superscripts + and − refer to the incident (upgoing) and reflected (downgoing) waves. By paralleling the previous procedure we have

$$\mathbf{Z}(-h_-) = \breve{C} \begin{pmatrix} \breve{\mathbf{A}}^+ \\ \breve{\mathbf{A}}^- \end{pmatrix}$$

with

$$\breve{\mathbf{A}}^+ = \left(\breve{\Phi}^+ \exp(-i\breve{\beta}_L h),\ \breve{\Psi}_x^+ \exp(-i\breve{\beta}_T h),\ \breve{\Psi}_y^+ \exp(-i\breve{\beta}_T h)\right)^\dagger,$$

$$\breve{\mathbf{A}}^- = \left(\breve{\Phi}^- \exp(i\breve{\beta}_L h),\ -\breve{\Psi}_x^- \exp(i\breve{\beta}_T h),\ -\breve{\Psi}_y^- \exp(i\breve{\beta}_T h)\right)^\dagger,$$

while

$$\breve{C} = \begin{pmatrix} \breve{C}_1 & \breve{C}_1 \\ \breve{C}_2 & -\breve{C}_2 \end{pmatrix}.$$

The field in the upper half-space has the form (5.1), (5.2), with both $\mathbf{\Phi}^-$ and $\mathbf{\Psi}^-$ vanishing. Hence we obtain

$$\mathbf{Z}(0_+) = \begin{pmatrix} C_1 \\ C_2 \end{pmatrix} \mathbf{A}^+.$$

The traction-displacement vector within the layer is written as in (5.5) with the corresponding related boundary values. Consider the propagator matrix P as before. In particular we can write

$$\mathbf{Z}(0_-) = \mathbf{a}, \qquad \mathbf{Z}(-h_+) = P_{-h}\mathbf{a}.$$

Elimination of $\mathbf{a}$ and comparison with the representations of $\mathbf{Z}(0_+)$ and $\mathbf{Z}(-h_-)$ leads to

$$P_{-h} \begin{pmatrix} C_1 \\ C_2 \end{pmatrix} \mathbf{A}^+ = \breve{C} \begin{pmatrix} \breve{\mathbf{A}}^+ \\ \breve{\mathbf{A}}^- \end{pmatrix}.$$

On multiplying both sides by $\breve{C}^{-1}$ we obtain

$$\begin{pmatrix} \Phi^+ \\ \Psi_x^+ \\ \Psi_y^+ \end{pmatrix} = T_u \begin{pmatrix} \breve{\Phi}^+ \\ \breve{\Psi}_x^+ \\ \breve{\Psi}_y^+ \end{pmatrix}$$

and

$$\begin{pmatrix} \breve{\Phi}^- \\ -\breve{\Psi}_x^- \\ -\breve{\Psi}_y^- \end{pmatrix} = R_u \begin{pmatrix} \breve{\Phi}^+ \\ \breve{\Psi}_x^+ \\ \breve{\Psi}_y^+ \end{pmatrix}$$

with

$$T_u^{-1} = \frac{1}{2}(\,\breve{C}_1^{-1},\quad \breve{C}_2^{-1}\,)\, P_{-h} \begin{pmatrix} C_1 \\ C_2 \end{pmatrix},$$

$$R_u = \frac{1}{2}(\,\breve{C}_1^{-1},\quad -\breve{C}_2^{-1}\,)\, P_{-h} \begin{pmatrix} C_1 \\ C_2 \end{pmatrix} T_u;$$

the subscript u is a reminder that the incident pair is upgoing. These constitute the analogues of (6.13) and (6.14).

6.7 Remarks about reflection and transmission matrices

Reflection and transmission of inhomogeneous waves by a continuously stratified layer has been discussed in terms of fundamental systems of solutions or, more precisely, in terms of the propagator matrix P. An analysis of the properties of P, namely its determination by a recursive algorithm, is given in [102] and can be extended directly to the dissipative media. By analogy with a result obtained in elasticity, we ask whether and how a more direct system of - possibly coupled - equations for $T_{d,u}$ and $R_{d,u}$ can be found.

By applying an invariant imbedding technique, Tromp and Snieder [163] have shown that the reflection and transmission matrices for plane longitudinal and transverse waves obey a system of 36 coupled, complex, first order non-linear differential Riccati equations. Here we are able to set up two systems of 18 linear differential equations which lead to a direct determination of the reflection and transmission matrices.

Consider the 6×6 matrix $M(z)$ defined as

$$M = \begin{pmatrix} M_1 & M_2 \\ M_3 & M_4 \end{pmatrix} = C^{-1} P^{-1}; \tag{7.1}$$

from the constancy of C and the definition of P it follows that

$$M(0) = C^{-1}, \qquad M(-h) = M_{-h} = C^{-1}(P_{-h})^{-1}.$$

To avoid frequent use of subscripts, for the time being we denote M_{-h} by M. Consider first the case of a downgoing incident pair. Comparison with the definition of M and (6.9), with W replaced by P, shows that

$$\begin{pmatrix} \mathbf{A}^+ \\ \mathbf{A}^- \end{pmatrix} = M \begin{pmatrix} \check{C}_1 \check{\mathbf{A}}^- \\ -\check{C}_2 \check{\mathbf{A}}^- \end{pmatrix}$$

that is

$$\mathbf{A}^+ = \left(M_1 \check{C}_1 - M_2 \check{C}_2 \right) \check{\mathbf{A}}^-, \qquad \mathbf{A}^- = \left(M_3 \check{C}_1 - M_4 \check{C}_2 \right) \check{\mathbf{A}}^-.$$

Of course $\mathbf{A}^-$ is regarded as known. Then the reflection and transmission matrices are easily recognized to be given by

$$T_d = \left(M_3 \check{C}_1 - M_4 \check{C}_2 \right)^{-1}, \tag{7.2}$$

$$R_d = \left(M_1 \check{C}_1 - M_2 \check{C}_2 \right) T_d. \tag{7.3}$$

Thus the knowledge of M leads to the determination of R_d and T_d through algebraic manipulations. Of course, (7.2) and (7.3) are equivalent to (6.13) and (6.14) via (7.1).

Now let the incident pair be upgoing, and coming from below. Multiplication of both sides of (4.9) by $C^{-1}(P_{-h})^{-1} = M_{-h}$ yields

$$C^{-1}\begin{pmatrix} C_1 \\ C_2 \end{pmatrix} \mathbf{A}^+ = M\check{C}\begin{pmatrix} \check{\mathbf{A}}^+ \\ \check{\mathbf{A}}^- \end{pmatrix}.$$

Comparison with (1.9) and substitution of the expression for C^{-1} shows that

$$C^{-1}\begin{pmatrix} C_1 \\ C_2 \end{pmatrix} = \begin{pmatrix} \mathbb{1} \\ 0 \end{pmatrix}.$$

Hence we find

$$\begin{pmatrix} \mathbf{A}^+ \\ 0 \end{pmatrix} = M\check{C}\begin{pmatrix} \check{\mathbf{A}}^+ \\ \check{\mathbf{A}}^- \end{pmatrix},$$

whence it follows

$$\mathbf{A}^+ = (M_1\check{C}_1 + M_2\check{C}_2)\check{\mathbf{A}}^+ + (M_1\check{C}_1 - M_2\check{C}_2)\check{\mathbf{A}}^-,$$

$$0 = (M_3\check{C}_1 + M_4\check{C}_2)\check{\mathbf{A}}^+ + (M_3\check{C}_1 - M_4\check{C}_2)\check{\mathbf{A}}^-.$$

Therefore, also in view of (7.2) and (7.3), we obtain

$$\check{\mathbf{A}}^- = R_u\check{\mathbf{A}}^+, \qquad \mathbf{A}^+ = T_u\check{\mathbf{A}}^+,$$

where

$$\begin{aligned} R_u &= -(M_3\check{C}_1 - M_4\check{C}_2)^{-1}(M_3\check{C}_1 + M_4\check{C}_2) \\ &= -T_d(M_3\check{C}_1 + M_4\check{C}_2), \end{aligned} \tag{7.4}$$

and

$$\begin{aligned} T_u &= (M_1\check{C}_1 + M_2\check{C}_2) + (M_1\check{C}_1 - M_2\check{C}_2)R_u \\ &= M_1\check{C}_1 + M_2\check{C}_2 + T_d^{-1}R_dR_u, \end{aligned} \tag{7.5}$$

This shows that also R_u and T_u are algebraically related to M_{-h}.

Now we again let M stand for $M(z)$, as $-h < z < 0$. An inspection of (7.2) and (7.4) shows that the matrices R_u and T_d, depend on the values of the matrices M_3 and M_4 at $z = -h$, besides on the given constant matrices $\check{C}_1$ and $\check{C}_2$. We prove in a moment that M_3 and M_4 are given as solutions to a reduced linear system of 18 coupled equations, while W_{-h} is involved in the system (6.3) of 36 equations. Of course, only the value of M at $z = -h$ is required if we are interested in the reflection and transmission matrices. If, rather, we look for the field inside the layer then we need the matrix $P(z)$, which is related to $M(z)$ through (7.1).

To find the differential equations for $M(z)$ we observe that, in view of (7.1), (6.3) and the relation $(P^{-1})' = -P^{-1}P'P^{-1}$,

$$\begin{aligned} M' &= C^{-1}\left(P^{-1}\right)' \\ &= -C^{-1}P^{-1}P'P^{-1} \\ &= -C^{-1}P^{-1}B, \end{aligned}$$

that is

$$M' = -MB. \tag{7.6}$$

Equation (7.6) is to be solved with the initial condition $M(0) = C^{-1}$ that follows from the definition (7.1) of M. The matrix C^{-1} is given by (1.9) by letting $s = 1$.

Owing to the structure of B that follows from (5.16), the linear system (7.6) can be split in the form

$$\begin{cases} M_1' = -M_2B_2, \\ M_2' = -M_1B_1, \end{cases} \qquad \begin{cases} M_3' = -M_4B_2, \\ M_4' = -M_3B_1. \end{cases}$$

The related initial conditions are

$$M_1(0) = M_3(0) = \frac{1}{2}C_1^{-1}, \qquad M_2(0) = -M_4(0) = \frac{1}{2}C_2^{-1}.$$

In particular, the 3×3 matrices M_3 and M_4 are obtained as solutions to a coupled linear system of two matrix equations, and this yields R_u and T_d. Then T_u and R_d follow directly by use of (7.5), (7.6) and the expressions of M_1, M_2.

We now consider the limit case when the incident wavelength is much longer than the thickness h of the layer and show that, as expected, the reflection and transmission matrices reduce to those characterizing the behaviour on inhomogemous waves at a plane interface between two homogeneous half-spaces. To this end we introduce a new variable $\nu = z/h$. The layer is described by $-1 < \nu < 0$ and $d/dz = (1/h)d/d\nu$. Now we define the parameter $\eta = h\beta_T$ and we model the condition of thin layer by letting $\eta \to 0$. Then we observe that the differential equation for the propagator matrix P can be rewritten in the form

$$\frac{dP(\nu)}{d\nu} = \frac{\eta}{\beta_T} B\,P. \tag{7.7}$$

We have to find a system of fundamental solutions to (7.7), satisfying the initial condition $P(0) = \mathbb{1}$, and to determine reflected and transmitted pairs in the limit $\eta \to 0$. To make the limit immediate, let P be parameterized by η and

$$P(\nu;\eta) = P(\nu;0) + \eta Q(\nu;\eta)$$

Q being bounded as $\eta \to 0$. Substitution in (7.7) requires that

$$\frac{dP(\nu;0)}{d\nu} = 0, \qquad P(0;0) = 1\!\!1. \tag{7.8}$$

The trivial solution to (7.8) is $P(\nu;0) = 1\!\!1$, $-1 < \nu < 0$, and hence, as $\eta \to 0$, we have $P_{-h} = 1\!\!1$.

Consider an incident downgoing wave. Substitution of the unit matrix for P_{-h} into (6.9) reduces the system for the determination of reflection and transmission matrices to the form

$$C\begin{pmatrix} \mathbf{A}^+ \\ \mathbf{A}^- \end{pmatrix} = \begin{pmatrix} \breve{C}_1\breve{\mathbf{A}}^- \\ -\breve{C}_2\breve{\mathbf{A}}^- \end{pmatrix}. \tag{7.9}$$

where, in the limit $\eta \to 0$, the expression of $\breve{\mathbf{A}}^-$ reads

$$\breve{\mathbf{A}}^- = (\breve{\Phi}^-, -\breve{\Psi}_x^-, -\breve{\Psi}_y^-)^\dagger.$$

The relation (7.9) states that the traction-displacement vectors of the upper and the lower half-space are equal, and this is just the continuity condition that is obtained in the case of two half-spaces in welded contact. Hence the reflected conjugate pair and the transmitted one coincide with those obtained in the case of a plane interface between two half-spaces. The explicit determination of the coefficients is given in Ch. 4.

Finally, as a simple application of the theory, suppose that the lower half-space is empty and an inhomogeneous pair impinges on the interface $z = 0$ from the upper half-space. At $z = 0$ the welded contact condition (6.7) must hold. The corresponding condition at $z = -h$ is that the limit of the traction vanishes, while no restriction is to be imposed on the limit of the displacement. In view of the definition of $\mathbf{Z}$ and the condition $\mathbf{Z}(-h_+) = P_{-h}\,\mathbf{a}$, the traction-free condition results in

$$(P_{-h})_{3j}a_j = (P_{-h})_{5j}a_j = (P_{-h})_{6j}a_j = 0. \tag{7.10}$$

Meanwhile, the continuity condition of the traction-displacement vector at $z = 0$ takes the form

$$C\begin{pmatrix} \mathbf{A}^+ \\ \mathbf{A}^- \end{pmatrix} = \mathbf{a} \tag{7.11}$$

that follows from (6.7). Equations (7.10) and (7.11) constitute a linear system of 9 equations for the 9 unknown components of $\mathbf{a}$ and $\mathbf{A}^+$. For example, once three components of the vector $\mathbf{a}$ have been found through (7.10), substitution into (7.11) yields the components of $\mathbf{A}^+$ and the remaining ones of $\mathbf{a}$. Other particular cases, such as, e.g., an upward propagating wave under the assumption that the upper half-space is filled either by a viscoelastic, or a viscous or an inviscid fluid can be dealt with straightforwardly.

7 SCATTERING BY OBSTACLES

In a typical scattering process a solid obstacle is immersed in a different solid material or a fluid and a time harmonic acoustic wave, which comes from infinity, interacts with the obstacle and is diffused to infinity. The direct scattering problem consists essentially in the determination of the transmitted field at infinity, in terms of the shape, the size and (possibly) the material properties of the scatterer. The inverse problem aims at the determination of the characteristic features of the obstacle in terms of the measured scattered wave. Such problems are of interest in many areas of scientific research and technology, such as non-destructive testing of materials, sonar exploration, analysis of living tissues, seismology and seismic exploration.

Within the framework of scattering theory, this chapter emphasizes some aspects related to inhomogeneous waves. The obstacle may be a cavity or an inclusion. To make the model more realistic, dissipative properties are incorporated for both the surrounding medium and the obstacle. In fact, only a lossy model can give a realistic representation of ultrasound propagation in living tissues and can take into account the observed attenuation and dispersion of seismic waves. The appropriate mathematical scheme that accounts for the presence of dissipation, though linear, requires the extension of the rather well established "scalar" theory of scattering to a more involved tensor version.

An integral representation is applied which involves a Green's tensor function and an appropriate radiation condition. The far-field is then shown to result from superposition of outgoing spherical, longitudinal and transverse, inhomogeneous waves. New aspects of dissipation are emphasized in connection with uniqueness properties, scattering cross section, and high-frequency behaviour.

7.1 The scalar theory of scattering

In this introductory section we consider a plane wave that comes from infinity and propagates in a homogeneous inviscid fluid. Start from the obvious observation that nothing of physical interest happens in the absence of inhomogeneities, while the wave is scattered, or diffracted, if an obstacle is placed in the fluid. This means that we can express the total field in the fluid as the sum of the "incident" wave and of a "scattered" wave originated at the boundary of the obstacle through a reflection phenomenon caused by the discontinuity in the material parameters. The behaviour of the scattered wave is thus determined by the nature of the surrounding fluid, the characteristics of the incident wave (namely wave vector and amplitude), the shape and the material properties of the obstacle. Our main interest is in the determination of influences of the obstacle on the scattered wave at large distances. The pertinent results are reviewed with the purpose of bringing into evidence those points that become crucial in performing the subsequent extension to the case of dissipative bodies. Proofs that are too technical and not strictly necessary to understand the contents of the chapter are omitted.

Assume that the obstacle occupies a convex, bounded, simply-connected, open domain D^- with regular boundary S, and that the open region D^+ exterior to S is occupied by the inviscid fluid. In the regions of the fluid - where no sources are acting - the excess pressure, the scalar potential, and the components of the velocity field of any time-harmonic wave, with frequency ω, obey the homogenous scalar Helmholtz equation

$$\Delta\phi + \kappa_L\phi = 0 \tag{1.1}$$

where $\kappa_L = k_L^2 = \omega^2/p_\rho$ is a positive constant. For definiteness we regard ϕ as a displacement potential; the associated velocity and pressure are then given by (3.3.11), (3.3.12). Consider the region D^+. In view of the linearity of the equations that govern wave propagation, (1.1) is to hold for the incident field ϕ^i, the scattered field ϕ and total field ϕ^+,

$$\phi^+ = \phi^i + \phi.$$

The incident wave is regarded as a datum. For definiteness, we take ϕ^i in the form of plane wave, that is

$$\phi^i = \Phi^i \exp[i(\mathbf{k}_L \cdot \mathbf{x} - \omega t)].$$

Therefore, we have to determine only one of the two fields ϕ^+, ϕ. Confine the attention on ϕ. Besides being a solution to the Helmholtz equation (1.1), the scattered field ϕ is required to satisfy suitable conditions at the surface S. Now, by (3.3.12), in a perfect fluid the pressure amplitude P is $\rho\omega^2$ times the potential amplitude Φ. Also, by (3.3.11), the velocity in the direction of the wave is $-i\omega$ times the derivative of ϕ in the direction of $\mathbf{n}$, viz $\mathbf{v}\cdot\mathbf{n} = -i\omega\partial\phi/\partial n$. In the limit case of a sound-soft obstacle the total pressure must vanish on the boundary and this results in

$$\phi = -\phi^i \qquad \text{on} \quad S$$

thus giving rise to a Dirichlet problem for the Helmholtz equation. If the obstacle is sound-hard the normal component of the velocity vanishes at S and hence

$$\frac{\partial\phi}{\partial n} = -\frac{\partial\phi^i}{\partial n} \qquad \text{on} \quad S.$$

In this case the scattered field is the solution to a Neumann problem. A more realistic treatment of the boundary conditions is thought to consist in the requirement that the normal velocity and the excess pressure are proportional at S, that is $\mathbf{v}\cdot\mathbf{n}+\chi\tilde{p}=0$. The coefficient χ is called the acoustic impedance of the obstacle and in general is a function defined on S. This approach gives rise to a boundary condition for ϕ^+ of the form

$$\frac{\partial\phi^+}{\partial n}+\alpha\phi^+=0 \qquad \text{on} \quad S.$$

Concerning the description of the scattered field, it is worth specifying that ϕ is generated by sources on S which model discontinuities of the material parameters. Accordingly the field ϕ is required to behave at infinity as an outgoing spherical wave or a radiation field; this is guaranteed by the well known Sommerfeld radiation condition

$$\frac{\partial\phi}{\partial R}-ik_L\phi=o(R^{-1}), \tag{1.2}$$

as $R:=|\mathbf{x}|\to\infty$, which is assumed to hold uniformly for all directions $\mathbf{x}/|\mathbf{x}|$. Really, the left side of (1.2) is identically zero in the case of spherical waves generated by a point source and it may be shown (cf. [159], p. 298) that the total outgoing energy flux through a sphere at infinity is positive if (1.2) holds.

Having characterized the relevant properties of the scattered field, we now proceed to the determination of an expression for ϕ as an integral over S involving the values of ϕ and its normal derivative. First we look for an integral representation of solutions to the Helmholtz equation (1.1), then we take into account the radiation condition (1.2) to find the required result for ϕ.

Denote by Ω a bounded connected domain in $\mathcal{E}^3$ with $\mathcal{C}^2$ boundary $\partial\Omega$. The second Green's theorem [126] states that for any pair of sufficiently regular functions $u(\mathbf{y})$ and $v(\mathbf{y})$ we have

$$\int_\Omega\{u\Delta v-v\Delta u\}dy=\int_{\partial\Omega}\left\{u\frac{\partial v}{\partial n}-v\frac{\partial u}{\partial n}\right\}da_y, \tag{1.3}$$

where $\mathbf{n}$ is the outward unit normal, dy and da_y denote the volume and surface elements. The theorem follows from the identity

$$u\Delta v-v\Delta u=\nabla\cdot(u\nabla v-v\nabla u)$$

through use of the divergence theorem. In the following we identify v with ϕ. Meanwhile the function u is identified with the Green's function g_L of the Helmholtz equation (1.1), namely the function satisfying

$$(\Delta+k_L^2)g_L=-4\pi\delta(\mathbf{y}-\mathbf{x}), \tag{1.4}$$

where $\delta(\mathbf{y}-\mathbf{x})$ is the three-dimensional Dirac's delta function. The explicit expression of g_L reads

$$g_L=\frac{\exp(ik_Lr)}{r} \tag{1.5}$$

where $r = |\mathbf{r}|$, $\mathbf{r} = \mathbf{y} - \mathbf{x}$. Substitute ϕ and g_L for v and u into (1.3). In view of (1.1) and (1.4) the volume integral reduces to integration of $4\pi\phi(\mathbf{y})\delta(\mathbf{y}-\mathbf{x})$, whence it follows that

$$\phi(\mathbf{x}) = \frac{1}{4\pi}\int_{\partial\Omega}\left\{g_L\frac{\partial\phi}{\partial n} - \phi\frac{\partial g_L}{\partial n}\right\}da_y \tag{1.6}$$

where $\mathbf{x}$ is interior to Ω and the notation da_y for the surface element reminds us that the integral is to be evaluated with respect to the vector variable $\mathbf{y}$.

An alternative proof of (1.6), not relying on the use of distributions, can be given as follows. Letting $\boldsymbol{\nu} = \mathbf{r}/r$ and taking the derivatives of g_L with respect to the variable $\mathbf{y}$ we have

$$\nabla g_L = \big(ik_L - \frac{1}{r}\big)g_L\boldsymbol{\nu}, \qquad \nabla\boldsymbol{\nu} = \frac{1}{r}(\mathbf{1} - \boldsymbol{\nu}\otimes\boldsymbol{\nu}). \tag{1.7}$$

Let $\nabla\otimes\nabla g_L$ be the tensor with components

$$(\nabla\otimes\nabla g_L)_{ij} = \partial_{y_i}\partial_{y_j}g_L;$$

the shorthand $\nabla\nabla g_l$ will be used for $\nabla\otimes\nabla g_l$ when no ambiguity arises. We have

$$\begin{aligned}\nabla\otimes\nabla g_L &= \frac{1}{r}\big(ik_L - \frac{1}{r}\big)g_L\,\mathbf{1} - \big(k_L^2 + \frac{3ik_L}{r} - \frac{3}{r^2}\big)\,g_L\,\boldsymbol{\nu}\otimes\boldsymbol{\nu}\\ &= -g_L\left[k_L^2\boldsymbol{\nu}\otimes\boldsymbol{\nu} + O(r^{-1})\right].\end{aligned} \tag{1.8}$$

It follows from (1.8) that g_L satisfies the Helmholtz equation provided $r \neq 0$, that is, $\mathbf{x} \neq \mathbf{y}$. For further reference we note that

$$\begin{aligned}&\nabla\otimes\nabla\otimes\nabla g_L\\ &= -g_L\big[\frac{3}{r}\big(k_L^2 + \frac{3ik_L}{r} - \frac{3}{r^2}\big)\,\mathrm{sym}(\boldsymbol{\nu}\otimes\mathbf{1}) + \big(ik_L^3 - \frac{6k_L^2}{r} - \frac{15ik_L}{r^2} + \frac{15}{r^3}\big)\boldsymbol{\nu}\otimes\boldsymbol{\nu}\otimes\boldsymbol{\nu}\big]\\ &\qquad = -g_L\left[ik_L^3\boldsymbol{\nu}\otimes\boldsymbol{\nu}\otimes\boldsymbol{\nu} + O(r^{-1})\right],\end{aligned} \tag{1.9}$$

where

$$[\mathrm{sym}(\boldsymbol{\nu}\otimes\mathbf{1})]_{ijk} = \frac{1}{3}(\delta_{ij}\nu_k + \delta_{jk}\nu_i + \delta_{ki}\nu_j).$$

Similarly, we find that

$$\nabla\otimes\nabla\otimes\nabla\otimes\nabla g_L = g_L\left[k_L^4\boldsymbol{\nu}\otimes\boldsymbol{\nu}\otimes\boldsymbol{\nu}\otimes\boldsymbol{\nu} + O(r^{-1})\right]. \tag{1.9'}$$

Now consider a constant $\epsilon > 0$ and let $\Omega_\epsilon = \{\mathbf{y} : |\mathbf{y}-\mathbf{x}| \leq \epsilon\}$ at a fixed interior point $\mathbf{x}$ of Ω; for a small enough ϵ we have $\Omega_\epsilon \subset \Omega$. Then apply the second Green's theorem in the form (1.3), where the domain Ω is to be replaced by $\Omega\setminus\Omega_\epsilon$ whereas v is identified with ϕ and u with g_L. The integrand in the volume integral is easily seen to vanish since ϕ and g_L obey Helmholtz equations (1.1) and (1.4); accordingly we are left with

$$\int_{\partial\Omega}\left\{g_L\frac{\partial\phi}{\partial n} - \phi\frac{\partial g_L}{\partial n}\right\}da_y + \int_{\partial\Omega_\epsilon}\left\{g_L\frac{\partial\phi}{\partial n} - \phi\frac{\partial g_L}{\partial n}\right\}da_y = 0.$$

On evaluating the normal derivative of g_L, taking the limit for $\epsilon \to 0$, and using the mean value theorem, we obtain that the integral over $\partial\Omega_\epsilon$ reduces to $-4\pi\phi(\mathbf{x})$, whence equation (1.6) follows.

Equation (1.6) shows that any solution to the Helmholtz equation can be represented as the combination of a single- and a double-layer surface potential defined on $\partial\Omega$ [52, 126]. In order to find a representation for the scattered field we regard Ω as the region bounded by S and by a spherical surface S_R of radius R centred at a point of D^- and containing the surface S. The analogue of (1.6) reads

$$\phi(\mathbf{x}) = \frac{1}{4\pi}\left(\int_S + \int_{S_R}\right)\left\{g_L\frac{\partial\phi}{\partial n} - \phi\frac{\partial g_L}{\partial n}\right\} da_y, \qquad \mathbf{x} \in \Omega,$$

$\mathbf{n}$ being the outward unit normal to Ω. Assuming that

$$\lim_{R\to\infty}\int_{S_R}\left\{g_L\frac{\partial\phi}{\partial n} - \phi\frac{\partial g_L}{\partial n}\right\} da_y = 0 \tag{1.10}$$

and reversing the orientation of the normal, so that $\mathbf{n}$ is now the outward normal to D^-, we find that ϕ is given by

$$\phi(\mathbf{x}) = \frac{1}{4\pi}\int_S\left(\phi\frac{\partial g_L}{\partial n} - g_L\frac{\partial\phi}{\partial n}\right) da_y. \tag{1.11}$$

The integral representation (1.11) holds for any $\mathbf{x}$ in the open domain D^+ exterior to S.

The limit (1.10) vanishes as a consequence of the radiation condition (1.2). To prove this we examine the asymptotic behaviour of the integrand, regarded as a function of $\mathbf{y}$, while $\mathbf{x}$ is kept fixed. First observe that, letting

$$\mathbf{y} = R\,\mathbf{n}$$

and $r = |\mathbf{y} - \mathbf{x}|$, we have

$$\frac{1}{r} = \frac{1}{R}[1 + O(R^{-1})] \qquad \frac{1}{r^m} = \frac{1}{R^m}[1 + O(R^{-1})], \tag{1.12}$$

where m denotes any positive integer. It follows that

$$\boldsymbol{\nu} = \frac{\mathbf{y} - \mathbf{x}}{r} = \frac{\mathbf{y} - \mathbf{x}}{R}\frac{R}{r} = \mathbf{n} + O(R^{-1}). \tag{1.13}$$

Substitution into the expression (1.5) provides

$$g_L = \frac{\exp(ik_L r)}{R}[1 + O(R^{-1})]. \tag{1.14}$$

Comparison with (1.7) yields

$$\frac{\partial g_L}{\partial R} = \frac{\partial g_L}{\partial n} = \mathbf{n}\cdot\nabla g_L = \big(ik_L - \frac{1}{r}\big)g_L\boldsymbol{\nu}\cdot\mathbf{n} = ik_L\frac{\exp(ik_L r)}{R}[1+O(R^{-1})].$$

Hence, by use of (1.14), it follows that

$$\frac{\partial g_L}{\partial R} - ik_L g_L = O(R^{-2}) \tag{1.15}$$

thus showing that g_L satisfies the radiation condition (1.2).

In connection with (1.10) we also need the asymptotic behaviour of ϕ, another quantity which enters the integral (1.10). Specifically, it follows from the radiation condition that

$$|\phi(\mathbf{y})| = O(R^{-1}). \tag{1.16}$$

To show this property we note that, because $\partial\phi/\partial R = \partial\phi/\partial n$, (1.2) yields

$$0 = \lim_{R\to\infty}\int_{S_R}\Big|\frac{\partial\phi}{\partial n} - ik_L\phi\Big|^2 da_y = \lim_{R\to\infty}\int_{S_R}\Big\{\Big|\frac{\partial\phi}{\partial n}\Big|^2 + k_L^2|\phi|^2 + ik_L\big(\phi^*\frac{\partial\phi}{\partial n} - \phi\frac{\partial\phi^*}{\partial n}\big)\Big\}da_y,$$

where * denotes the complex conjugate. Now we evaluate the contribution in braces through the use of (1.3), where Ω is identified with the domain $D^+(R)$ included between S and S_R, while u and v are replaced by ϕ^* and ϕ. In this way we find that

$$\Big(\int_S + \int_{S_R}\Big)\Big\{\phi^*\frac{\partial\phi}{\partial n} - \phi\frac{\partial\phi^*}{\partial n}\Big\}da_y = \int_{D^+(R)}\{\phi^*\Delta\phi - \phi\Delta\phi^*\}dy = 0,$$

where the last equality follows from the fact that both ϕ^* and ϕ satisfy (1.1) with a real κ_L. This in turn allows us to write

$$\begin{aligned}\lim_{R\to\infty}\Big(\int_{S_R}\Big|\frac{\partial\phi}{\partial n}\Big|^2 da_y + \int_{S_R}k_L^2|\phi|^2 da_y\Big) &= -ik_L\int_{S_R}\Big\{\phi^*\frac{\partial\phi}{\partial n} - \phi\frac{\partial\phi^*}{\partial n}\Big\}da_y\\ &= ik_L\int_S\Big\{\phi^*\frac{\partial\phi}{\partial n} - \phi\frac{\partial\phi^*}{\partial n}\Big\}da_y\end{aligned}$$

The integral over S is a finite quantity and so must be the limit of $\int_{S_R}(|\partial\phi/\partial n|^2 + k_L^2|\phi|^2)da_y$ as $R\to\infty$. Since the two integrals $\int_{S_R}|\partial\phi/\partial n|^2 da_y$, $\int_{S_R}k_L^2|\phi|^2 da_y$ are non-negative, each of them must be bounded as $R\to\infty$ and this means that (1.16) holds.

Coming back to the proof of (1.10), we rewrite the integrand as

$$g_L\frac{\partial\phi}{\partial n} - \phi\frac{\partial g_L}{\partial n} = g_L\big(\frac{\partial\phi}{\partial R} - ik_L\phi\big) - \phi\big(\frac{\partial g_L}{\partial R} - ik_L g_L\big).$$

Then comparison with (1.14), (1.16), and the radiation conditions (1.2) and (1.15) for ϕ and g_L shows that, in spherical coordinates θ and φ,

$$\lim_{R\to\infty}\int_{S_R}\Big\{g_L\frac{\partial\phi}{\partial n} - \phi\frac{\partial g_L}{\partial n}\Big\}R^2 d\theta d\varphi = 0,$$

which is the desired result.

The representation (1.11) is of fundamental importance in the analysis of the scalar theory of scattering. Here, though, we are content with quoting a few results that follow from (1.11) [52], without going into the details of the proofs. Slight changes in the proofs allow an extension of these conclusions to the case when the wavenumber k_L is complex-valued.

For any given exterior domain, solutions to the Helmholtz equation satisfying the Sommerfeld radiation condition are analytic functions of their independent variables and admit an expansion of the form

$$\phi(\mathbf{x}) = \frac{\exp ik_L R}{R} \sum_{n=0}^{\infty} \frac{F_n(\theta, \varphi)}{R^n}, \qquad R = |\mathbf{x}|,$$

which is absolutely and uniformly convergent with respect to the usual spherical coordinates. The representation is valid for $R \geq R_0$, where R_0 is the radius of a spherical surface containing the domain D of the obstacle.

As an immediate consequence it follows that the asymptotic behaviour of ϕ is given by

$$\phi(\mathbf{x}) = \frac{\exp ik_L R}{R} F_0(\theta, \varphi) + O(R^{-2})$$

where F_0 is called the far-field pattern or radiation pattern of ϕ. Moreover, substitution of the development in terms of the radial variable into the Helmholtz equation (1.1) shows that the coefficients F_n are determined in terms of the far field through the recurrence formula

$$2ik_L n F_n = n(n-1)F_{n-1} + \left[\frac{1}{\sin\theta}\frac{\partial}{\partial\theta}\left(\sin\theta\frac{\partial}{\partial\theta}\right) + \frac{1}{\sin^2\theta}\frac{\partial^2}{\partial\varphi^2}\right] F_{n-1}, \qquad n = 1, 2, \ldots.$$

This proves that the far-field pattern does completely determine solutions to the Helmholtz equation satisfying the radiation condition.

7.2 Integral representations for displacement

The procedures examined in the previous section are now extended to the case when the obstacle is immersed in a dissipative medium. The obstacle, whose properties and shape are known, occupies a domain D^- with boundary S. As before D^+ denotes the exterior domain of the solid matrix. The scattered field and the incident one, as well as their sum, obey the linearized equations of motion for a viscoelastic solid in D^+. Motivated by the results of the previous section, we look for an integral representation of the scattered field which in turn allows a detailed analysis of the imprint left by the obstacle on the field transmitted at infinity. This requires a preliminary discussion of integral representations for a general regular time-harmonic viscoelastic displacement fields.

The viscoelastic obstacle and the surrounding medium are taken as *homogeneous* and isotropic. The incoming wave perturbs an unstressed equilibrium placement which is chosen as reference. Accordingly, the body force is disregarded. The motion is governed by (the linear approximation)

$$\rho \ddot{\mathbf{u}} = \nabla \cdot \mathbf{T}$$

where ρ is the reference mass density and

$$\mathbf{T} = \lambda_0 \operatorname{tr}\mathbf{E}\,\mathbf{1} + 2\mu_0 \mathbf{E} + \int_0^\infty [\lambda' (\operatorname{tr}\mathbf{E}^t)\mathbf{1} + 2\mu' \mathbf{E}^t](s)\, ds\ .$$

Once the displacement $\mathbf{u}(\mathbf{x}, t)$ is determined, the actual mass density $\tilde{\rho}$ is given by $\tilde{\rho} = \rho(1 - \nabla \cdot \mathbf{u})$.

Suppose that $\mathbf{u}$ is time-harmonic, namely

$$\mathbf{u}(\mathbf{x}, t) = \mathbf{U}(\mathbf{x}) \exp(-i\omega t)$$

In this case the equation of motion takes the simplified form

$$\nabla \cdot \mathbf{T}[\mathbf{U}] + \rho\omega^2 \mathbf{U} = 0 \tag{2.1}$$

where the *stress operator* $\mathbf{T}[\mathbf{U}]$ is defined as a mapping of the (displacement) vector $\mathbf{U}$ into the (stress) tensor

$$\mathbf{T}[\mathbf{U}] = \lambda(\nabla \cdot \mathbf{U})\mathbf{1} + \mu(\nabla \mathbf{U} + \nabla \mathbf{U}^\dagger), \tag{2.2}$$

where λ and μ are the usual complex moduli. Whenever possible, we omit writing the common factor $\exp(-i\omega t)$. So $\mathbf{T}[\mathbf{U}]$ is the stress determined by $\mathbf{U}$, or by $\mathbf{u}$ if the factor $\exp(-i\omega t)$ is considered. In this sense the equation of motion can be written as

$$\Delta^\diamond \mathbf{U} + \rho\omega^2 \mathbf{U} = 0 \tag{2.3}$$

where $\Delta^\diamond$ is the operator given by

$$\Delta^\diamond \mathbf{U} := \nabla \cdot \mathbf{T}[\mathbf{U}] = \mu \Delta \mathbf{U} + (\mu + \lambda)\nabla(\nabla \cdot \mathbf{U}) \tag{2.4}$$

for any vector $\mathbf{U}$. Comparison with (1.1) shows that $\Delta^\diamond$ plays the same role as the Laplacian Δ in the scalar (perfect fluid) theory. This interpretation is enforced by the observation that $\Delta^\diamond$ reduces to Δ for $\lambda = -\mu = -1$. Notice also that equations (2.1)-(2.3) hold for time-harmonic waves in a viscoelastic fluid, provided λ and μ in (2.4) are replaced by $-i\omega\mu$ and $\rho p_\rho - i\omega\lambda$, cf. §3.3. The particular case of the elastic solid is obtained by merely letting λ and μ be real and independent of ω.

We now address the general problem of finding an integral representation for the solution to (2.1) or (2.3) in a homogenous domain. For the sake of generality and for future convenience we discuss the mathematical problem corresponding to a nonvanishing right-hand side, that is we consider the equation

$$\Delta^{\circ}\mathbf{U} + \rho\omega^2\mathbf{U} = -\rho\mathbf{f}; \tag{2.5}$$

$\mathbf{f}$ represents a "force term" and is given by a known vector function of the position $\mathbf{x}$ with compact support, possibly parameterized by ω. Paralleling the approach of the scalar theory, we have to find the analogue of the second Green's theorem, in the form of the so-called Betti's third formula [110, 111, 67, 55]. Then we need fundamental solutions of (2.5), namely the Green's displacement tensor [140]. The required integral representation will follow through a combination of these results.

For any regular vector field $\mathbf{V}(\mathbf{y})$ a direct check shows that the identity

$$\{\nabla\cdot\mathbf{T}[\mathbf{U}]\}\cdot\mathbf{V} = \nabla\cdot\{\mathbf{T}[\mathbf{U}]\,\mathbf{V}\} - \mathbf{T}[\mathbf{U}]\cdot\nabla\mathbf{V}$$

holds. Subtraction of the expression obtained by interchange of $\mathbf{U}$ with $\mathbf{V}$, and comparison with the definition (2.2) of the stress operator $\mathbf{T}[\cdot]$ yields

$$\{\nabla\cdot\mathbf{T}[\mathbf{U}]\}\cdot\mathbf{V} - \{\nabla\cdot\mathbf{T}[\mathbf{V}]\}\cdot\mathbf{U} = \nabla\cdot\{\mathbf{T}[\mathbf{U}]\,\mathbf{V}\} - \mathbf{T}[\mathbf{V}]\,\mathbf{U}\}.$$

Use of the Gauss theorem leads to the integral identity

$$\int_{\Omega}\{(\nabla\cdot\mathbf{T}[\mathbf{U}])\cdot\mathbf{V} - (\nabla\cdot\mathbf{T}[\mathbf{V}])\cdot\mathbf{U}\}dy = \int_{\partial\Omega}\{\mathbf{V}\cdot(\mathbf{T}[\mathbf{U}]\,\mathbf{n}) - \mathbf{U}\cdot(\mathbf{T}[\mathbf{V}]\,\mathbf{n})\}\,da_y,$$

that holds for λ and μ being dependent on the position. If, further, λ and μ are constant then we can write

$$\int_{\Omega}(\mathbf{V}\cdot\Delta^{\circ}\mathbf{U} - \mathbf{U}\cdot\Delta^{\circ}\mathbf{V})dy = \int_{\partial\Omega}\{\mathbf{V}\cdot(\mathbf{T}[\mathbf{U}]\,\mathbf{n}) - \mathbf{U}\cdot(\mathbf{T}[\mathbf{V}]\,\mathbf{n})\}\,da_y, \tag{2.6}$$

where Ω is a bounded connected domain with C^2 boundary $\partial\Omega$, $\mathbf{n}$ is the outward unit normal to $\partial\Omega$. To obtain the desired integral representation for the displacement field we have merely to identify the vector $\mathbf{U}$ with the viscoelastic displacement field satisfying (2.5), while $\mathbf{V}$ is to be replaced by a fundamental solution to (2.5), say $\mathcal{G}$, which constitutes the analogue of the Green's function g_L defined by (1.5). Thus, the determination of the integral representation is to be delayed until the explicit form of $\mathcal{G}$ has been discussed.

For any two points $\mathbf{x}, \mathbf{y} \in \mathcal{E}^3$, consider the *Green's displacement tensor* $\mathcal{G}$ [140], with values in Sym, defined as a solution to the differential equation

$$\Delta^{\circ}\mathcal{G}(\mathbf{y},\mathbf{x}) + \rho\omega^2\mathcal{G}(\mathbf{y},\mathbf{x}) = -\delta(\mathbf{y}-\mathbf{x})\,\mathbf{1}, \tag{2.7}$$

where $\mathbf{y}$ is the current position and $\mathbf{x}$ is fixed. Really, since Δ° operates on vectors, we should write (2.7) more properly as

$$\Delta^\circ \mathcal{G}\mathbf{e} + \rho\omega^2 \mathcal{G}\mathbf{e} = -\delta(\mathbf{y} - \mathbf{x})\,\mathbf{e},$$

for any constant vector $\mathbf{e}$. Letting $\mathbf{e}$ be the j-th unit vector we obtain the component form

$$\mu\Delta\mathcal{G}_{ij}(\mathbf{y},\mathbf{x}) + (\lambda + \mu)\partial_{y_i}\partial_{y_h}\mathcal{G}_{hj}(\mathbf{y},\mathbf{x}) + \rho\omega^2\mathcal{G}_{ij}(\mathbf{y},\mathbf{x}) = -\delta(\mathbf{y} - \mathbf{x})\,\delta_{ij}$$

whereby $\mathcal{G}_{ij}(\mathbf{y},\mathbf{x})$ represents the i-th component of the (complex) displacement field which arises at the point $\mathbf{y}$ of an infinite homogeneous isotropic body when it is acted upon by a unit "force" concentrated at the point $\mathbf{x}$ and parallel to the j-th direction.

The solution to (2.7) in the unbounded space $\mathcal{E}^3$ is well known when μ and λ are real-valued. It is expressed as a linear combination of the free-space Green's functions for longitudinal and transverse waves and of their derivatives. We observe that formally replacing the real parameters μ_0, λ_0 of the elastic case with μ, λ provides the corresponding results for the viscoelastic case. Yet, on ascertaining that this is really so, we deduce useful relations for later developments.

By analogy with (1.4) and (1.5) we say that the free-space Green's functions $g_L(r)$, $g_T(r)$ are defined as solutions to the Helmholtz equation with point-like sources,

$$(\Delta + \kappa_L)g_L = -4\pi\delta(\mathbf{y} - \mathbf{x}), \qquad (\Delta + \kappa_T)g_T = -4\pi\delta(\mathbf{y} - \mathbf{x}), \tag{2.8}$$

namely

$$g_L = \frac{\exp(ik_L r)}{r}, \qquad g_T = \frac{\exp(ik_T r)}{r}, \tag{2.9}$$

where $k_L^2 = \kappa_L$, $k_T^2 = \kappa_T$ and $r = |\mathbf{y} - \mathbf{x}|$. We observe that, κ_L and κ_T are complex quantities, parameterized by the frequency ω and the material parameters of the medium, while the constant entering (1.4) is real. Roughly, k_L and k_T are square roots of κ_L and κ_T and then are defined to within a factor -1. To resolve this ambiguity we have to consider the physical meaning of g_L and g_T. To fix ideas, let us consider g_L. Since the time dependence is through the factor $\exp(-i\omega t)$, whatever the choice of the sign for k_L, the Green's function g_L describes a scalar spherical wave centred at $\mathbf{x}$ of the form

$$g_L \exp(-i\omega t) = \frac{\exp[i(k_L r - \omega t)]}{r} = \exp(-\mathrm{Im}\, k_L\, r)\frac{\exp[i(\mathrm{Re}\, k_L\, r - \omega t)]}{r}$$

since the corresponding wave equation reduces to the Helmholtz equation for g_L. Clearly, the wave is outgoing if $\mathrm{Re}\, k_L > 0$ and is incoming otherwise. In scattering problems we are concerned with the representation of outgoing waves and henceforth we assume $\mathrm{Re}\, k_L = \mathrm{Re}\sqrt{\kappa_L} > 0$. The real and imaginary parts of κ_L,

$$\frac{\rho\omega^2}{|2\mu + \lambda|^2}[2(\mu_0 + \mu_c') + \lambda_0 + \lambda_c'], \qquad -\frac{\rho\omega^2}{|2\mu + \lambda|^2}[2\mu_s' + \lambda_s'],$$

are both positive (cf. (3.2.8) and (3.2.9)). Accordingly $\operatorname{Im} k_L > 0$ and this means that the amplitude decays as r grows, namely as the wave propagates. Conversely, if we make the choice $\operatorname{Re} k_L < 0$ then also $\operatorname{Im} k_L < 0$ and the amplitude of the related wave decreases as the wave comes from infinity, and propagates toward the origin. The same conclusion holds for the potential g_T. Similar remarks also hold for the displacement field associated with a scalar potential proportional to g_L.

We are now in a position to show that the tensor function

$$\mathcal{G}(\mathbf{y}, \mathbf{x}) = \frac{1}{4\pi \rho\omega^2}[k_T^2\, g_T\, \mathbf{1} - \nabla \otimes \nabla(g_L - g_T)], \tag{2.10}$$

with components

$$\mathcal{G}_{ij} = \frac{1}{4\pi\rho\omega^2}[k_T^2\, g_T\, \delta_{ij} - \partial_{y_i}\partial_{y_j}(g_L - g_T)]$$

is a solution to equation (2.7). To this end we observe that, upon substitution of (2.10), the left-hand side of (2.7) may be written as

$$\begin{aligned}\Delta^\circ \mathcal{G} + \rho\omega^2\mathcal{G} = \frac{1}{4\pi\rho\omega^2}\{&\mu k_T^2\, \Delta g_T\, \mathbf{1} - \mu\nabla \otimes \nabla(\Delta g_L - \Delta g_T) \\ &+ (\mu+\lambda)k_T^2 \nabla \otimes \nabla g_T - (\mu + \lambda)\nabla \otimes \nabla(\Delta g_L - \Delta g_T) \\ &+ \rho\omega^2[k_T^2\, \mathbf{1}\, g_T - \nabla \otimes \nabla(g_L - g_T)]\}.\end{aligned}$$

In view of (2.8) we find that

$$\begin{aligned}\Delta^\circ \mathcal{G} + \rho\omega^2\mathcal{G} = \frac{1}{4\pi\rho\omega^2}\{&[k_T^2(\rho\omega^2 - \mu k_T^2)\, g_T - 4\pi\mu k_T^2\, \delta(\mathbf{y}-\mathbf{x})]\,\mathbf{1} \\ &+ \nabla \otimes \nabla[((\lambda + 2\mu)k_L^2 - \rho\omega^2)\, g_L - (\mu k_T^2 - \rho\omega^2)\, g_T]\},\end{aligned}$$

which shows that the two conditions

$$k_T^2 = \frac{\rho\omega^2}{\mu}, \qquad k_L^2 = \frac{\rho\omega^2}{2\mu + \lambda}$$

are necessary and sufficient for $\mathcal{G}$ to be a solution to (2.7). The final conclusion is that the Green's tensor for the elastic (complex) displacement vector is given by (2.10), with g_L and g_T defined in (2.9). Incidentally, as is evident from the definition, $\mathcal{G}$ maps the vectors $\mathbf{y} - \mathbf{x}$ into the set of symmetric tensors.

On the basis of Betti's formula (2.6) and the properties of Green's tensor $\mathcal{G}$ we can obtain the integral representation for the displacement field $\mathbf{U}$. Actually, substitution of $\mathcal{G}$ for $\mathbf{V}$ into (2.6) yields, in component form,

$$\int_\Omega \Big\{\mathcal{G}_{ij}\, \Delta^\circ U_i - U_i\, \Delta^\circ \mathcal{G}_{ij}\Big\}\, dy = \int_{\partial\Omega} \Big\{\mathcal{G}_{ij}\, (\mathbf{T}[\mathbf{U}])_{ih}\, n_h - U_i\, (\mathbf{T}[\mathcal{G}])_{ihj}\, n_h\Big\}\, da_y$$

where

$$(\mathbf{T}[\mathcal{G}])_{ihj} = \lambda \partial_{y_k} \mathcal{G}_{kj}\, \delta_{ih} + \mu(\partial_{y_i} \mathcal{G}_{hj} + \partial_{y_h} \mathcal{G}_{ij}),$$

whence it follows that $\mathbf{T}[\mathcal{G}]$ is symmetric in the first two indices. On assuming that $\mathbf{U}$ is a solution of (2.5) and inserting (2.7) into the above identity we obtain

$$\int_\Omega \{ -\mathcal{G}_{ij}\rho f_i + U_i \delta(\mathbf{y}-\mathbf{x})\delta_{ij} \} dy = \int_{\partial\Omega} \{ \mathcal{G}_{ij}\,(\mathbf{T}[\mathbf{U}])_{ih}\, n_h - U_i\,(\mathbf{T}[\mathcal{G}])_{ihj}\, n_h \} da_y.$$

Using the symmetry of $\mathcal{G}$ we can express this result in vector form as

$$\int_\Omega \mathbf{U}(\mathbf{y})\,\delta(\mathbf{y}-\mathbf{x})\,dy = \int_\Omega \rho \mathcal{G} \mathbf{f}\, dy + \int_{\partial\Omega} \Big\{ \mathcal{G}\mathbf{t} - \lambda(\mathbf{U}\cdot\mathbf{n})\nabla\cdot\mathcal{G} - 2\mu\,[\mathrm{sym}(\mathbf{U}\otimes\mathbf{n})\nabla]\mathcal{G} \Big\}\, da_y, \quad (2.11)$$

where $\mathbf{t}$ is the surface traction that corresponds to the stress associated with the displacement $\mathbf{U}$, namely

$$\mathbf{t} = \mathbf{T}[\mathbf{U}]\mathbf{n}, \quad (2.12)$$

and use has been made of the relation

$$U_i\,(\mathbf{T}[\mathcal{G}])_{ihj}\, n_h = \lambda U_h n_h \partial_{y_k} \mathcal{G}_{kj} + \mu U_i n_h (\partial_{y_i} \mathcal{G}_{hj} + \partial_{y_h} \mathcal{G}_{ij})$$

or

$$\mathbf{U}\mathbf{T}[\mathcal{G}]\mathbf{n} = \lambda(\mathbf{U}\cdot\mathbf{n})\nabla\cdot\mathcal{G} + 2\mu\,[\mathrm{sym}(\mathbf{U}\otimes\mathbf{n})\nabla]\mathcal{G} \quad (2.13)$$

where

$$[\mathrm{sym}(\mathbf{U}\otimes\mathbf{n})\nabla]_k = \mathrm{sym}(\mathbf{U}\otimes\mathbf{n})_{ki}\partial_{y_i},$$

$$\big\{[\mathrm{sym}(\mathbf{U}\otimes\mathbf{n})\nabla]\mathcal{G}\big\}_j = [\mathrm{sym}(\mathbf{U}\otimes\mathbf{n})\nabla]_k \mathcal{G}_{kj}.$$

By analogy with the standard notation for the scalar theory, we identify Ω with the region $D^+(R)$ bounded internally by a closed surface S and externally by the spherical surface S_R, of radius R, centred at the origin. We observe that, as $R \to \infty$, $D^+(R) \to D^+$. For the time being, the behaviour of $\mathbf{t}$ and $\mathbf{U}$ at infinity is taken to satisfy

$$\lim_{R\to\infty} \int_{S_R} \Big\{ \mathcal{G}\mathbf{t} - \lambda(\mathbf{U}\cdot\mathbf{n})\nabla\cdot\mathcal{G} - 2\mu\,[\mathrm{sym}(\mathbf{U}\otimes\mathbf{n})\nabla]\mathcal{G} \Big\} da_y = 0, \quad (2.14)$$

which is the analogue of the radiation condition (1.2) for the scalar theory. Sufficient conditions for the validity of (2.14) are examined in the next section. Taking the limit of (2.11) as $R \to \infty$ we find that

$$\int_S \Big\{ \lambda(\mathbf{U}\cdot\mathbf{n})\nabla\cdot\mathcal{G} + 2\mu\,[\mathrm{sym}(\mathbf{U}\otimes\mathbf{n})\nabla]\mathcal{G} - \mathcal{G}\mathbf{t} \Big\}\, da_y + \int_{D^+} \rho\mathcal{G}\mathbf{f}\, dy = \begin{cases} \mathbf{U}(\mathbf{x}), & \mathbf{x}\in D^+ \\ 0, & \mathbf{x}\in D^- \end{cases}, \quad (2.15)$$

n being the outward normal to S. It is understood that the values of **U** and **t** at S are evaluated by taking the exterior limits. Alternative representations of **U** by means of an integral over S are possible [99], but will not be examined here.

By means of (2.11) and identifying Ω with the domain interior to the surface S we obtain

$$\int_S \left\{ \mathcal{G}\mathbf{t} - \lambda(\mathbf{U}\cdot\mathbf{n})\nabla\cdot\mathcal{G} - 2\mu\,[\mathrm{sym}(\mathbf{U}\otimes\mathbf{n})\nabla]\mathcal{G} \right\} da_y + \int_{D^-} \rho\mathcal{G}\mathbf{f}\, dy = \begin{cases} \mathbf{U}(\mathbf{x}), & \mathbf{x}\in D^- \\ 0, & \mathbf{x}\in D^+. \end{cases} \tag{2.16}$$

We have thus shown that the displacement field is completely determined by the values assumed by the traction and the displacement at the surface S, and by the force density **f**. Of course, the representations (2.15) and (2.16) also hold for solutions to the homogeneous equation (2.3), provided that we set $\mathbf{f} = 0$. In that case the solution has continuous partial derivatives of any order at points not belonging to S.

It is worth observing that the exponentials in the definitions (2.9) of g_L and g_T could be chosen with a minus sign, namely $g_L = \exp(-ik_L r)/r$. Under these conditions the previous considerations can be repeated, *mutatis mutandis*, for incoming waves.

7.3 Radiation condition

In this section we investigate sufficient conditions for the validity of (2.14). In this regard we restrict attention to a particular class of displacement fields as far as the behaviour at infinity is concerned. On the one hand the conditions to be imposed should be as weak as possible, that is, mild enough to be compatible with the existence of a measurable flux of outgoing radiation at infinity. On the other, they should be so chosen as to guarantee existence and uniqueness of the solution to the resulting boundary value problem. Starting from the standard radiation condition for the elastic displacement field we develop an analysis that provides useful relations for the asymptotic behaviour of displacement and traction. This analysis applies formally to both elasticity and viscoelasticity.

As in the previous sections, we consider a displacement field **U** in the region exterior to a connected domain D^- with boundary S. To state the radiation condition we need a well known representation theorem which shows that any regular elastic displacement field belonging to a monochromatic wave may be represented as the sum of an irrotational and a solenoidal vector, both of them obeying the Helmholtz equation and corresponding to a longitudinal and a transverse wave, respectively [111]. The radiation condition for elastic bodies will then be expressed in terms of these fields. For future reference the theorem is extended to viscoelasticity.

Any displacement field **U** may be expressed as

$$\mathbf{U} = \mathbf{U}_L + \mathbf{U}_T \tag{3.1}$$

where the irrotational and solenoidal fields $\mathbf{U}_L$ and $\mathbf{U}_T$ satisfy

$$\begin{cases} \Delta \mathbf{U}_L + \kappa_L \mathbf{U}_L = 0, & \nabla \times \mathbf{U}_L = 0 \\ \Delta \mathbf{U}_T + \kappa_T \mathbf{U}_T = 0, & \nabla \cdot \mathbf{U}_T = 0. \end{cases} \tag{3.2}$$

A constructive proof of this result is obtained by defining

$$\mathbf{U}_L = \frac{1}{\kappa_T - \kappa_L}(\Delta + \kappa_T)\mathbf{U}, \qquad \mathbf{U}_T = \frac{1}{\kappa_L - \kappa_T}(\Delta + \kappa_L)\mathbf{U}.$$

It is immediately seen by direct evaluation that (3.1) holds identically. To show that $\nabla \cdot \mathbf{U}_T = 0$ and $\nabla \times \mathbf{U}_L = 0$ it suffices to recall that $\mathbf{U}$ is a solution of (2.3), that is $\Delta^\diamond \mathbf{U} + \rho\omega^2 \mathbf{U} = 0$. Application of the divergence operator to (2.3) yields

$$\begin{aligned} 0 &= \nabla \cdot (\Delta^\diamond \mathbf{U} + \rho\omega^2 \mathbf{U}) = \nabla \cdot \left[\mu \, \Delta \mathbf{U} + (\lambda + \mu) \nabla(\nabla \cdot \mathbf{U}) + \rho\omega^2 \mathbf{U}\right] \\ &= (\lambda + 2\mu)\Delta(\nabla \cdot \mathbf{U}) + \rho\omega^2(\nabla \cdot \mathbf{U}) = (\lambda + 2\mu)\nabla \cdot \left[(\Delta + \kappa_L)\mathbf{U}\right] \\ &= (\lambda + 2\mu)(\kappa_L - \kappa_T)(\nabla \cdot \mathbf{U}_T) \end{aligned}$$

and hence $\mathbf{U}_T$ is a solenoidal field. Similarly application of the curl operator to (2.3) gives

$$\begin{aligned} 0 &= \nabla \times (\Delta^\diamond \mathbf{U} + \rho\omega^2 \mathbf{U}) = \mu \nabla \times [(\Delta + \kappa_T)\mathbf{U}] \\ &= \mu(\kappa_T - \kappa_L) \nabla \times \mathbf{U}_L, \end{aligned}$$

whence the irrotationality of $\mathbf{U}_L$. Also the Helmholtz equations (3.2) follow as a consequence of (2.3). By applying the operator $\Delta + \kappa_L$ to (2.3) and using the definition of $\mathbf{U}_T$ we obtain

$$\begin{aligned} 0 &= (\Delta + \kappa_L)(\Delta^\diamond \mathbf{U} + \rho\omega^2 \mathbf{U}) \\ &= (\Delta + \kappa_L)\left[\mu \, \Delta \mathbf{U} + (\lambda + \mu) \nabla(\nabla \cdot \mathbf{U}) + \rho\omega^2 \mathbf{U}\right] \\ &= \mu(\Delta + \kappa_L)(\Delta + \kappa_T)\mathbf{U} + (\lambda + \mu) \nabla\left[\nabla \cdot (\Delta + \kappa_L)\mathbf{U}\right] \\ &= \mu(\Delta + \kappa_L)(\Delta + \kappa_T)\mathbf{U} + (\lambda + \mu)(\kappa_L - \kappa_T)\nabla(\nabla \cdot \mathbf{U}_T). \end{aligned}$$

We already know that $\nabla \cdot \mathbf{U}_T = 0$. Moreover, by definition,

$$(\Delta + \kappa_T)\mathbf{U} = (\kappa_T - \kappa_L)\mathbf{U}_L, \qquad (\Delta + \kappa_L)\mathbf{U} = (\kappa_L - \kappa_T)\mathbf{U}_T.$$

Then $\mu(\Delta + \kappa_L)(\Delta + \kappa_T)\mathbf{U} = 0$ yields

$$\mu(\kappa_T - \kappa_L)(\Delta + \kappa_L)\mathbf{U}_L = 0 = \mu(\kappa_L - \kappa_T)(\Delta + \kappa_T)\mathbf{U}_T,$$

which completes the proof of (3.2). An alternative, non-constructive proof follows from the representation (3.1.11) provided that $\mathbf{U}_L$ and $\mathbf{U}_T$ are identified with ∇M and $\nabla \times \mathbf{W}$, respectively.

The following procedure is somewhat customary for elastic bodies in that it parallels that for the scalar theory of scattering (cf. [110]). By straightforward changes the procedure applies also to viscoelasticity. The detailed analysis provides useful relations about the asymptotic behaviour of displacement and traction. Further, it shows how the radiation condition is in a sense unrestrictive in the context of viscoelasticity.

Consider an exterior domain D^+ and let the vector $\mathbf{U}$ be decomposed into an irrotational and a solenoidal part, viz $\mathbf{U} = \mathbf{U}_L + \mathbf{U}_T$. If $\mathbf{U}$ describes the displacement of an elastic body then the radiation condition is expressed by the asymptotic behaviour of $\mathbf{U}_L$ and $\mathbf{U}_T$ as

$$\frac{\partial \mathbf{U}_L(\mathbf{x})}{\partial R} - ik_L \mathbf{U}_L(\mathbf{x}) = o(R^{-1}), \tag{3.3a}$$

$$\frac{\partial \mathbf{U}_T(\mathbf{x})}{\partial R} - ik_T \mathbf{U}_T(\mathbf{x}) = o(R^{-1}), \tag{3.3b}$$

where $R = |\mathbf{x}|$, uniformly for all directions. Of course, in elasticity k_L and k_T are real.

Really, in [111] it is also assumed that

$$\lim_{R\to\infty} \mathbf{U}_L(\mathbf{x}) = 0, \qquad \lim_{R\to\infty} \mathbf{U}_T(\mathbf{x}) = 0,$$

but these conditions seem redundant in that they follow from (3.3). Actually, each component of $\mathbf{U}_L$ and $\mathbf{U}_T$ is a solution of the scalar Helmholtz equation in the exterior domain D^+, as follows from projection of equations (3.2) on the coordinate axes, and satisfies a radiation condition of the form (1.2), which is obtained by projection of (3.3). Accordingly the results pertaining to the scalar theory apply to each component of $\mathbf{U}_L$ and $\mathbf{U}_T$. Our interest is in the integral representation (1.11) and the asymptotic behaviour (1.16); when expressed in compact notation they read

$$\mathbf{U}_L(\mathbf{x}) = O(R^{-1}), \qquad \mathbf{U}_L(\mathbf{x}) = \frac{1}{4\pi}\int_S \left\{ \mathbf{U}_L(\mathbf{y})\, \frac{\partial g_L(r)}{\partial n} - g_L(r)\, \frac{\partial \mathbf{U}_L(\mathbf{y})}{\partial n} \right\} da_y, \tag{3.4a}$$

$$\mathbf{U}_T(\mathbf{x}) = O(R^{-1}), \qquad \mathbf{U}_T(\mathbf{x}) = \frac{1}{4\pi}\int_S \left\{ \mathbf{U}_T(\mathbf{y})\, \frac{\partial g_T(r)}{\partial n} - g_T(r)\, \frac{\partial \mathbf{U}_T(\mathbf{y})}{\partial n} \right\} da_y. \tag{3.4b}$$

Hence $\mathbf{U}_L$ and $\mathbf{U}_T$ vanish as $R \to \infty$.

Henceforth the dependences on $\mathbf{y}$ and r are not indicated, it being understood that the integrands in (3.4) are evaluated at S; the dependence on $\mathbf{x}$ is through the argument r of g_L and g_T. As shown by (3.4) the radiation condition gives us information on the representations and the asymptotic properties of the irrotational and solenoidal parts of $\mathbf{U}$. We also analyse the asymptotic behaviour of the gradients of $\mathbf{U}_L$ and $\mathbf{U}_T$. Combining these results we find the expression of the traction $\mathbf{t}$ over the sphere at infinity and this allows us to understand better the limit (2.14).

For definiteness we investigate the irrotational field $\mathbf{U}_L$ only; the pertinent results can be extended to $\mathbf{U}_T$ and then to $\mathbf{U}$. It is apparent from the representation (3.4a) that $\mathbf{U}_L$

depends on $\mathbf{x}$ through the Green's function g_L and its normal derivative $\partial g_L/\partial n = \mathbf{n}\cdot\nabla_{\mathbf{y}}\, g_L$. Since $g_L = g_L(r) = g_L(|\mathbf{y}-\mathbf{x}|)$ a suffix $\mathbf{y}$ or $\mathbf{x}$ is appended in the evaluation of the derivatives to indicate which of the two variables is being operated on. Obviously, the dependence on r implies that $\nabla_{\mathbf{x}}\, g_L(r) = -\nabla_{\mathbf{y}}\, g_L(r)$.

The analysis of the asymptotic properties has many points in common with that performed for the scalar theory in §7.1. A trivial difference is that here we have $|\mathbf{x}| = R \to \infty$, instead of $|\mathbf{y}| = R \to \infty$. Accordingly, when no ambiguity arises, ∇ stands for $\nabla_{\mathbf{x}}$. For example,

$$\nabla R = \mathbf{x}/|\mathbf{x}| =: \hat{\mathbf{x}}.$$

By the asymptotic expansions (1.12) we see that $1/r^n$ equals $1/R^n$ plus higher order terms in $1/R$. Equation (1.13) is then changed to

$$\boldsymbol{\nu} = \frac{\mathbf{y}-\mathbf{x}}{R}\frac{R}{r} = -\hat{\mathbf{x}} + O(R^{-1}).$$

By (1.14), the leading term at infinity in the expression of g_L is $\exp(ik_L r)/R$. As regards the derivatives of g_L, comparison of (1.7) and (1.8) and account of the previous observations show that

$$\nabla_{\mathbf{x}}\, g_L = ik_L \frac{\exp(ik_L r)}{R}\hat{\mathbf{x}} + O(R^{-2}),$$

$$\nabla_{\mathbf{x}} \otimes \nabla_{\mathbf{y}}\, g_L = k_L^2 \frac{\exp(ik_L r)}{R}\hat{\mathbf{x}}\otimes\hat{\mathbf{x}} + O(R^{-2}).$$

We now come to the asymptotic analysis of $\mathbf{U}_L$ and its derivatives. Substitution of $\nabla_{\mathbf{x}} g_L$ into the representation (3.4a) of $\mathbf{U}_L$ yields

$$\mathbf{U}_L(\mathbf{x}) = -\frac{1}{4\pi R}\int_S \left\{ ik_L\,[\hat{\mathbf{x}}\cdot\mathbf{n}]\,\mathbf{U}_L + \frac{\partial \mathbf{U}_L}{\partial n}\right\} \exp(ik_L r)\, da_y + O(R^{-2}). \tag{3.5}$$

Similarly, substitution of $\nabla_{\mathbf{x}} g_L$ and $\nabla_{\mathbf{x}} \otimes \nabla_{\mathbf{y}} g_L$ into (3.4a) gives

$$\partial_{x_k} \mathbf{U}_L(\mathbf{x}) = \frac{1}{4\pi}\int_S \left\{ \mathbf{U}_L\, n_j\, \partial_{x_k}\partial_{y_j} g_L - \partial_{x_k} g_L \frac{\partial \mathbf{U}_L}{\partial n}\right\} da_y.$$

Then we can write the asymptotic expression

$$\nabla\otimes\mathbf{U}_L(\mathbf{x}) = \frac{1}{4\pi R}\int_S \left\{ k_L^2\,(\hat{\mathbf{x}}\cdot\mathbf{n})\,\hat{\mathbf{x}}\otimes\mathbf{U}_L - ik_L\,\hat{\mathbf{x}}\otimes\frac{\partial\mathbf{U}_L}{\partial n}\right\} \exp(ik_L r)\, da_y + O(R^{-2}).$$

Accordingly, comparison with the asymptotic form of $\mathbf{U}_L$ and some rearrangement yield

$$\nabla\otimes\mathbf{U}_L = ik_L\,\hat{\mathbf{x}}\otimes\mathbf{U}_L + O(R^{-2}). \tag{3.6}$$

On observing that $\partial\mathbf{U}_L/\partial R = (\hat{\mathbf{x}}\cdot\nabla_{\mathbf{x}})\mathbf{U}_L$ we also find from (3.6)

$$\frac{\partial\mathbf{U}_L(\mathbf{x})}{\partial R} - ik_L\,\mathbf{U}_L(\mathbf{x}) = O(R^{-2})\,. \tag{3.7}$$

On the basis of the representation (3.6) for the gradient of $\mathbf{U}_L$ it follows that the condition $\nabla \times \mathbf{U}_L = 0$ leads to

$$\hat{\mathbf{x}} \times \mathbf{U}_L = O(R^{-2}), \tag{3.8}$$

thus showing that the component of $\mathbf{U}_L$, normal to the radial direction, falls off at infinity as $O(R^{-2})$ although $\mathbf{U}_L$ is $O(R^{-1})$. Moreover, by (3.6), a direct calculation yields

$$\nabla \cdot \mathbf{U}_L - ik_L(\hat{\mathbf{x}} \cdot \mathbf{U}_L) = O(R^{-2}). \tag{3.9}$$

We now establish the analogous results for $\mathbf{U}_T$. On paralleling the above procedure we find that

$$\nabla \otimes \mathbf{U}_T = ik_T\, \hat{\mathbf{x}} \otimes \mathbf{U}_T + O(R^{-2}), \tag{3.10}$$

$$\frac{\partial \mathbf{U}_T(\mathbf{x})}{\partial R} - ik_T \mathbf{U}_T(\mathbf{x}) = O(R^{-2})\,. \tag{3.11}$$

Then the condition $\nabla \cdot \mathbf{U}_T = 0$ and the representation (3.10) imply that

$$\hat{\mathbf{x}} \cdot \mathbf{U}_T = O(R^{-2}) \tag{3.12}$$

showing that at large distances the radial component of $\mathbf{U}_T$ decreases as $O(R^{-2})$ while $\mathbf{U}_T$ is $O(R^{-1})$. In essence this means that $\mathbf{U}_T$ is asymptotically tangent to the spherical surfaces centred at the origin. Finally, evaluation of the curl of $\mathbf{U}_T$ and comparison with (3.10) gives

$$\nabla \times \mathbf{U}_T - ik_T(\hat{\mathbf{x}} \times \mathbf{U}_T) = O(R^{-2}), \tag{3.13}$$

A more detailed discussion of the asymptotic relations between $\mathbf{U}_L$, $\mathbf{U}_T$, and $\mathbf{U}$ can be given as follows. Recall the vector identity

$$\mathbf{A} = (\hat{\mathbf{x}} \cdot \mathbf{A})\hat{\mathbf{x}} + (\hat{\mathbf{x}} \times \mathbf{A}) \times \hat{\mathbf{x}},$$

for any vector $\mathbf{A}$ and unit vector $\hat{\mathbf{x}}$. The choice $\mathbf{A} = \mathbf{U}_L$ and use of (3.8) lead to

$$\mathbf{U}_L = (\mathbf{U}_L \cdot \hat{\mathbf{x}})\hat{\mathbf{x}} + O(R^{-2}), \tag{3.14}$$

showing that $\mathbf{U}_L$ is asymptotically radial. Similarly, on setting $\mathbf{A} = \mathbf{U}_T$ and comparing with (3.12) we find

$$\mathbf{U}_T = (\hat{\mathbf{x}} \times \mathbf{U}_T) \times \hat{\mathbf{x}} + O(R^{-2}), \tag{3.15}$$

which means that, at large distances from the origin, $\mathbf{U}_T$ is orthogonal to the radial direction. Finally, substitution of $\mathbf{U}$ for $\mathbf{A}$ in the left-hand side and of $\mathbf{U}_L + \mathbf{U}_T$ in the right side leads to

$$\mathbf{U} = (\hat{\mathbf{x}} \cdot \mathbf{U}_L)\hat{\mathbf{x}} + (\hat{\mathbf{x}} \times \mathbf{U}_T) \times \hat{\mathbf{x}} + O(R^{-2}). \tag{3.16}$$

This shows that the irrotational and solenoidal fields $\mathbf{U}_L$ and $\mathbf{U}_T$ determine the leading contributions to the radial and transverse components of $\mathbf{U}$, respectively.

In order to evaluate the asymptotic expression of the traction we now consider the surface S_R of the sphere at infinity. In the present notation the outward normal is simply the radial vector $\hat{\mathbf{x}}$ and then we have

$$\mathbf{t} = \mathbf{T}[\mathbf{U}]\hat{\mathbf{x}} = \mathbf{T}[\mathbf{U}_L]\hat{\mathbf{x}} + \mathbf{T}[\mathbf{U}_T]\hat{\mathbf{x}} =: \mathbf{t}_L + \mathbf{t}_T,$$

with obvious meanings of the symbols. By using well-known identities we can write the stress operator $\mathbf{T}[\mathbf{U}]$, defined in (2.2), as

$$\mathbf{T}[\mathbf{U}] = \lambda(\nabla \cdot \mathbf{U})\mathbf{1} + 2\mu\nabla\mathbf{U}^\dagger + \mu\nabla \times \mathbf{U},$$

whence it follows the general form of the traction as

$$\mathbf{t} = \lambda(\nabla \cdot \mathbf{U})\,\hat{\mathbf{x}} + 2\mu\,\partial\mathbf{U}/\partial R + \mu\,\hat{\mathbf{x}} \times (\nabla \times \mathbf{U}).$$

By specializing to $\mathbf{t}_L$ and recalling that $\nabla \times \mathbf{U}_L = 0$ we find

$$\mathbf{t}_L = \lambda(\nabla \cdot \mathbf{U}_L)\,\hat{\mathbf{x}} + 2\mu\,\partial\mathbf{U}_L/\partial R.$$

Then we observe that, in view of (3.9) and (3.14),

$$(\nabla \cdot \mathbf{U}_L)\hat{\mathbf{x}} = ik_L(\hat{\mathbf{x}} \cdot \mathbf{U}_L)\hat{\mathbf{x}} + O(R^{-2}) = ik_L\mathbf{U}_L + O(R^{-2}).$$

Substitution into the expression of $\mathbf{t}_L$ and account of (3.7) leads to

$$\mathbf{t}_L = ik_L(\lambda + 2\mu)\mathbf{U}_L + O(R^{-2}).$$

Similarly, because $\nabla \cdot \mathbf{U}_T = 0$, we find

$$\mathbf{t}_T = 2\mu\,\partial\mathbf{U}_T/\partial R + \mu\,\hat{\mathbf{x}} \times (\nabla \times \mathbf{U}_T).$$

In view of (3.12) and (3.13) we have

$$\hat{\mathbf{x}} \times (\nabla \times \mathbf{U}_T) = ik_T\,\hat{\mathbf{x}} \times (\hat{\mathbf{x}} \times \mathbf{U}_T) + O(R^{-2}) = -ik_T\mathbf{U}_T + O(R^{-2}).$$

Therefore, use of (3.11) leads to

$$\mathbf{t}_T = ik_T\mu\mathbf{U}_T + O(R^{-2}).$$

Since $\mathbf{t} = \mathbf{t}_L + \mathbf{t}_T$, we obtain the asymptotic expression

$$\mathbf{t} = ik_L(\lambda + 2\mu)\mathbf{U}_L + ik_T\mu\mathbf{U}_T + O(R^{-2}). \tag{3.17}$$

The results (3.16) and (3.17) hold for elastic displacements obeying the radiation condition.

To complete the scheme we have to determine the asymptotic behaviour of $\mathcal{G}$ and $\nabla\mathcal{G}$. To this end it is convenient to consider representations that bring into evidence the dependence on $r = |\mathbf{y} - \mathbf{x}|$, r being the variable that eventually is taken to go to infinity. For later reference we let k_L and k_T be complex-valued. We start from the definition (2.10) of $\mathcal{G}$, namely

$$\mathcal{G}(\mathbf{y},\mathbf{x}) = \frac{1}{4\pi\rho\omega^2}[k_T^2\, g_T\, \mathbf{1} - \nabla\otimes\nabla(g_L - g_T)].$$

In view of the asymptotic behaviour (1.8) for $\nabla\otimes\nabla g_L$ we have

$$\mathcal{G}(\mathbf{y},\mathbf{x}) = \frac{1}{4\pi\rho\omega^2}\left\{g_T\, k_T^2\left[\mathbf{1} - \boldsymbol{\nu}\otimes\boldsymbol{\nu} + O(r^{-1})\right] + g_L k_L^2\left[\boldsymbol{\nu}\otimes\boldsymbol{\nu} + O(r^{-1})\right]\right\}. \tag{3.18}$$

Consider again the definition of $\mathcal{G}$ and evaluate $\nabla\mathcal{G}$, $\nabla\otimes\nabla\mathcal{G}$ to obtain

$$\nabla\mathcal{G}(\mathbf{y},\mathbf{x}) = \frac{1}{4\pi\rho\omega^2}[k_T^2\,\nabla g_T\otimes\mathbf{1} - \nabla\otimes\nabla\otimes\nabla(g_L - g_T)],$$

$$\nabla\otimes\nabla\mathcal{G}(\mathbf{y},\mathbf{x}) = \frac{1}{4\pi\rho\omega^2}[k_T^2\,\nabla\otimes\nabla g_T\otimes\mathbf{1} - \nabla\otimes\nabla\otimes\nabla\otimes\nabla(g_L - g_T)].$$

Comparison with (1.9) and (1.9') yields, asymptotically,

$$\nabla\mathcal{G} = \frac{1}{4\pi\rho\omega^2}\left\{ik_T^3 g_T[\boldsymbol{\nu}\otimes\mathbf{1} + O(r^{-1})] + i(k_L^3 g_L - k_T^3 g_T)[\boldsymbol{\nu}\otimes\boldsymbol{\nu}\otimes\boldsymbol{\nu} + O(r^{-1})]\right\} \tag{3.19}$$

$$\nabla\otimes\nabla\mathcal{G} = \frac{1}{4\pi\rho\omega^2}\left\{-k_T^4 g_T[\boldsymbol{\nu}\otimes\boldsymbol{\nu}\otimes\mathbf{1} + O(r^{-1})] + (k_T^4 g_T - k_L^4 g_L)[\boldsymbol{\nu}\otimes\boldsymbol{\nu}\otimes\boldsymbol{\nu}\otimes\boldsymbol{\nu} + O(r^{-1})]\right\}. \tag{3.20}$$

Now that the asymptotic behaviour of $\mathbf{U}$, $\mathbf{t}$, $\mathcal{G}$ and $\nabla\otimes\mathcal{G}$ is known, we address attention to the proof of the condition

$$\lim_{R\to\infty}\int_{S_R}\{\mathcal{G}\mathbf{t} - \lambda(\mathbf{U}\cdot\mathbf{n})\nabla\cdot\mathcal{G} - 2\mu[\mathrm{sym}(\mathbf{U}\otimes\mathbf{n})\nabla]\mathcal{G}\}\, da_y = 0 \tag{3.21}$$

in the elastic case; here S_R has the usual meaning while $\mathbf{n}$ is the outward unit normal.

To prove the validity of (3.21) we show that the integrand is $O(R^{-3})$; since the surface element da_y is proportional to R^2, it follows that if the integrand is $O(R^{-3})$ then the integral over S_R tends to zero as $R\to\infty$. Now we let $\mathbf{y}$ belong to the spherical surface which is allowed to go to infinity, while $\mathbf{x}$ is kept fixed. Therefore we have $|\mathbf{y}| = R$ at S_R and the correspondent of the unit vector $\hat{\mathbf{x}}$ is the radial normal $\mathbf{n} = \mathbf{y}/R$. For example, equation (3.14) changes to

$$\mathbf{U}_L - (\mathbf{U}_L\cdot\mathbf{n})\mathbf{n} = O(R^{-2}),$$

while (3.12) now reads $\mathbf{U}_T\cdot\mathbf{n} = O(R^{-2})$. Further we have $\mathbf{U}_L = O(R^{-1})$ and $\mathbf{U}_T = O(R^{-1})$ whence $\mathbf{U} = \mathbf{U}_L + \mathbf{U}_T = O(R^{-1})$ and $\mathbf{t} = O(R^{-1})$, the last relation being a consequence of (3.17).

The asymptotic expressions (3.18) and (3.19) of $\mathcal{G}$ and $\nabla\mathcal{G}$, are given in terms of the variable $r = |\mathbf{y} - \mathbf{x}|$; a more convenient form should involve $R = |\mathbf{y}|$. In this regard we observe that the relations (1.12) to (1.14) apply and, in particular, $\boldsymbol{\nu} = \mathbf{n} + O(R^{-1})$. Therefore we have

$$\mathcal{G} = \frac{1}{4\pi\rho\omega^2 R}\left\{\exp(ik_T r)\, k_T^2 [\mathbf{1} - \mathbf{n}\otimes\mathbf{n} + O(R^{-1})] + \exp(ik_L r) k_L^2 \left[\mathbf{n}\otimes\mathbf{n} + O(R^{-1})\right]\right\}, \tag{3.22}$$

$$\begin{aligned}\nabla\mathcal{G} = \frac{1}{4\pi\rho\omega^2 R}\{&ik_T^3 \exp(ik_T r)\left[\mathbf{n}\otimes\mathbf{1} + O(R^{-1})\right] \\ &+ i[k_L^3 \exp(ik_L r) - k_T^3 \exp(ik_T r)]\left[\mathbf{n}\otimes\mathbf{n}\otimes\mathbf{n} + O(R^{-1})\right]\},\end{aligned} \tag{3.23}$$

that hold also for complex k_L and k_T. Back to elastic bodies, (3.17) and the canonical splitting of $\mathbf{U}$ allow us to find that

$$\mathcal{G}\mathbf{t} = \frac{1}{4\pi\rho\omega^2 R}\left[i\mu k_T^3 \exp(ik_T r)\mathbf{U}_T + i(\lambda + 2\mu)k_L^3 \exp(ik_L r)(\mathbf{U}_L\cdot\mathbf{n})\mathbf{n}\right] + O(R^{-3}),$$

$$\lambda(\mathbf{U}\cdot\mathbf{n})\nabla\cdot\mathcal{G} = \frac{i\lambda}{4\pi\rho\omega^2 R} k_L^3 \exp(ik_L r)(\mathbf{U}_L\cdot\mathbf{n})\mathbf{n} + O(R^{-3}),$$

$$\begin{aligned}2\mu[\mathrm{sym}(\mathbf{U}\otimes\mathbf{n})\nabla]\mathcal{G} &= \frac{\mu}{4\pi\rho\omega^2 R}\{ik_T^3 \exp(ik_T r)[(\mathbf{U}_L\cdot\mathbf{n})\mathbf{n} + \mathbf{U}_L + \mathbf{U}_T] \\ &\qquad + 2i[k_L^3 \exp(ik_L r) - k_T^3 \exp(ik_T r)](\mathbf{U}_L\cdot\mathbf{n})\mathbf{n}\} + O(R^{-3}) \\ &= \frac{\mu}{4\pi\rho\omega^2 R}\{ik_T^3 \exp(ik_T r)[\mathbf{U}_L - (\mathbf{U}_L\cdot\mathbf{n})\mathbf{n}] \\ &\qquad + ik_T^3 \exp(ik_T r)\mathbf{U}_T + 2ik_L^3 \exp(ik_L r)(\mathbf{U}_L\cdot\mathbf{n})\mathbf{n}\} + O(R^{-3}).\end{aligned}$$

Combining these three expressions we obtain the desired result

$$\mathcal{G}\mathbf{t} - \lambda(\mathbf{U}\cdot\mathbf{n})\nabla\cdot\mathcal{G} - 2\mu[\mathrm{sym}(\mathbf{U}\otimes\mathbf{n})\nabla]\mathcal{G} = O(R^{-3})$$

whence (3.21) follows.

Also to establish a connection with other formulations, it is worth considering an alternative approach that yields sufficient conditions for the vanishing of the flux over the sphere at infinity. Substitute the asymptotic expressions (3.22) and (3.23) for $\mathcal{G}$ and $\nabla\mathcal{G}$ into the integrand in (3.21), where $\mathbf{U}$ and $\mathbf{t}$ are regarded as given, but without specifying their asymptotic dependence on R. Explicit calculations show that

$$\mathcal{G}\mathbf{t} \simeq \frac{1}{4\pi\rho\omega^2 R}\{k_T^2 \exp(ik_T r)[\mathbf{t} - (\mathbf{t}\cdot\mathbf{n})\mathbf{n}] + k_L^2 \exp(ik_L r)(\mathbf{t}\cdot\mathbf{n})\mathbf{n}\}$$

$$\nabla\cdot\mathcal{G} \simeq \frac{ik_L^3}{4\pi\rho\omega^2 R}\exp(ik_L r)\mathbf{n},$$

$$[\text{sym}(\mathbf{U} \otimes \mathbf{n})\nabla]\boldsymbol{\mathcal{G}} \simeq \frac{1}{4\pi\rho\omega^2 R}\{ik_T^3 \exp(ik_T r)[(\mathbf{U}\cdot\mathbf{n})\mathbf{n} + \mathbf{U}] + i[k_L^3 \exp(ik_L r) - k_T^3 \exp(ik_T r)]2(\mathbf{U}\cdot\mathbf{n})\mathbf{n}\};$$

here the symbol $\simeq$ means equality to within terms in $\mathbf{t}$, $\mathbf{U}$ times $O(R^{-2})$. It follows that

$$\boldsymbol{\mathcal{G}}\mathbf{t} - \lambda(\mathbf{U}\cdot\mathbf{n})\nabla\cdot\boldsymbol{\mathcal{G}} - 2\mu[\text{sym}(\mathbf{U}\otimes\mathbf{n})\nabla]\boldsymbol{\mathcal{G}} \simeq \frac{1}{4\pi\omega^2 R}\{k_T^2 \exp(ik_T r)\big[\mathbf{t} - i\mu k_T\mathbf{U} - \mathbf{n}\cdot(\mathbf{t} - i\mu k_T\mathbf{U})\,\mathbf{n}\big] + k_L^2 \exp(ik_L r)\,\mathbf{n}\cdot\big[\mathbf{t} - i(\lambda+2\mu)k_L\mathbf{U}\big]\;\mathbf{n}\}. \quad (3.24)$$

If the right side of (3.24) is $o(R^{-2})$, that is the expression in braces is $o(R^{-1})$, then the integral over S_R is to be disregarded. A sufficient condition for the vanishing of the limit (3.21) is given by

$$\mathbf{t} - i\mu k_T\mathbf{U} - \mathbf{n}\cdot(\mathbf{t} - i\mu k_T\mathbf{U})\,\mathbf{n} = o(R^{-1}), \quad (3.25)$$

$$\mathbf{n}\cdot\big[\mathbf{t} - i(\lambda+2\mu)k_L\,\mathbf{U})\big]\;\mathbf{n} = o(R^{-1}), \quad (3.26)$$

along with $\mathbf{t}, \mathbf{U} = O(R^{-1})$. By analogy with [98] it follows that the boundedness of $R\mathbf{t}$ and $R\mathbf{U}$ is a consequence of (3.25) and (3.26). Then, apart from the change in sign $ik_T \to -ik_T$ and $ik_L \to -ik_L$, which is due to the choice $\exp(-i\omega t)$ in the time-harmonic dependence, addition of (3.25) and (3.26) yields the radiation condition employed by Jones [98]. However, it is worth remarking that the two vectors (3.25) and (3.26) are perpendicular and parallel to the normal $\mathbf{n}$, respectively, that is, they are perpendicular to each other. Thus the simultaneous validity of (3.25) and (3.26) is equivalent to the statement that the limit of their sum vanishes. But when we take the sum the contributions $(\mathbf{t}\cdot\mathbf{n})\mathbf{n}$ cancel and we obtain Jones' results.

Look specifically at the viscoelastic scattering. The conditions (3.25) and (3.26) are too restrictive. For, in terms of spherical coordinates R, θ, ϕ, the leading contribution to the integrand of the integral over the sphere at infinity is given by the right side of (3.24) times $R^2\,d\theta\,d\phi$. Its asymptotic behaviour is governed by the factors

$$R\exp(ik_L r), \qquad R\exp(ik_T r).$$

Now, the thermodynamic restrictions (3.2.8), (3.2.9) and the fact that the scattered wave is outgoing imply that $\text{Im}\,k_L$, $\text{Im}\,k_T > 0$ and then

$$R\exp(ik_L r) = \exp(i\,\text{Re}k_L r)\frac{\exp(-\text{Im}k_L r)}{1/R} \to 0 \quad \text{as } R\to\infty,$$

and the same for $\exp(ik_T r)$. Along with the fact that exponentials can be factorized also in the higher-order terms, this proves that, if the surrounding body is viscoelastic, the limit (3.21) holds provided only that $\mathbf{U}$ and $\mathbf{t}$ are bounded, not necessarily infinitesimal.

Henceforth it is understood that the radiation condition is given by (3.3), or (3.25) and (3.26), if the solid is elastic, and by the boundedness at infinity of $\mathbf{U}$ and $\mathbf{t}$ if the solid is viscoelastic.

7.4 The scattered field

Upon understanding that the time dependence is $\exp(-i\omega t)$, we consider an incident field $\mathbf{U}^i(\mathbf{x})$ impinging on the obstacle which is bounded by the closed surface S. We denote by $\mathbf{U}$ the scattered field and by

$$\mathbf{U}^+ = \mathbf{U}^i + \mathbf{U}$$

the total (scattered plus incident) exterior field. Properties of $\mathbf{U}$ are now determined through the integral representation for the exterior field.

As usual, the incident field $\mathbf{U}^i$ is regarded as a solution to the balance equation (2.1) for viscoelastic displacements generated by sources at infinity. The field $\mathbf{U}^i$ is also taken to satisfy the relation

$$\int_S \{\boldsymbol{\mathcal{G}}\mathbf{t}^i - \lambda(\mathbf{U}^i\cdot\mathbf{n})\nabla\cdot\boldsymbol{\mathcal{G}} - 2\mu[\mathrm{sym}(\mathbf{U}^i\otimes\mathbf{n})\nabla]\boldsymbol{\mathcal{G}}\}\, da_y = 0,$$

at S. This condition follows from the assumption that the incident field exists for all times as though the obstacle were not in place. This statement is not obvious and sometimes (cf. [15]) is regarded as "in some sense unphysical". Actually, we admit that $\mathbf{U}^i$ may be extended to D^- by regarding D^- as filled by the same material as is D^+ and letting $\mathbf{U}^i$ be continuous across S. Then the integral identity follows by applying (2.11) to the field $\mathbf{U}^i$, provided D^- coincides with Ω, the point $\mathbf{x}$ belongs to D^+, and the force density $\mathbf{f}$ is set equal to zero. The last condition is simply a restatement of the fact that the sources of $\mathbf{U}^i$ are placed at infinity.

Consider again (2.11), but now identify Ω with the domain $D^+(R)$ which is comprised between the boundary S of the obstacle and the sphere at infinity S_R. Suppose $\mathbf{x}\in D^+(R)$ and observe that in the absence of sources the field $\mathbf{U}^i$ is given by

$$\mathbf{U}^i(\mathbf{x}) = \int_{S_R} \{\boldsymbol{\mathcal{G}}\mathbf{t}^i - \lambda(\mathbf{U}^i\cdot\mathbf{n})\nabla\cdot\boldsymbol{\mathcal{G}} - 2\mu[\mathrm{sym}(\mathbf{U}^i\otimes\mathbf{n})\nabla]\boldsymbol{\mathcal{G}}\}\, da_y,$$

the contribution of the integral over S being zero.

Suppose now that these geometric conditions are maintained, but consider the scattered field $\mathbf{U}$ which obviously obeys (2.1) in D^+. No sources for $\mathbf{U}$ are allowed to occur in D^+ and hence the scattered field is assumed to meet the radiation condition. As shown in the previous section, this requirement represents the most natural counterpart of the fact that $\mathbf{U}$ is completely originated from S and behaves as an outgoing radiation field.

From the mathematical viewpoint the radiation condition is needed to ensure uniqueness of the scattered field. Under these assumptions we find an integral representation for the scattered field $\mathbf{U}$ of the form (2.11), namely

$$\mathbf{U}(\mathbf{x}) = \int_S \{ \lambda(\mathbf{U}\cdot\mathbf{n})\nabla\cdot\mathcal{G} + 2\mu\,[\mathrm{sym}(\mathbf{U}\otimes\mathbf{n})\,\nabla]\mathcal{G} - \mathcal{G}\mathbf{t} \}\, da_y. \tag{4.1}$$

Sometimes it is convenient to represent $\mathbf{U}$ in the equivalent form

$$\mathbf{U}(\mathbf{x}) = \int_S \{ \lambda(\mathbf{U}^+\cdot\mathbf{n})\nabla\cdot\mathcal{G} + 2\mu[\mathrm{sym}(\mathbf{U}^+\otimes\mathbf{n})\,\nabla]\mathcal{G} - \mathcal{G}\mathbf{t}^+ \}\, da_y, \tag{4.1'}$$

where $\mathbf{U}$ has been replaced by $\mathbf{U}^+$. There are two ways of arriving at (4.1'). Either we can subtract from (4.1) the vanishing integral involving $\mathbf{U}^i$ and then collect together $\mathbf{U}^i + \mathbf{U}$ as $\mathbf{U}^+$ as well as the corresponding tractions, or we can apply (2.11) to the total field to find

$$\begin{aligned}\mathbf{U}^+ = \mathbf{U}^i + \mathbf{U} = &\int_S \{ \lambda(\mathbf{U}^+\cdot\mathbf{n})\nabla\cdot\mathcal{G} + 2\mu[\mathrm{sym}(\mathbf{U}^+\otimes\mathbf{n})\,\nabla]\mathcal{G} - \mathcal{G}\mathbf{t}^+ \}\, da_y \\ &+ \int_{S_R} \{\mathcal{G}\mathbf{t}^i - \lambda(\mathbf{U}^i\cdot\mathbf{n})\nabla\cdot\mathcal{G} - 2\mu[\mathrm{sym}(\mathbf{U}^i\otimes\mathbf{n})\nabla]\mathcal{G}\}\, da_y,\end{aligned}$$

the contribution of $\mathbf{U}$ at infinity being zero because of the radiation condition; then (4.1') follows through comparison with the integral representation of $\mathbf{U}^i$ in the exterior domain.

A typical situation when the representation (4.1') of $\mathbf{U}$ proves particularly useful occurs when the scattering is due to the action of an empty inclusion. Then the traction $\mathbf{t}^+$ of $\mathbf{U}^+$ at S vanishes and (4.1') simplifies to

$$\mathbf{U}(\mathbf{x}) = \int_S \{ \lambda(\mathbf{U}^+\cdot\mathbf{n})\nabla\cdot\mathcal{G} + 2\mu[\mathrm{sym}(\mathbf{U}^+\otimes\mathbf{n})\nabla]\mathcal{G}\}\, da_y,$$

while, on the contrary, the value of the traction $\mathbf{t}$ of $\mathbf{U}$ at S is unknown and no natural simplification occurs in (4.1). For definiteness, throughout this section we refer to the representation (4.1).

In the previous section, the vector $\mathbf{U}$ has ultimately been expressed in terms of $\mathbf{U}_L$, $\mathbf{U}_T$, and their derivatives at S. Although their values are still unspecified, we have been able to develop an asymptotic analysis of $\mathbf{U}$ and to see that, in the integral representation of $\mathbf{U}$, the integral over the sphere at infinity vanishes. Accordingly we have proved (4.1). As a further step we now determine the irrotational and solenoidal parts $\mathbf{U}_L$, $\mathbf{U}_T$ of $\mathbf{U}$ in terms of the values of $\mathbf{U}$ and $\mathbf{t}$ at S.

Substitution of the expression (2.10) of $\mathcal{G}$ into (4.1), some rearrangement and use of (2.8) yield the canonical splitting

$$\mathbf{U} = \mathbf{U}_L + \mathbf{U}_T$$

where

$$\mathbf{U}_L(\mathbf{x}) = -\frac{1}{4\pi\rho\omega^2}\int_S \{-\lambda(\mathbf{U}\cdot\mathbf{n})\,k_L^2\nabla g_L + 2\mu\,(\mathbf{U}\otimes\mathbf{n})\cdot(\nabla\otimes\nabla\otimes\nabla g_L) - (\nabla\otimes\nabla g_L)\mathbf{t}\}da_y, \quad (4.2)$$

$$\mathbf{U}_T(\mathbf{x}) = \frac{1}{4\pi\rho\omega^2}\int_S \{\mu\,(\mathbf{U}\otimes\mathbf{n}+\mathbf{n}\otimes\mathbf{U})\cdot(k_T^2\,\nabla g_T\otimes\mathbf{1}+\nabla\otimes\nabla\otimes\nabla g_T) - (k_T^2 g_T\mathbf{1}+\nabla\otimes\nabla g_T)\mathbf{t}\}da_y, \quad (4.3)$$

$\mathbf{n}$ denoting the outward normal to S and the operator ∇ standing for $\nabla_{\mathbf{y}}$.

That $\mathbf{U}_L$ and $\mathbf{U}_T$ represent the irrotational and solenoidal parts of $\mathbf{U}$ is easily ascertained. The position $\mathbf{x}$ belongs to the open domain D^+ exterior to S. Then, as follows from (2.8), g_L and g_T satisfy the scalar Helmholtz equation and this implies that

$$\Delta\mathbf{U}_L + k_L^2\mathbf{U}_L = 0, \qquad \Delta\mathbf{U}_T + k_T^2\mathbf{U}_T = 0.$$

The observation that $\nabla_{\mathbf{y}}\, g_L = -\nabla_{\mathbf{x}}\, g_L$ allows $\mathbf{U}_L$ to be written as

$$\mathbf{U}_L(\mathbf{x}) = \frac{1}{4\pi\rho\omega^2}\nabla_{\mathbf{x}}\int_S \{-\lambda(\mathbf{U}\cdot\mathbf{n})\,k_L^2 g_L + 2\mu\,(\mathbf{U}\otimes\mathbf{n})\cdot(\nabla\otimes\nabla g_L) - \nabla g_L\cdot\mathbf{t}\}da_y.$$

This shows that $\mathbf{U}_L$ is a gradient and hence its curl vanishes identically, as required by (3.2). It remains to prove that $\nabla\cdot\mathbf{U}_T = 0$. To this end it is convenient to use the indicial notation (in Cartesian coordinates). Also, formally a change in sign occurs by replacing $\nabla_{\mathbf{x}}$ with $\nabla_{\mathbf{y}}$. Then by (4.3) we find that

$$\nabla\cdot\mathbf{U}_T = -\frac{1}{4\pi\rho\omega^2}\int_S \{\mu\,(U_i n_j + n_i U_j)(k_T^2\,\partial_i\partial_j g_T + \partial_i\partial_j\Delta g_T) - (k_T^2\,\partial_j g_T + \partial_j\Delta g_T)t_j\}da_y,$$

where ∂_j stands for $\partial/\partial y_j$ and $\Delta = \partial_j\partial_j$. Because $\Delta g_T = -k_T^2 g_T$ in D^+, we have the desired result $\nabla\cdot\mathbf{U}_T = 0$.

We now come to the determination of the asymptotic form of $\mathbf{U}_L$ and $\mathbf{U}_T$, and hence of $\mathbf{U}$. Observe that $r = |\mathbf{y}-\mathbf{x}|$, $R = |\mathbf{x}|$ and $\boldsymbol{\nu} = (\mathbf{y}-\mathbf{x})/r$ satisfy

$$\nabla R = \mathbf{x}/R = \hat{\mathbf{x}}, \qquad \boldsymbol{\nu} = -\hat{\mathbf{x}} + O(R^{-1}),$$

$$r = R - \hat{\mathbf{x}}\cdot\mathbf{y} + O(R^{-1}), \qquad 1/r = 1/R + O(R^{-2}).$$

Substitution into (1.5) and (1.7)-(1.9') yields

$$\begin{aligned}
g_L = \frac{\exp(ik_L r)}{r} &= \frac{\exp(ik_L R)}{R}\exp(-ik_L\,\hat{\mathbf{x}}\cdot\mathbf{y})\big[1+O(R^{-1})\big],\\
\nabla g_L &= -g_L\big[ik_L\hat{\mathbf{x}} + O(R^{-1})\big],\\
\nabla\otimes\nabla g_L &= -g_L\big[k_L^2\hat{\mathbf{x}}\otimes\hat{\mathbf{x}} + O(R^{-1})\big],\\
\nabla\otimes\nabla\otimes\nabla g_L &= g_L\big[ik_L^3\hat{\mathbf{x}}\otimes\hat{\mathbf{x}}\otimes\hat{\mathbf{x}} + O(R^{-1})\big],\\
\nabla\otimes\nabla\otimes\nabla\otimes\nabla g_L &= g_L\big[k_L^4\hat{\mathbf{x}}\otimes\hat{\mathbf{x}}\otimes\hat{\mathbf{x}}\otimes\hat{\mathbf{x}} + O(R^{-1})\big]
\end{aligned}$$

By (2.9) similar relations hold for g_T.

Substitution into the expression (4.2) of $\mathbf{U}_L$ yields the representation

$$\mathbf{U}_L(\mathbf{x}) = -\frac{k_L^2}{4\pi\rho\omega^2}\int_S \{[ik_L\lambda(\mathbf{U}\cdot\mathbf{n})\hat{\mathbf{x}} + 2ik_L\mu(\mathbf{U}\cdot\hat{\mathbf{x}})(\mathbf{n}\cdot\hat{\mathbf{x}})\hat{\mathbf{x}} + (\mathbf{t}\cdot\hat{\mathbf{x}})\hat{\mathbf{x}} + O(R^{-1})]g_L\}da_y.$$

In view of the asymptotic form of g_L we conclude that

$$\mathbf{U}_L(\mathbf{x}) = \frac{\exp(ik_L R)}{R}\,[\mathbf{U}_L^0 + O(R^{-1})] \tag{4.4}$$

where

$$\mathbf{U}_L^0(\hat{\mathbf{x}}) = -\frac{k_L^2}{4\pi\rho\omega^2}\,\hat{\mathbf{x}}\int_S \{ik_L\,[\lambda(\mathbf{U}\cdot\mathbf{n}) + 2\mu(\mathbf{U}\cdot\hat{\mathbf{x}})(\mathbf{n}\cdot\hat{\mathbf{x}})] + \mathbf{t}\cdot\hat{\mathbf{x}}\}\exp(-ik_L\,\hat{\mathbf{x}}\cdot\mathbf{y})\,da_y, \tag{4.5}$$

which shows, in particular, that $\mathbf{U}_L^0$ is parallel to $\hat{\mathbf{x}}$. As regards $\mathbf{U}_T$, substitution into (4.3) of the asymptotic expressions leads to

$$\mathbf{U}_T = -\frac{k_T^2}{4\pi\rho\omega^2}\int_S \{ik_T\mu\,[(\mathbf{U}\cdot\hat{\mathbf{x}})\mathbf{n} + (\mathbf{n}\cdot\hat{\mathbf{x}})\mathbf{U} - 2(\mathbf{U}\cdot\hat{\mathbf{x}})(\mathbf{n}\cdot\hat{\mathbf{x}})\,\hat{\mathbf{x}}] + [\mathbf{t} - (\mathbf{t}\cdot\hat{\mathbf{x}})\hat{\mathbf{x}}] + O(R^{-1})\}\,g_T\,da_y.$$

On setting

$$\mathbf{B}(\hat{\mathbf{x}}) = -\frac{k_T^2}{4\pi\rho\omega^2}\int_S \{ik_T\mu\,[(\mathbf{U}\cdot\hat{\mathbf{x}})\mathbf{n} + (\mathbf{n}\cdot\hat{\mathbf{x}})\mathbf{U}] + \mathbf{t}\}\exp(-ik_T\,\hat{\mathbf{x}}\cdot\mathbf{y})\,da_y$$

and

$$\mathbf{U}_T^0(\hat{\mathbf{x}}) = \hat{\mathbf{x}}\times(\mathbf{B}\times\hat{\mathbf{x}}) \tag{4.6}$$

we conclude that

$$\mathbf{U}_T = \frac{\exp(ik_T R)}{R}\,[\mathbf{U}_T^0 + O(R^{-1})]. \tag{4.7}$$

Summation of (4.4) and (4.7) provides the asymptotic behaviour of the displacement $\mathbf{U}$ in the form

$$\mathbf{U} = \frac{\exp(ik_L R)}{R}\,[\mathbf{U}_L^0 + O(R^{-1})] + \frac{\exp(ik_T R)}{R}\,[\mathbf{U}_T^0 + O(R^{-1})], \tag{4.8}$$

thus showing that $\mathbf{U}_L^0$ and $\mathbf{U}_T^0$ represent the radial and transverse parts, respectively; in other words, very far from the obstacle the displacement may be regarded as the superposition of a radial term, generated by a spherical longitudinal wave, and a tangent term (perpendicular to the radius), generated by a spherical transverse wave [148]. This corroborates the results of the previous section (cf. (3.16)). The fields $\mathbf{U}_L^0$ and $\mathbf{U}_T^0$ define the longitudinal and transverse constituents of the *far-field* pattern of $\mathbf{U}$; essentially, they are given by integrals over the scattering surface S, where the integrands depend on the values of $\mathbf{t}$ and $\mathbf{U}$ at S, the material parameters and the geometry of S. In the elastic case $\mathbf{U}_L^0$

and $\mathbf{U}_T^0$ reduce to the longitudinal and transverse normalized scattering amplitudes of [55, 56].

In the remaining part of this section we show that, by analogy with the scalar theory, to each far-field pattern there corresponds a unique displacement field satisfying the radiation condition. A preliminary step is the proof that both $\mathbf{U}_L$ and $\mathbf{U}_T$ obey the radiation condition in the form (3.3). To this end we need the asymptotic limits of the derivatives of $\mathbf{U}_L$ and $\mathbf{U}_T$. Once they are known, we can also determine the asymptotic form of the stress tensor, which is useful in the proof of uniqueness results.

To evaluate the gradient of $\mathbf{U}_L$ we apply $\nabla_{\mathbf{x}}$ to both sides of (4.2). The right-hand side depends on $\mathbf{x}$ through $g_L(r)$ and hence we can make use of the identity $\nabla_{\mathbf{x}} = -\nabla_{\mathbf{y}}$. This allows the validity of the asymptotic representations for the derivatives and leads to

$$\nabla_{\mathbf{x}}\otimes\mathbf{U}_L(\mathbf{x}) = \frac{1}{4\pi\rho\omega^2}\int_S \{\,\lambda k_L^4\,(\mathbf{U}\cdot\mathbf{n}) + 2\mu k_L^4\,(\mathbf{U}\cdot\hat{\mathbf{x}})(\mathbf{n}\cdot\hat{\mathbf{x}}) - ik_L^3(\mathbf{t}\cdot\hat{\mathbf{x}}) + O(R^{-1})\}g_L\, da_y\,\hat{\mathbf{x}}\otimes\hat{\mathbf{x}}.$$

Comparison with (4.5) shows that

$$\nabla_{\mathbf{x}}\otimes\mathbf{U}_L = \frac{\exp(ik_L R)}{R}\left[ik_L\,\hat{\mathbf{x}}\otimes\mathbf{U}_L^0 + O(R^{-1})\right]. \tag{4.9}$$

Through the same procedure, (4.3) yields

$$\nabla_{\mathbf{x}}\otimes\mathbf{U}_T = \frac{\exp(ik_T R)}{R}\left[ik_T\,\hat{\mathbf{x}}\otimes\mathbf{U}_T^0 + O(R^{-1})\right]. \tag{4.10}$$

Incidentally, $\mathbf{U}_L$ and $\mathbf{U}_T$ satisfy the radiation condition. For, since $\partial/\partial R = \hat{\mathbf{x}}\cdot\nabla_{\mathbf{x}}$, we conclude from (4.9) that

$$\frac{\partial\mathbf{U}_L}{\partial R} - ik_L\mathbf{U}_L = \frac{\exp(ik_L R)}{R}\left[ik_L\mathbf{U}_L^0 - ik_L\mathbf{U}_L^0 + O(R^{-1})\right],$$

uniformly for all directions, and hence the radiation condition (3.3a) is valid for $\mathbf{U}_L$. Similar considerations hold for $\mathbf{U}_T$.

As an aside we find also the asymptotic form of the stress tensor. The general expression for the stress is given by (2.2). Because $\mathbf{U}_T^0$ is perpendicular to $\hat{\mathbf{x}}$, it follows from (4.9) and (4.10) that

$$\nabla\cdot\mathbf{U} = \frac{\exp(ik_L R)}{R}\left[ik_L\,\hat{\mathbf{x}}\cdot\mathbf{U}_L^0 + O(R^{-1})\right].$$

Substitution into the expression of $\mathbf{T}$ and comparison with (4.9) and (4.10) yields

$$\begin{aligned}\mathbf{T} =& ik_L\frac{\exp(ik_L R)}{R}\left[\lambda(\hat{\mathbf{x}}\cdot\mathbf{U}_L^0)\mathbf{1} + 2\mu\hat{\mathbf{x}}\otimes\mathbf{U}_L^0 + O(R^{-1})\right]\\ &+ ik_T\mu\frac{\exp(ik_T R)}{R}\left[\hat{\mathbf{x}}\otimes\mathbf{U}_T^0 + \mathbf{U}_T^0\otimes\hat{\mathbf{x}} + O(R^{-1})\right].\end{aligned} \tag{4.11}$$

As an ancillary result we derive a series representation of $\mathbf{U}_L$. In a Cartesian coordinate system each component U_{Lj} of $\mathbf{U}_L$ obeys the scalar Helmholtz equation

$$\Delta U_{Lj} + k_L^2 U_{Lj} = 0$$

and the radiation condition. We know (cf. §7.1 and [52]) that U_{Lj} may be represented through the expansion

$$U_{Lj}(\mathbf{x}) = \frac{\exp(ik_L R)}{R} \sum_{n=0}^{\infty} \frac{f_n(\theta, \phi)}{R^n}.$$

Then a similar representation theorem for $\mathbf{U}_L$ holds, viz

$$\mathbf{U}_L(\mathbf{x}) = \frac{\exp(ik_L R)}{R} \sum_{n=0}^{\infty} \frac{\mathbf{U}_L^n(\theta, \phi)}{R^n}. \tag{4.12}$$

Here R, θ, ϕ denote the spherical coordinates of the point $\mathbf{x}$; the expansion is valid for all $R \geq R_0$, where R_0 is such that S is completely enclosed in the sphere with centre at the origin and radius R_0. The series and their derivatives converge absolutely and uniformly with respect to the variables R, θ, ϕ. As our notation might have suggested, comparison with (4.4) shows that $\mathbf{U}_L^0$ coincides with the longitudinal constituent of the far field.

We now show that $\mathbf{U}_L$ is completely determined by the corresponding far field $\mathbf{U}_L^0$, in the sense that the coefficients $\mathbf{U}_L^n$ are ultimately determined by $\mathbf{U}_L^0$ via a recurrence algorithm. To this end we observe that the expansion (4.12) has to satisfy the vector Helmholtz equation. Then apply the Laplacian operator (in spherical coordinates)

$$\Delta f(R, \theta, \phi) = \frac{1}{R^2} \frac{\partial}{\partial R}\left(R^2 \frac{\partial f}{\partial R}\right) + \frac{1}{R^2 \sin\theta} \frac{\partial}{\partial \theta}\left(\sin\theta \frac{\partial f}{\partial \theta}\right) + \frac{1}{R^2 \sin^2\theta} \frac{\partial^2 f}{\partial \phi^2},$$

to each component U_{Lj} of $\mathbf{U}_L$ and substitute into the Helmholtz equation. The requirement that the coefficients of the powers of $1/R$ in the resulting series vanish gives the recurrence formula

$$2(n+1)ik_L \mathbf{U}_L^{n+1} = n(n+1)\mathbf{U}_L^n + \frac{1}{\sin\theta} \frac{\partial}{\partial \theta}\left(\sin\theta \frac{\partial \mathbf{U}_L^n}{\partial \theta}\right) + \frac{1}{\sin^2\theta} \frac{\partial^2 \mathbf{U}_L^n}{\partial \phi^2}. \tag{4.13}$$

As a consequence of (4.12) and (4.13) the condition $\mathbf{U}_L^0 = 0$ implies that $\mathbf{U}_L^n = 0$ for every value of n. Thus $\mathbf{U}_L$ is completely determined by its asymptotic values. The same result can be extended to the solenoidal field $\mathbf{U}_T$. We have thus established a one-to-one correspondence between viscoelastic displacement fields in exterior domains and their far-field pattern. Actually, the asymptotic expression of $\mathbf{U}$ shows that the asymptotic expressions of $\mathbf{U}_L$ and $\mathbf{U}_T$, namely $\mathbf{U}_L^0$ and $\mathbf{U}_T^0$, are the radial and transverse components. Then recurrence formulae yield $\mathbf{U}_L$ and $\mathbf{U}_T$, whence $\mathbf{U}$ follows.

7.5 Uniqueness theorems

This section is devoted to the uniqueness of the solution to direct scattering problems for penetrable and impenetrable obstacles. Both the obstacle and the surrounding medium are regarded as viscoelastic solids. In general a scattered wave and the associated transmitted wave originate at the boundary of the obstacle, where suitable continuity conditions are assumed to hold. The total exterior field in D^+ is the superposition of the incident and the scattered wave. The scattered wave is taken to satisfy the radiation condition. The consistency of the mathematical model demands that the exterior and interior displacement fields be unique; of course the interior field is non-trivial if the obstacle is penetrable.

There are essentially three types of boundary conditions that are supposed to hold at the surface S of the obstacle, and give rise to rather different mathematical problems. When an interior transmitted wave occurs inside the obstacle it is usual to impose that the two media are in welded contact at S, and this means that the displacement and the traction are continuous at S. We refer to this case as the transmission problem for a penetrable obstacle. The other two possibilities correspond to what is usually known as an impenetrable obstacle. In one case the scatterer is modelled as perfect rigid body and this means that the total exterior displacement field vanishes at S while no condition is imposed on the traction. In the other case no force is taken to occur at S and the surface deforms until the total exterior surface traction vanishes. This corresponds, for example, to scattering by an empty inclusion inside a given body.

We first consider a penetrable obstacle. The linearity of the pertinent equations reduces the proof of uniqueness to the proof that the difference of two possible solutions vanishes everywhere. Of course the difference of two solutions corresponds to a vanishing incident wave and obeys the radiation condition.

We recall that D^- is the domain occupied by the obstacle, D^+ is the exterior domain, $\mathbf{n}$ is the outward normal to the boundary S, S_R is a sphere of radius R centred at the origin; R_0 is such that $D^- \subset S_{R_0}$, $D^+(R)$ is the region between S and S_R, so that D^+ is the limit of $D^+(R)$ as $R \to \infty$. To avoid ambiguities a superscript $-$ is added to quantities pertaining to the domain D^-, while a $+$ or no superscript refers to D^+. We assume that

$$\lambda^- = \lambda^-(\mathbf{x}), \quad \mu^- = \mu^-(\mathbf{x}), \quad \rho^- = \rho^-(\mathbf{x}), \qquad \text{in } D^-;$$
$$\lambda^+ = \lambda^+(\mathbf{x}), \quad \mu^+ = \mu^+(\mathbf{x}), \quad \rho^+ = \rho^+(\mathbf{x}), \qquad \text{in } D^+ \cap S_{R_0};$$

$$\lambda^+,\ \mu^+,\ \rho^+ \text{ constant} \quad \text{for } |\mathbf{x}| = R > R_0;$$

the material parameters are taken to be discontinuous at S but continuous along with their derivatives both inside and outside S, and asymptotically constant. The total interior and exterior displacement fields are denoted by $\mathbf{U}^-$ and $\mathbf{U}^+$, respectively. It is supposed that

$$\mathbf{U}^+ = \mathbf{U}^s + \mathbf{U}^i,$$

where $\mathbf{U}^i$, $\mathbf{U}^s$ denote the incident and scattered fields. The incident field obeys the linearized equations of motion

$$\rho^+\omega^2\mathbf{U}^i + \nabla\cdot\mathbf{T}^+[\mathbf{U}^i] = 0 \qquad \text{in } D^+,$$

the stress $\mathbf{T}^+[\mathbf{U}^i]$ being defined by (2.2).

The *transmission problem* consists in finding $\mathbf{U}^s \in \mathcal{C}^2(D^+)\cap\mathcal{C}^1(\overline{D^+})$ and $\mathbf{U}^- \in \mathcal{C}^2(D^-)\cap\mathcal{C}^1(\overline{D^-})$ such that

$$\rho^+\omega^2\mathbf{U}^s + \nabla\cdot\mathbf{T}^+[\mathbf{U}^s] = 0, \qquad \text{in } D^+,$$

$$\rho^-\omega^2\mathbf{U}^- + \nabla\cdot\mathbf{T}^-[\mathbf{U}^-] = 0, \qquad \text{in } D^-.$$

The conditions of welded contact are

$$(\mathbf{U}^-)_- = (\mathbf{U}^+)_+, \qquad (\mathbf{T}^-[\mathbf{U}^-]\mathbf{n})_- = (\mathbf{T}^+[\mathbf{U}^+]\mathbf{n})_+ \qquad \text{at } S,$$

the symbol $(\ )_\pm$ denoting the limit values in $D^\pm$. The dependence on the datum $\mathbf{U}^i$ is made evident by writing the boundary conditions in the alternative form

$$(\mathbf{U}^-)_- - (\mathbf{U}^s)_+ = (\mathbf{U}^i)_+, \qquad (\mathbf{T}^-[\mathbf{U}^-]\mathbf{n})_- - (\mathbf{T}^+[\mathbf{U}^s]\mathbf{n})_+ = (\mathbf{T}^+[\mathbf{U}^i]\mathbf{n})_+.$$

Of course the field $\mathbf{U}^s$ is required to satisfy the radiation condition.

To establish the uniqueness of the solution to the transmission problem we consider two solutions $\mathbf{U}_1^-$, $\mathbf{U}_1^s$, and $\mathbf{U}_2^-$, $\mathbf{U}_2^s$ corresponding to the same incident field $\mathbf{U}^i$. Set

$$\mathbf{U} = \mathbf{U}_2^- - \mathbf{U}_1^-, \qquad \text{in } D^-,$$

$$\mathbf{U} = \mathbf{U}_2^s - \mathbf{U}_1^s = \mathbf{U}_2^+ - \mathbf{U}_1^+, \qquad \text{in } D^+.$$

Obviously $\mathbf{U}$ satisfies the radiation condition and, in D^+ and D^-, obeys the field equations

$$\rho^\pm\omega^2\mathbf{U} + \nabla\cdot\mathbf{T}^\pm[\mathbf{U}] = 0 \tag{5.1}$$

for viscoelastic bodies. Moreover $\mathbf{U}$ meets the boundary conditions

$$(\mathbf{U})_- = (\mathbf{U})_+, \qquad (\mathbf{T}^-[\mathbf{U}]\mathbf{n})_- = (\mathbf{T}^+[\mathbf{U}]\mathbf{n})_+ \qquad \text{at } S. \tag{5.2}$$

We have to prove that, as a consequence of (5.1) and (5.2), the field $\mathbf{U}$ vanishes identically.

To this end we need a few preliminary results concerning the behaviour of time-harmonic waves in heterogeneous viscoelastic media. We omit momentarily the superscripts + and − and observe that, upon inner multiplication by the complex conjugate $\mathbf{U}^*$ of $\mathbf{U}$ and straightforward calculations, eq. (5.1) gives

$$\nabla\cdot(\mathbf{T}\mathbf{U}^*) = \mathbf{T}\cdot\mathbf{E}^* - \rho\omega^2|\mathbf{U}|^2,$$

where $\mathbf{T}$ stands for $\mathbf{T}[\mathbf{U}]$ and $\mathbf{E} = \frac{1}{2}(\nabla\mathbf{U} + \nabla\mathbf{U}^\dagger)$. Meanwhile, by comparison with (2.2), we can write

$$\mathbf{T} = \lambda(\operatorname{tr}\mathbf{E})\mathbf{1} + 2\mu\,\mathbf{E}. \tag{5.3}$$

Accordingly we have

$$\mathbf{T}\cdot\mathbf{E}^* = \lambda\,|\mathrm{tr}\mathbf{E}|^2 + 2\mu\,\mathbf{E}\cdot\mathbf{E}^*,$$

and the divergence of $\mathbf{T}\mathbf{U}^*$ may be written as

$$\nabla\cdot(\mathbf{T}\mathbf{U}^*) = \lambda\,|\operatorname{tr}\mathbf{E}|^2 + 2\mu\,\mathbf{E}\cdot\mathbf{E}^* - \rho\omega^2|\mathbf{U}|^2. \tag{5.4}$$

Consider (5.4) in the exterior domain. Integration of both sides over $D^+(R)$ and use of Gauss theorem yield

$$-\int_S \left(\mathbf{U}^*\cdot\mathbf{t}\right)_+ da + \int_{S_R} \mathbf{U}^*\cdot\mathbf{t}\,da = \int_{D^+(R)} \{\lambda\,|\mathrm{tr}\mathbf{E}|^2 + 2\mu\,\mathbf{E}\cdot\mathbf{E}^* - \rho\omega^2|\mathbf{U}|^2\}dx \tag{5.5}$$

where $\mathbf{t} = \mathbf{T}\mathbf{n}$ and the minus sign in front of the integral over S occurs because $\mathbf{n}$ is the outward normal to S and then inward normal with respect to $D^+(R)$. Now integrate (5.4) in the domain D^- to find

$$\int_S \left(\mathbf{U}^*\cdot\mathbf{t}\right)_- da = \int_{D^-} \{\lambda^-\,|\mathrm{tr}\mathbf{E}|^2 + 2\mu^-\,\mathbf{E}\cdot\mathbf{E}^* - \rho^-\omega^2|\mathbf{U}|^2\}dV.$$

Summation of the last two integrals and account of the continuity assumptions (5.2) give

$$\begin{aligned}\int_{S_R} \mathbf{U}^*\cdot\mathbf{t}\,da = &\int_{D^+(R)} \{\lambda\,|\mathrm{tr}\mathbf{E}|^2 + 2\mu\,\mathbf{E}\cdot\mathbf{E}^* - \rho\omega^2|\mathbf{U}|^2\}dx \\ &+ \int_{D^-} \{\lambda^-\,|\mathrm{tr}\mathbf{E}|^2 + 2\mu^-\,\mathbf{E}\cdot\mathbf{E}^* - \rho^-\omega^2|\mathbf{U}|^2\}dx.\end{aligned} \tag{5.6}$$

The next step is to show that the limit of the left side for $R\to\infty$ vanishes. To prove that it is so we recall that the displacement field $\mathbf{U}$ is a solution of (5.1) and consider a spherical surface $S_{\hat{R}}$, $\hat{R} > R_0$, so that the material parameters are constant for $|\mathbf{x}| > \hat{R}$. Accordingly we obtain for $\mathbf{U}(\mathbf{x})$, $|\mathbf{x}| > \hat{R}$, an integral representation of the form (4.1) with integration performed over $S_{\hat{R}}$. Hence we may define a far-field pattern and (4.8) applies. We thus obtain

$$\mathbf{U}^* = \frac{\exp(-ik_L^* R)}{R}\left[(\mathbf{U}_L^0)^* + O(R^{-1})\right] + \frac{\exp(-ik_T^* R)}{R}\left[(\mathbf{U}_T^0)^* + O(R^{-1})\right].$$

It is worth remarking that, in view of (4.5) and (4.6), both $(\mathbf{U}_L^0)^*$ and $\mathbf{U}_L^0$ are parallel to $\hat{\mathbf{x}}$, while $(\mathbf{U}_T^0)^*$ and $\mathbf{U}_T^0$ are perpendicular to $\hat{\mathbf{x}}$. Accordingly we have

$$(\mathbf{U}_L^0)^*\cdot\mathbf{U}_T^0 = (\mathbf{U}_T^0)^*\cdot\mathbf{U}_L^0 = 0.$$

Moreover, since $\mathbf{n}$ coincides with $\hat{\mathbf{x}}$ on S_R, we deduce from (4.11) that

$$\mathbf{t} = \mathbf{T}\hat{\mathbf{x}} = ik_L(\lambda + 2\mu)\frac{\exp(ik_L R)}{R}\left[\mathbf{U}_L^0 + O(R^{-1})\right] + ik_T\mu\frac{\exp(ik_T R)}{R}\left[\mathbf{U}_T^0 + O(R^{-1})\right].$$

As a consequence we can write

$$\int_{S_R} \mathbf{U}^* \cdot \mathbf{t}\, da = \int_{S_R} \Big\{ ik_L(\lambda + 2\mu)\frac{|\exp(ik_L R)|^2}{R^2}\left[|\mathbf{U}_L^0|^2 + O(R^{-1})\right] + ik_T\mu\frac{|\exp(ik_T R)|^2}{R^2}\left[|\mathbf{U}_T^0|^2 + O(R^{-1})\right]\Big\} da. \quad (5.7)$$

Of course, $|\exp(ik_L R)|^2 = \exp(-2\,\mathrm{Im}k_L\, R)$ and $|\exp(ik_T R)|^2 = \exp(-2\,\mathrm{Im}k_T\, R)$ and, because of (3.2.8), (3.2.9) and the fact that the waves are outgoing, we have $\mathrm{Im}\, k_L > 0$ and $\mathrm{Im}\, k_T > 0$. Since da is proportional to R^2, we obtain

$$\lim_{R\to\infty} \int_{S_R} \mathbf{U}^* \cdot \mathbf{t}\, da = 0.$$

Then the asymptotic limit of (5.6) yields

$$\int_{D^+} \{\lambda\, |\mathrm{tr}\mathbf{E}|^2 + 2\mu\, \mathbf{E}\cdot\mathbf{E}^* - \rho\omega^2|\mathbf{U}|^2\} dx + \int_{D^-} \{\lambda^-\, |\mathrm{tr}\mathbf{E}|^2 + 2\mu^-\, \mathbf{E}\cdot\mathbf{E}^* - \rho^-\omega^2|\mathbf{U}|^2\} dx = 0. \quad (5.8)$$

As the last step we show that (5.8) yields the vanishing of $\mathbf{E}$ in the whole space. Consider the first integral on D^+ and denote by $\mathbf{E}_1$ and $\mathbf{E}_2$ the real and imaginary parts of $\mathbf{E}$. Letting a superposed ring stand for the trace-free part we represent the complex-valued tensor $\mathbf{E}$ as

$$\mathbf{E} = \mathbf{E}_1 + i\mathbf{E}_2 = \mathring{\mathbf{E}}_1 + \tfrac{1}{3}\mathrm{tr}\mathbf{E}_1\, \mathbf{1} + i(\,\mathring{\mathbf{E}}_2 + \tfrac{1}{3}\mathrm{tr}\mathbf{E}_2\, \mathbf{1}).$$

It follows that

$$|\mathrm{tr}\mathbf{E}|^2 = (\mathrm{tr}\mathbf{E}_1)^2 + (\mathrm{tr}\mathbf{E}_2)^2,$$

$$\mathbf{E}\cdot\mathbf{E}^* = \mathring{\mathbf{E}}_1 \cdot \mathring{\mathbf{E}}_1 + \tfrac{1}{3}(\mathrm{tr}\mathbf{E}_1)^2 + \mathring{\mathbf{E}}_2 \cdot \mathring{\mathbf{E}}_2 + \tfrac{1}{3}(\mathrm{tr}\mathbf{E}_2)^2,$$

whence

$$\lambda\, |\mathrm{tr}\mathbf{E}|^2 + 2\mu\, \mathbf{E}\cdot\mathbf{E}^* = (\lambda + \tfrac{2}{3}\mu)\left[(\mathrm{tr}\mathbf{E}_1)^2 + (\mathrm{tr}\mathbf{E}_2)^2\right] + 2\,\mu\,(\mathring{\mathbf{E}}_1 \cdot \mathring{\mathbf{E}}_1 + \mathring{\mathbf{E}}_2 \cdot \mathring{\mathbf{E}}_2).$$

As a consequence of the thermodynamic inequalities (2.4.7) we have

$$\mathrm{Im}\lambda\, |\mathrm{tr}\mathbf{E}|^2 + \mathrm{Im}(2\mu)\, \mathbf{E}\cdot\mathbf{E}^* = \mathrm{Im}(\lambda + \tfrac{2}{3}\mu)\left[(\mathrm{tr}\mathbf{E}_1)^2 + (\mathrm{tr}\mathbf{E}_2)^2\right] + 2\,\mathrm{Im}\mu\,(\mathring{\mathbf{E}}_1 \cdot \mathring{\mathbf{E}}_1 + \mathring{\mathbf{E}}_2 \cdot \mathring{\mathbf{E}}_2) < 0 \quad (5.9)$$

for any pair of tensor fields $\mathbf{E}_1, \mathbf{E}_2$ not simultaneously vanishing. With the change $\mu \to \mu^-$, $\lambda \to \lambda^-$, $\rho \to \rho^-$, the inequality (5.9) holds for D^- too. Now consider the imaginary part of the integral relation (5.8). In view of (5.9) and the continuity of $\mathbf{E}$ in D^-, D^+, it follows that $\mathbf{E} = 0$ in D^+ and D^-. The stress $\mathbf{T}$ vanishes too, because of (5.3), and then (5.1) shows that $\mathbf{U} = 0$ in D^+ and D^-. This completes the proof of uniqueness.

As already remarked, quite often the scheme of impenetrable obstacles is adopted in that the field in the interior of the obstacle is neglected. In such a case suitable conditions are needed at the boundary S for the determination of the exterior field. Their explicit formulation depends on the characteristic properties of the obstacle. Letting $S_1 \cup S_2 = S$ and $S_1 \cap S_2 = \emptyset$, we can write the most customary types of conditions as

- (D) the displacement is given at S;
- (T) the traction is given at S;
- (M) the displacement is given at S_1 and the traction is given at S_2.

Apart from the fact that now we are dealing with exterior problems only, the remaining conditions coincide with those examined above in the study of penetrable obstacles. The material parameters λ, μ, and ρ are constant outside the sphere S_{R_0} and vary continuously from point to point in the region between S and S_{R_0}. An incident field $\mathbf{U}^i$ comes from infinity and is diffused at S; the total field $\mathbf{U}^+ = \mathbf{U}^i + \mathbf{U}^s$ satisfies the field equations (5.1) in D^+ and one of the boundary conditions (D), (T), and (M) at S. The scattered field $\mathbf{U}^s$, satisfying the radiation condition, proves to be uniquely determined.

To fix ideas we prove that $\mathbf{U}^s$ is unique in the case when (D) holds. Consider two scattered fields corresponding to the same incident field $\mathbf{U}^i$. By linearity, the difference $\mathbf{U}$ of the total external fields obeys the equations of motion (5.1) and the radiation condition, since $\mathbf{U}^i$ cancels. At S we have $\mathbf{U} = 0$ because (D) holds for both total fields. We remark in passing that the two scattered fields coincide at S in that they are determined as the difference between the assigned total field and the (common) incident one. By paralleling the analysis developed for penetrable obstacles we may write an equation of the form (5.5). The integral over S vanishes as a consequence of the boundary conditions; similarly, that over S_R does not give any contribution because of dissipation and the radiation condition. Thus, as $R \to \infty$ we are left with

$$\int_{D^+} (\lambda |\mathrm{tr}\,\mathbf{E}|^2 + 2\mu \mathbf{E} \cdot \mathbf{E}^* - \rho\omega^2 |\mathbf{U}|^2) dx = 0$$

and hence we reach the conclusion that $\mathbf{U} = 0$.

The procedure can be modified very easily to deal with boundary conditions of the form (T) or (M). It is also possible to find similar results concerning both penetrable and impenetrable obstacles when either the external medium or the obstacle itself or both of them are modelled as viscous or viscoelastic fluids [36]. Furthermore, the proof of uniqueness developed in the analysis of the penetrable obstacle can be extended to multilayered bodies. However, we are not pursuing these points here. Rather, we observe that this part improves somewhat the analysis of [98] for viscoelastic solids and that an alternative approach has been recently developed in [169].

Although elasticity may be viewed as a particular case of viscoelasticity the absence of dissipation requires some qualitative change in the proof of uniqueness. For definiteness, we consider a penetrable obstacle under the standard assumptions except that now λ and μ, and hence k_L and k_T, are real quantities.

We start from the observation that (5.6) holds also for elastic solids. The left side is given by (5.7) where, however, $|\exp(ik_L R)|^2 = |\exp(ik_T R)|^2 = 1$. Accordingly we find that

$$\int_{S_R} \mathbf{U}^* \cdot \mathbf{t}\, da = i \int_{S_R} \{k_L(\lambda + 2\mu)\,|\mathbf{U}_L^0|^2 + k_T \mu\,|\mathbf{U}_T^0|^2\} \frac{da}{R^2} + O(R^{-1}),$$

the leading contribution being of course purely imaginary. As to the right side of (5.6), comparison with (5.9) shows that it may be written as

$$\begin{aligned}
&\int_{D^+(R)} \{\lambda\,|\mathrm{tr}\mathbf{E}|^2 + 2\mu\,\mathbf{E}\cdot\mathbf{E}^* - \rho\omega^2|\mathbf{U}|^2\}dx \\
&\qquad + \int_{D^-} \{\lambda^-\,|\mathrm{tr}\mathbf{E}|^2 + 2\mu^-\,\mathbf{E}\cdot\mathbf{E}^* - \rho^-\omega^2|\mathbf{U}|^2\}dx \\
&= \int_{D^+(R)} \{(\lambda + 2\mu)\,[(\mathrm{tr}\mathbf{E}_1)^2 + (\mathrm{tr}\mathbf{E}_2)^2] + 2\mu\,(\mathring{\mathbf{E}}_1 \cdot \mathring{\mathbf{E}}_1 + \mathring{\mathbf{E}}_2 \cdot \mathring{\mathbf{E}}_2) - \rho\omega^2|\mathbf{U}|^2\}dx \\
&+ \int_{D^-} \{(\lambda^- + 2\mu^-)\,[(\mathrm{tr}\mathbf{E}_1)^2 + (\mathrm{tr}\mathbf{E}_2)^2] + 2\,\mu^-(\mathring{\mathbf{E}}_1 \cdot \mathring{\mathbf{E}}_1 + \mathring{\mathbf{E}}_2 \cdot \mathring{\mathbf{E}}_2) - \rho\omega^2|\mathbf{U}|^2\}dx,
\end{aligned}$$

and hence is a real quantity. It follows that the integral over S_R vanishes, and this means that $\mathbf{U}_L^0 = 0$ and $\mathbf{U}_T^0 = 0$. According to (4.13) $\mathbf{U}_L$ and $\mathbf{U}_T$ vanish in the region where the material parameters are constant. Following [169] we conclude that $\mathbf{U} = 0$ in the region exterior to S.

7.6 Scattering cross section

The presence of an obstacle modifies the power distribution of the incident wave. It is of interest to predict the amount of power, relative to the incident one, that is scattered into a given direction. Comparison with experimental measurements may serve to gain information on the properties of the obstacle, if these are unknown. Since the asymptotic representations of the stress and the displacement have already been determined as integrals over the boundary S of the obstacle, what remains to do is to insert these integrals into the asymptotic expression for the energy flux intensity and to find the total and differential cross sections.

The *scattering cross section* is defined as the ratio of the time average rate at which energy is scattered by the obstacle to the corresponding time average rate at which the energy of the incident wave crosses a unit area normal to the direction of incidence [9,

55, 56]. Here the energy scattered by the obstacle is the energy of the scattered field $\mathbf{U}$ transmitted across a sphere S_R, $R \to \infty$. The *differential cross section* is a measure of the fraction of incident power scattered into a particular direction [77], relative to the differential element of solid angle. Since the scattering cross section is a total energy measured in units of energy per unit area, it has the dimensions of an area. Indeed, it represents an area over the plane surface, orthogonal to the incident wave, which receives a time average rate of energy equal to that scattered by the obstacle [56].

These definitions correspond to the customary ones adopted in acoustics and electromagnetism. In the classical theory of elasticity the above unit of energy is defined unambiguously in that it is independent of the spatial position of the surface element at which the energy rate of the incident wave is evaluated. But when we specialize to viscoelastic bodies the definition of the time average rate of the incident energy is not invariant under translations along the direction of phase propagation, due to the (exponential) decay of the amplitude of the incident wave. Hence the choice of a unit of energy is not that obvious and deserves some attention. Here we determine the asymptotic expression of the time average rate of the energy of the scattered field, which is the superposition of a longitudinal and a transverse expanding spherical wave; the surface element at an arbitrary direction is taken to be orthogonal to the corresponding radial vector.

According to the analysis of Ch. 3, the expression of the time average of the energy flux intensity is given by (3.5.1), namely

$$\langle \mathcal{I} \rangle = \tfrac{1}{2}\omega \, \mathrm{Im}[\,(\mathbf{T}\mathbf{n}_1)\cdot \mathbf{u}^*],$$

$\mathbf{n}_1$ being the direction of propagation of the phase, that is the unit vector of the real part of $\mathbf{k}$. Here we consider the scattered field $\mathbf{U}$ on the spherical surface S_R centred at the origin and far enough from the obstacle. According to (4.8) we have

$$\mathbf{U}(\mathbf{x}) = \frac{\exp(ik_L R)}{R}\,[\,\mathbf{U}_L^0(\hat{\mathbf{x}}) + O(R^{-1})] + \frac{\exp(ik_T R)}{R}\,[\,\mathbf{U}_T^0(\hat{\mathbf{x}}) + O(R^{-1})],$$

which shows that $\mathbf{U}$ is the superposition of a longitudinal and a transverse outgoing spherical wave with direction-dependent amplitudes. Thus we take as $\mathbf{n}_1$ the radial outward unit vector, namely $\mathbf{n}_1 = \hat{\mathbf{x}}$. Hence, by comparison with (4.11), we find that

$$\begin{aligned}(\mathbf{T}\mathbf{n}_1)\cdot \mathbf{u}^* &= \{\exp(-i\omega t)\,\mathbf{T}[\mathbf{U}]\,\hat{\mathbf{x}}\}\cdot \mathbf{U}^*\,\exp(i\omega t)\\ &= ik_L(\lambda + 2\mu)\frac{|\exp(ik_L R)|^2}{R^2}\,[\,|\mathbf{U}_L^0|^2 + O(R^{-1})]\\ &\quad + ik_T\mu\frac{|\exp(ik_T R)|^2}{R^2}\,[\,|\mathbf{U}_T^0|^2 + O(R^{-1})].\end{aligned}$$

To simplify the notation, the dependence on $R\,\hat{\mathbf{x}}$ is henceforth understood and not written. Upon substitution into the expression of $\langle \mathcal{I} \rangle$ we find that the time average energy flux intensity across an element of area tangent to the sphere S_R at $\mathbf{x} = R\hat{\mathbf{x}}$ is given by

$$\begin{aligned}\langle \mathcal{I} \rangle = \tfrac{1}{2}\omega\{\mathrm{Re}\big[k_L(\lambda + 2\mu)\big] \frac{|\exp(ik_L R)|^2}{R^2}\big[\,|\mathbf{U}_L^0|^2 + O(R^{-1})\big] \\ + \mathrm{Re}(k_T\mu)\frac{|\exp(ik_T R)|^2}{R^2}\big[\,|\mathbf{U}_T^0|^2 + O(R^{-1})\big]\}.\end{aligned} \tag{6.1}$$

Incidentally, if the solid is elastic then k_L, k_T, λ, μ are real and $|\exp(ik_L R)| = |\exp(ik_T R)| = 1$. As a consequence, (6.1) reduces to

$$\langle \mathcal{I} \rangle = \frac{1}{2}\frac{\omega}{R^2}\big[k_L(\lambda + 2\mu)\,|\mathbf{U}_L^0|^2 + k_T\mu\,|\mathbf{U}_T^0|^2\big] + O(R^{-3}), \tag{6.1'}$$

a result already known in the literature [55], though in a different form. To find the differential and total cross sections we also need the average time rate of the energy density carried by the incident wave. If the incident wave is longitudinal, the energy flux intensity is given by

$$\langle \mathcal{I}_L^i \rangle = \tfrac{1}{2}\rho\omega^3|\mathbf{U}^i|^2/k_L,$$

(cf. §3.5), while, for an incident transverse wave, we have

$$\langle \mathcal{I}_T^i \rangle = \tfrac{1}{2}\rho\omega^3|\mathbf{U}^i|^2/k_T.$$

If we denote by $d\Omega$ the element of solid angle, the differential cross section $dP/d\Omega$ is defined as

$$\frac{dP_{L,T}}{d\Omega} = \frac{R^2\langle \mathcal{I} \rangle}{\langle \mathcal{I}_{L,T}^i \rangle} \tag{6.2}$$

while the cross section P is obtained as

$$P_{L,T} = \int_\Omega \frac{dP_{L,T}}{d\Omega}\,d\Omega; \tag{6.3}$$

here the subscript L or T is relative to the incident wave, Ω denotes the unit sphere, and the dependence on R is considered in the limit $R \to \infty$. By applying these definitions in the framework of elastic scattering we find

$$\frac{dP_{L,T}}{d\Omega} = \frac{k_{L,T}}{|\mathbf{U}^i|^2}\Big(\frac{|\mathbf{U}_L^0|^2}{k_L} + \frac{|\mathbf{U}_T^0|^2}{k_T}\Big), \tag{6.4}$$

$$P_{L,T} = \int_\Omega \frac{k_{L,T}}{|\mathbf{U}^i|^2}\Big[\frac{|\mathbf{U}_L^0|^2}{k_L} + \frac{|\mathbf{U}_T^0|^2}{k_T}\Big]\,d\Omega,$$

Notice that P is positive, thus showing that the radiation condition causes the energy flux to be outgoing.

When wave scattering occurs in dissipative bodies changes in the distribution of the mean energy flux intensity may be ascribed to two different causes, namely the presence of the scatterer, which determines the diffusion of the incident energy, and the attenuation that occurs as a consequence of dissipation. Thus we have to take into account these two effects, and possibly to distinguish between them. Preliminarily we need an expression for the intensity of the incident energy.

For definiteness, let the incident wave be longitudinal. We know (cf. §3.5) that the mean energy flux intensity of an inhomogeneous longitudinal wave takes the form

$$\begin{aligned} \mathcal{I}_L^i &= \tfrac{1}{2}\omega \exp(-2\,\mathbf{k}_{2L}^i \cdot \mathbf{x})(|\mathbf{U}^i|/|\mathbf{k}_L^i|)^2 \left[\rho\omega^2 k_{1L}^i + 4\mu_1 k_{1L}^i (k_{2L}^i)^2 \sin^2\gamma\right], \\ &= \tfrac{1}{2}\omega \exp(-2\,k_{2L}^i R\,\mathbf{n}_{2L}^i \cdot \hat{\mathbf{x}})\, M_L \end{aligned} \tag{6.5}$$

where

$$M_L = (|\mathbf{U}^i|/|\mathbf{k}_L^i|)^2 \left[\rho\omega^2 k_{1L}^i + 4\mu_1 k_{1L}^i (k_{2L}^i)^2 \sin^2\gamma\right],$$

and $\mathbf{k}_L^i = \mathbf{k}_{1L}^i + i\mathbf{k}_{2L}^i = k_{1L}^i\mathbf{n}_1^i + k_{2L}^i\mathbf{n}_2^i$, γ is the angle between $\mathbf{k}_{1L}^i$ and $\mathbf{k}_{2L}^i$, μ_1 is the real part of μ. Unlike the elastic case, $\langle\mathcal{I}_L^i\rangle$ depends on the position. More precisely, $\mathcal{I}_L^i$ is constant on planes perpendicular to $\mathbf{k}_{2L}^i$, approaches zero as $\mathbf{x}$ increases in the direction of $\mathbf{k}_{2L}^i$, and approaches infinity as $\mathbf{x}$ increases in the direction opposite to $\mathbf{k}_{2L}^i$. In particular, if $\mathbf{k}_{1L}^i$ and $\mathbf{k}_{2L}^i$ are parallel, we can say that when the point $\mathbf{x}$ is sufficiently far from the origin, opposite to the direction of the incident wave, then $\mathcal{I}_L^i \to \infty$, while $\mathcal{I}_L^i \to 0$ in the direction of the incident wave. In other words, a wave that comes from infinity and carries a finite energy density at the boundary of an obstacle should have been generated with an infinite energy. Quite analogously, for an inhomogeneous transverse incident wave we can write (cf. §3.5)

$$\mathcal{I}_T^i = \tfrac{1}{2}\omega \exp(-2\,k_{2T}^i R\,\mathbf{n}_{2T}^i \cdot \hat{\mathbf{x}})\, M_T.$$

Consider now the definition (6.2) of the differential scattering cross section. The numerator is to be replaced by (6.1). In view of the relation

$$|\exp(ik_L R)|^2 - \exp(-2k_{2L}R)|\exp(ik_{1L}R)|^2 = \exp(-2k_{2L}R)$$

and the analogous one for the transverse component of the asymptotic field we can write the leading term of the numerator as

$$R^2\langle\mathcal{I}\rangle = \tfrac{1}{2}\omega\big\{\exp(-2k_{2L}R)\,|\mathbf{U}_L^0|^2\,\mathrm{Re}\big[k_L(\lambda+2\mu)\big] + \exp(-2k_{2T}R)\,|\mathbf{U}_T^0|^2\,\mathrm{Re}(k_T\mu)\big\}.$$

The values of μ, λ, k_L, k_T determine which of the two asymptotic waves, the longitudinal and the transverse one, is predominant in the scattered field, far from the obstacle.

The conceptual difficulty, inherent in the dissipativity, is that the denominator (6.5) depends on $\mathbf{x}$. A way of overcoming this difficulty might be as follows. Let $\hat{\mathbf{x}}$ be the direction of the differential cross section. Since $\langle\mathcal{I}\rangle$ is a function of R and $\hat{\mathbf{x}}$, it seems natural to compare $R^2\langle\mathcal{I}\rangle$ with the values assumed by $\langle\mathcal{I}_L^i\rangle$ or $\langle\mathcal{I}_T^i\rangle$ at the point $R\hat{\mathbf{x}}$, that is with the time rate average energy of the incident wave at the same point. Upon substitution of the pertinent expressions into the definition (6.2) of the differential cross section we find that

$$\frac{dP_L}{d\Omega} = \frac{1}{M_L}\{|\mathbf{U}_L^0|^2 \operatorname{Re}[k_L(\lambda + 2\mu)] \exp[-2R(k_{2L} - k_{2L}^i \, \mathbf{n}_{2L}^i \cdot \hat{\mathbf{x}})] + |\mathbf{U}_T^0|^2 \operatorname{Re}(k_T\mu) \exp[-2R(k_{2T} - k_{2L}^i \, \mathbf{n}_{2L}^i \cdot \hat{\mathbf{x}})]\}, \tag{6.6}$$

$$\frac{dP_T}{d\Omega} = \frac{1}{M_T}\{|\mathbf{U}_L^0|^2 \operatorname{Re}[k_L(\lambda + 2\mu)] \exp[-2R(k_{2L} - k_{2T}^i \, \mathbf{n}_{2T}^i \cdot \hat{\mathbf{x}})] + |\mathbf{U}_T^0|^2 \operatorname{Re}(k_T\mu) \exp[-2R(k_{2T} - k_{2T}^i \, \mathbf{n}_{2T}^i \cdot \hat{\mathbf{x}})]\}. \tag{6.7}$$

Because of (3.2.14) - with $\alpha = 1$ - k_{2L} depends only on the material parameters. Meanwhile k_{2L}^i depends also on the geometry of the incident wave through $\alpha = \mathbf{n}_1^i \cdot \mathbf{n}_2^i$. Hence in general $k_{2L} \neq k_{2L}^i$ unless $\alpha = 1$. The same holds for k_T. Now observe that, in view of (2.4.7) and the assumptions $\operatorname{Re}\mu > 0$, $\operatorname{Re}(\lambda + 2\mu/3) > 0$ we have

$$\operatorname{Re}[k_L(\lambda + 2\mu)] = k_1(\lambda_1 + 2\mu_1) - k_2(\lambda_2 + 2\mu_2) > 0.$$

This shows that for any $\hat{\mathbf{x}}$ at any R we find a positive differential scattering cross section. If $\mathbf{n}_{2L}^i \cdot \hat{\mathbf{x}} < 0$ or $\mathbf{n}_{2T}^i \cdot \hat{\mathbf{x}} < 0$ the two expressions of the differential cross section approach zero as $R \to \infty$, as expected. Consider, instead, the case when these inner products are positive. For simplicity evaluate $dP_{L,T}/d\Omega$ in the direction of $\mathbf{n}_2^i$, that is assume that $\mathbf{n}_{2L}^i \cdot \hat{\mathbf{x}} = \mathbf{n}_{2T}^i \cdot \hat{\mathbf{x}} = 1$. Suppose also that $\mathbf{k}_1^i$ is parallel to $\mathbf{k}_2^i$ so that $k_{2L}^i = k_{2L}$ and $k_{2T}^i = k_{2T}$. The two expressions (6.6) and (6.7) for the differential scattering cross sections reduce to

$$\frac{dP_L}{d\Omega} = \frac{1}{M_L}\{|\mathbf{U}_L^0|^2 \operatorname{Re}[k_L(\lambda + 2\mu)] + |\mathbf{U}_T^0|^2 \operatorname{Re}(k_T\mu) \exp[-2R(k_{2T} - k_{2L}]\},$$

$$\frac{dP_T}{d\Omega} = \frac{1}{M_T}\{|\mathbf{U}_L^0|^2 \operatorname{Re}[k_L(\lambda + 2\mu)] \exp[-2R(k_{2L} - k_{2T}] + |\mathbf{U}_T^0|^2 \operatorname{Re}(k_T\mu)\},$$

thus showing that one of them is necessarily unbounded as $R \to \infty$. As a comment, we can say that the numerator and the denominator of the pertinent fraction (6.2) approach zero as $R \to \infty$. The vanishing of the numerator is related to the decay of the scattered wave while that of the denominator corresponds to the decay of the incident wave. Since we are comparing rates of decay of different waves, it should come as no surprise that one limit value turns out to be infinity.

In a sense this difficulty rules out what could be regarded as a reasonable choice for $\langle \mathcal{I}_L^i \rangle$ and $\langle \mathcal{I}_T^i \rangle$. A seemingly alternative possibility has been realized in a recent paper [57] on thermoelastic scattering. In fact, as an ultimate consequence of the thermal effects, the allowed incident and scattered fields exhibit essentially the same behaviour at infinity as displacement fields in viscoelasticity. There, $\langle \mathcal{I}_L^i \rangle$ and $\langle \mathcal{I}_T^i \rangle$ are identified with the minimum value assumed on the sphere S_R by the time average rate of the corresponding incident wave.

With reference to (6.5), in the case of an incident longitudinal wave, this corresponds to the choice $\mathbf{n}^i_{2L} \cdot \hat{\mathbf{x}} = 1$, which leads to quite paradoxical results. On the contrary, as already observed, some difficulties are avoided if we choose for $\langle \mathcal{I}^i_L \rangle$ the maximum value of (6.5) on the sphere S_R, which corresponds to setting $\mathbf{n}^i_{2L} \cdot \hat{\mathbf{x}} = -1$. Nevertheless, the measure of the scattering intensity so obtained also seems meaningless, because the numerator and the denominator tend very rapidly to zero and infinity, respectively, as $R \to \infty$. Hence this ratio is not suitable to characterize the scattering efficiency of the obstacle.

It might seem more convenient to choose a measure of the incident rate of energy $\langle \mathcal{I}^i_L \rangle$ or $\langle \mathcal{I}^i_T \rangle$ independent of R. Since we are interested in the determination of the intrinsic scattering efficiency of the obstacle with a minimal influence of the attenuation effects, a realistic estimate of the relative amount of power reflected at a given direction might be given as follows. Consider the sphere of smallest radius, say $\bar{R}$, enclosing the surface S. Choose as $\langle \mathcal{I}^i_L \rangle$ and $\langle \mathcal{I}^i_T \rangle$ the maximum values assumed by these quantities on the sphere $S_{\bar{R}}$; they are taken when $\hat{\mathbf{x}} = -\mathbf{n}^i$ and read

$$\langle \mathcal{I}^i_L \rangle = \tfrac{1}{2}\omega \exp(2k^i_{2L}\bar{R}) M_L, \qquad \langle \mathcal{I}^i_T \rangle = \tfrac{1}{2}\omega \exp(2k^i_{2T}\bar{R}) M_T.$$

Accordingly we find that detailed information about the distribution of the energy of incident waves is given by the differential scattering cross sections

$$\begin{aligned} \frac{dP_L}{d\Omega} = \frac{1}{M_L}\{ &|\mathbf{U}^0_L|^2 \operatorname{Re}[k_L(\lambda + 2\mu)] \exp[-2\bar{R}(k_{2L} + k^i_{2L})] \\ &+ |\mathbf{U}^0_T|^2 \operatorname{Re}(k_T\mu) \exp[-2\bar{R}(k_{2T} + k^i_{2L})]\}, \end{aligned} \tag{6.8}$$

$$\begin{aligned} \frac{dP_T}{d\Omega} = \frac{1}{M_T}\{ &|\mathbf{U}^0_L|^2 \operatorname{Re}[k_L(\lambda + 2\mu)] \exp[-2\bar{R}(k_{2L} + k^i_{2T})] \\ &+ |\mathbf{U}^0_T|^2 \operatorname{Re}(k_T\mu) \exp[-2\bar{R}(k_{2T} + k^i_{2T})]\}. \end{aligned} \tag{6.9}$$

Altogether, these conceptual difficulties seem to indicate that the scattering cross section may be rightly considered only when the surrounding medium allows disregarding the decay of the waves namely when it can be regarded as elastic.

7.7 High-frequency far field and curvature effects

In the last two sections we have examined a few results that follow from the asymptotic expression of the displacement field, as superposition of a longitudinal and a transverse wave. Such waves have been described in terms of the fields $\mathbf{U}^0_L$ and $\mathbf{U}^0_T$, which in turn are evaluated through integrals over the surface S. Here we investigate the high-frequency limit of these integrals and determine the scattered field at infinity. This is performed by looking for approximate expressions of

$$\mathbf{U}^0_L(\hat{\mathbf{x}}) = U^0_L\, \hat{\mathbf{x}}$$

and

$$\mathbf{U}_T^0(\hat{\mathbf{x}}) = \hat{\mathbf{x}} \times (\mathbf{B} \times \hat{\mathbf{x}}),$$

where $\mathbf{U}_L^0$ and $\mathbf{B}$ depend on $\hat{\mathbf{x}}$ as

$$U_L^0 = -\frac{k_L^2}{4\pi\rho\omega^2}\int_S \{ik_L[\lambda(\mathbf{U}\cdot\mathbf{n}) + 2\mu(\mathbf{U}\cdot\hat{\mathbf{x}})(\mathbf{n}\cdot\hat{\mathbf{x}})] + \mathbf{t}\cdot\hat{\mathbf{x}}\}\exp(-ik_L\,\hat{\mathbf{x}}\cdot\mathbf{y})\,da \tag{7.1}$$

and

$$\mathbf{B} = -\frac{k_T^2}{4\pi\rho\omega^2}\int_S \{ik_T\mu[(\mathbf{U}\cdot\hat{\mathbf{x}})\mathbf{n} + (\mathbf{n}\cdot\hat{\mathbf{x}})\mathbf{U}] + \mathbf{t}\}\exp(-ik_T\,\hat{\mathbf{x}}\cdot\mathbf{y})\,da. \tag{7.2}$$

Evaluation of these double integrals through the method of stationary phase shows that the behaviour of the far field at high frequencies ω is affected by the curvature of S.

The integrands in (7.1) and (7.2) are regarded as functions of $\mathbf{y}$ parameterized by ω through k_L, k_T, λ, and μ. For conciseness we evaluate only the limit of (7.1).

By (3.2.17), the asymptotic dependence of λ and μ on ω can be written as

$$\lambda = \lambda_0 + i\lambda_0'\omega^{-1} + o(\omega^{-1}), \qquad \mu = \mu_0 + i\mu_0'\omega^{-1} + o(\omega^{-1}). \tag{7.3}$$

Then, upon substitution and keeping only the leading terms, we have

$$k_L^2 \simeq \rho\omega^2\frac{\lambda_0 + 2\mu_0 - i(\lambda_0' + 2\mu_0')\omega^{-1}}{(\lambda_0 + 2\mu_0)^2}.$$

Evaluation of k_{1L} and k_{2L} through (3.2.14), in the case $\alpha = 1$, and taking only the leading terms as $\omega \to \infty$ yield the limit functions

$$k_{1L}^\infty = \sqrt{\frac{\rho}{\lambda_0 + 2\mu_0}}\,\omega, \qquad k_{2L}^\infty = -\sqrt{\frac{\rho}{\lambda_0 + 2\mu_0}}\,\frac{\lambda_0' + 2\mu_0'}{2(\lambda_0 + 2\mu_0)}. \tag{7.4}$$

Asymptotically, the real part of $\mathbf{k}_L$ is a linear function of ω while the imaginary part is constant. The change $\lambda + 2\mu \to \mu$ provides the analogous result for k_T. Of course, by (7.3),

$$\lambda \to \lambda_0, \qquad \mu \to \mu_0$$

as $\omega \to \infty$.

Consider the exponential in (7.1) and write

$$\exp(-ik_L\,\hat{\mathbf{x}}\cdot\mathbf{y}) = \exp(k_{2L}\,\hat{\mathbf{x}}\cdot\mathbf{y})\exp(-ik_{1L}\,\hat{\mathbf{x}}\cdot\mathbf{y}).$$

The fact that $k_{2L} \to k_{2L}^\infty$ as $\omega \to \infty$ and $k_{1L} \simeq k_{1L}^\infty \propto \omega$ as $\omega \to \infty$ suggests a natural way of applying the method of stationary phase to the integral (7.1).

According to the method of stationary phase (cf. Appendix, Corollary A.1)

$$\int_{\mathbb{R}^n} f(x,\lambda)\exp[i\lambda\varphi(x)]dx$$
$$\simeq \exp[i\lambda\varphi(x_0)]\frac{f_\infty(x_0)(2\pi)^{n/2}}{\sqrt{|\det\varphi''(x_0)|}}\exp[i\tfrac{\pi}{4}\mathrm{sgn}\,\varphi''(x_0)]\,\lambda^{-n/2} + O(\lambda^{-(1+n/2)}), \tag{7.5}$$

as $\lambda \to \infty$, provided x_0 is the only point of stationary phase φ in the domain D where f is non-vanishing, it is non-degenerate and lies strictly inside D. Here φ'' denotes the Hessian matrix and the writing sgn, applied to a matrix, means the difference between the number of positive eigenvalues and the number of negative ones of the matrix. The parameter λ occurring in (7.5) is naturally identified with k_{1L} and the phase factor is identified with $\exp(-ik_{1L}\,\hat{\mathbf{x}}\cdot\mathbf{y})$.

To ascertain the applicability of (7.5) it is convenient to split the smooth connected surface S into two parts S_+ and S_- defined by

$$S_+ = \{\mathbf{y} \in S, \quad \hat{\mathbf{x}}\cdot\mathbf{y} \geq 0\},$$

$$S_- = \{\mathbf{y} \in S, \quad \hat{\mathbf{x}}\cdot\mathbf{y} \leq 0\}.$$

It is assumed that S_+ and S_- intersect at a common line and that they are connected. If we consider a light source placed at infinity in the direction of $\hat{\mathbf{x}}$ then S_+ may be regarded as the illuminated region and S_- the shadow region. Then we look at (7.1) as a sum of integrals over the two subsets of S and we assume that S_+ is described by the equation $\sigma(\mathbf{y}) = 0$. For definiteness we consider the integral (7.1) on S_+. The results can be extended straightforwardly to S_-.

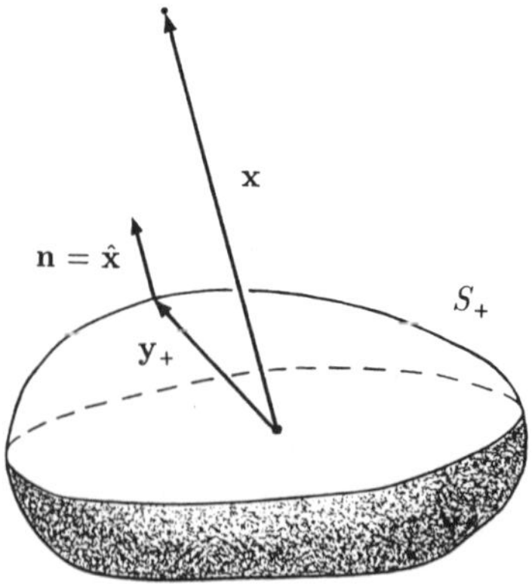

Fig. 7.1 Scattering by the illuminated region.

We identify the phase φ with the scalar product $-\hat{\mathbf{x}}\cdot\mathbf{y}$, where $\mathbf{y}$ satisfies the condition $\sigma(\mathbf{y}) = 0$. The function φ has a stationary point if there exists a Lagrange multiplier τ such that

$$\nabla\big[-\hat{\mathbf{x}}\cdot\mathbf{y} + \tau\sigma(\mathbf{y})\big] = 0, \qquad \sigma(\mathbf{y}) = 0,$$

whence

$$\hat{\mathbf{x}} - \tau \nabla \sigma(\mathbf{y}) = 0.$$

Since in general $\nabla \sigma(\mathbf{y})$ is parallel to the normal at $\mathbf{y}$, we conclude that $\mathbf{y}_+$ is a stationary point if and only if the normal $\mathbf{n}_+$ at $\mathbf{y}_+$ is parallel to $\hat{\mathbf{x}}$. If S_+ is convex then the existence of such a $\mathbf{y}_+$ is guaranteed [145]. Now we take advantage of the parallelism between $\mathbf{n}_+$ and $\hat{\mathbf{x}}$ and choose Cartesian axes such that the 2-dimensional domain S_+ is represented in the form

$$\zeta - h(\xi, \eta) = 0,$$

and the ζ-axis coincides with $\mathbf{n}_+$. Consequently the ζ-axis is also parallel to $\hat{\mathbf{x}}$ and the (ξ, η)-plane is parallel to the tangent plane to S_+ at $\mathbf{y}_+$. It follows that the coordinate pair (ξ, η) of an arbitrary point of S_+ may be identified with the projection on the tangent plane, and the domain of the integral (7.1) is the projection of S_+ on the (ξ, η)-plane. In particular $n = 2$ and $\mathbf{y}_+$ corresponds to the pair $(0,0)$. The phase function φ is taken to be

$$\varphi = -\hat{\mathbf{x}} \cdot \mathbf{y} = -h(\xi, \eta).$$

The stationary value of φ occurs at $(0,0)$ and then

$$\nabla h(0,0) = 0.$$

Of course $\nabla \otimes \nabla \varphi(\xi, \eta) = -\nabla \otimes \nabla h(\xi, \eta)$ at any point of S_+.

If the domain is strictly convex then $\mathbf{y}_+$ is the unique stationary point of φ and is non-degenerate. If, instead, the domain is convex but not strictly convex we can have more (possibly infinitely many) stationary points and they may be degenerate [145]. We assume strict convexity and then $h \leq 0$ in S_+ while, at $\mathbf{y}_+$,

$$\operatorname{sgn}(\nabla \otimes \nabla \varphi) = -\operatorname{sgn}(\nabla \otimes \nabla h) = 2,$$

$$\det(\nabla \otimes \nabla \varphi) = K_+,$$

K_+ being the (positive) Gaussian curvature.

Once the surface element is written in the convenient form $da = \sqrt{1 + (\nabla h)^2} d\xi d\eta$ we find that the integrand of (7.1) can be written in the form

$$F(\xi, \eta) \exp(ik\varphi),$$

where

$$F(\xi, \eta) = -\frac{k_L^2}{4\pi \rho \omega^2} \{ik_L [\lambda(\mathbf{U} \cdot \mathbf{n}) + 2\mu(\mathbf{U} \cdot \hat{\mathbf{x}})(\mathbf{n} \cdot \hat{\mathbf{x}})] + \mathbf{t} \cdot \hat{\mathbf{x}}\} \exp(k_{2L} \hat{\mathbf{x}} \cdot \mathbf{y}) \sqrt{1 + (\nabla h)^2}.$$

and $\mathbf{y} = (\xi, \eta, h(\xi, \eta))$. Let

$$\tilde{\mathbf{t}}_+ = \lim_{\omega \to \infty} \mathbf{t}_+ / \omega.$$

the suffix + denoting the value of the pertinent quantity at $\mathbf{y}_+$. Here we regard $\mathbf{U}$ and $\mathbf{t}$ as given vectors on S. In the case they are known as functions of the parameter ω then $\mathbf{U}_+$ and $\mathbf{t}_+$ are taken as the limit expressions as $\omega \to \infty$. In this sense, since the stress behaves as the gradient of $\mathbf{U}$ then $\tilde{\mathbf{t}}_+$ behaves as $\mathbf{U}$ when $\omega \to \infty$. In view of (7.5) we obtain

$$\int_{S_+} F(\xi,\eta) \exp\left(-ik_{1L}\, \hat{\mathbf{x}} \cdot \mathbf{y}\right) da \simeq \frac{\exp(-ik_L^{\infty}\, \hat{\mathbf{x}} \cdot \mathbf{y}_+)}{2\sqrt{K_+}}\, \hat{\mathbf{x}} \cdot \left[\mathbf{U}_+ - \frac{i}{\sqrt{\rho(\lambda_0 + 2\mu_0)}} \tilde{\mathbf{t}}_+\right] \tag{7.6}$$

to within $O(\omega^{-1})$. Notice also the expression in the right side of (7.6) is independent of the choice of coordinates.

The approximate expression for $\mathbf{B}$ follows by proceeding along the same lines. We only point out that the phase is still given by $-\hat{\mathbf{x}} \cdot \mathbf{y}$ and then the stationary point at S_+ is $\mathbf{y}_+$, as before. Skipping over the details, we write the final result in the form

$$\mathbf{B} \simeq \frac{\exp(-ik_T^{\infty}\, \hat{\mathbf{x}} \cdot \mathbf{y}_+)}{2\sqrt{K_+}} \left[(\mathbf{U}_+ \cdot \hat{\mathbf{x}})\, \hat{\mathbf{x}} + \mathbf{U}_+ - \frac{i}{\sqrt{\rho\mu_0}} \tilde{\mathbf{t}}_+\right].$$

Substitution into the expression of $\mathbf{U}_T^0$ yields

$$\mathbf{U}_T^0(\hat{\mathbf{x}}) \simeq \frac{\exp(-ik_T^{\infty}\, \hat{\mathbf{x}} \cdot \mathbf{y}_+)}{2\sqrt{K_+}}\, \hat{\mathbf{x}} \times \left[(\mathbf{U}_+ - \frac{i}{\sqrt{\rho\mu_0}} \tilde{\mathbf{t}}_+) \times \hat{\mathbf{x}}\right]. \tag{7.7}$$

By (4.8) and (7.6), (7.7) we obtain the high-frequency limit of the displacement field at infinity in the form

$$\begin{aligned} \mathbf{U}(\mathbf{x}) = {} & \frac{\exp(ik_L^{\infty} R)}{R} \left\{ \frac{\exp(-ik_L^{\infty}\, \hat{\mathbf{x}} \cdot \mathbf{y}_+)}{2\sqrt{K_+}}\, \hat{\mathbf{x}} \cdot \left[\mathbf{U}_+ - \frac{i}{\sqrt{\rho(\lambda_0 + 2\mu_0)}} \tilde{\mathbf{t}}_+\right] \right\} \hat{\mathbf{x}} \\ & + \frac{\exp(ik_T^{\infty} R)}{R} \left\{ \frac{\exp(-ik_T^{\infty}\, \hat{\mathbf{x}} \cdot \mathbf{y}_+)}{2\sqrt{K_+}}\, \hat{\mathbf{x}} \times \left(\mathbf{U}_+ - \frac{i}{\sqrt{\rho\mu_0}} \tilde{\mathbf{t}}_+\right) \right\} \times \hat{\mathbf{x}} + O(R^{-2}). \end{aligned} \tag{7.8}$$

These results show that the asymptotic field results from three effects. One is obviously due to the boundary values of the displacement $\mathbf{U}$ and the traction $\mathbf{t}$, at the point $\mathbf{y}_+$. In accordance with §7.4, the scattered field and the related traction at S may be replaced by the corresponding values of the total exterior field. The second one is due to dissipation and shows up through the radial damping factors $\exp(-k_2^{\infty} R)$ but also through the anisotropic factors $\exp(k_2^{\infty} \hat{\mathbf{x}} \cdot \mathbf{y}_+)$. The third one is of geometric origin and is due to the term $1/\sqrt{K_+}$. With equal radii of curvature, the curvature K and the radius of curvature r are related by $K = 1/r^2$ [139]. Then it follows that the amplitude of the displacement field at infinity goes as the radius of curvature at the stationary points. This agrees with the intuitive view that we expect a higher value for the amplitude when the portion of the surface yielding the leading contribution is nearly flat.

The dependence on the inverse square root of the curvature obtained in (7.8) or (7.6) and (7.7) agrees qualitatively with analogous results which hold for the scalar theory (cf.

[145], §2.2). Really, the analysis of [145] is confined to the case when the field vanishes at the boundary of the obstacle. Moreover the normal derivative of the field at the obstacle is found through an application of the so-called Kirchhoff approximation, which means in particular that the field and its normal derivative are assumed to vanish at the shadow boundary. That is why here we have disregarded the contribution of S_-.

7.8 Boundary integral equations

Through the previous analysis we have found an integral representation for the scattered field in the open exterior domain D^+ in terms of the values assumed at the boundary S of the obstacle. On this basis the asymptotic expression of the scattered field has been determined explicitly. A natural question then arises as to whether the given data on S turn out to be the limit of the corresponding quantities as given by the integral representations. The related answer provides the basic framework for the determination of the boundary data to be inserted into the integral representations.

If the obstacle is rigid then the displacement vanishes at S while the surface traction is unknown; if the obstacle is in fact a cavity then the surface traction vanishes at S and the displacement is unknown. In both cases one of the two vectors $\mathbf{U}$ or $\mathbf{t}$ is regarded as given at S and the other one is unknown. The investigation of uniqueness has shown that specification of one datum at S, either $\mathbf{U}$ or $\mathbf{t}$, together with the radiation condition, determines uniquely the scattered field. This means that the data entering the integral representation cannot be chosen independently if continuity of the scattered field with the data is required. This in turn leads to a mathematical problem expressed by an integral equation where the value at S is the unknown.

To evaluate the limit of the integral representation for the displacement when the point $\mathbf{x}$ tends to a point of S we assume that S is a Liapunov surface [126, 110, 155]. This means essentially that S admits a tangent plane at each point, that the tangent plane varies continuously when the point is moved along the surface, and that at each point S may be locally described in the form $\zeta = \zeta(\xi, \eta)$, where the ξ- and η-axis are in the tangent plane at the given point while ζ is taken along the outward normal.

Consider a vector "density" $\mathbf{w}$ defined on S with components satisfying Holder's condition with index not greater than 1. We define the *single-layer potential* as

$$\mathcal{S}\mathbf{w} = \int_S \boldsymbol{\mathcal{G}} \mathbf{w} da_y \tag{8.1}$$

and the *double-layer potential* as

$$\mathcal{D}\mathbf{w} = \int_S \{ \lambda(\mathbf{w} \cdot \mathbf{n}) \nabla \cdot \boldsymbol{\mathcal{G}} + 2\mu \, [\mathrm{sym}(\mathbf{w} \otimes \mathbf{n}) \nabla] \boldsymbol{\mathcal{G}} \} \, da_y, \tag{8.2}$$

at the point $\mathbf{x}$, the variable of integration being $\mathbf{y}$. The single- and double-layer potentials are regarded as linear operators that map functions of S into analytic functions defined over the open sets D^+ or D^-, according to the choice of $\mathbf{x}$. Later we show the connection of the integrals (8.1) and (8.2) with those for single- and double-layer potentials in the theory of harmonic functions [52, 104, 126]. The analogy can also be pushed further. Actually, replacement of the definitions (8.1) and (8.2) into the integral expression (4.1) of the exterior displacement field yields

$$\mathbf{U}(\mathbf{x}) = \mathcal{D}\mathbf{U}(\mathbf{x}) - \mathcal{S}\mathbf{t}(\mathbf{x}), \tag{8.3}$$

provided that only the radiation condition holds. The above integral representation is strictly similar to those holding for harmonic functions and for solutions of the Helmholtz equation [52, 104, 126]. Inside S we have

$$\mathbf{U}(\mathbf{x}) = \mathcal{S}\mathbf{t}(\mathbf{x}) - \mathcal{D}\mathbf{U}(\mathbf{x}). \tag{8.4}$$

Here $\mathbf{U}$ and $\mathbf{t}$ have the usual meaning of displacement and traction at S. For the time being we regard the whole space as occupied by a homogeneous isotropic medium while the surface S has a purely geometric meaning. Scattering problems and heterogeneities will be considered after the introduction of the pertinent mathematical tools.

We now quote without proof a few results concerning the behaviour of single- and double- layer potentials at S. The highly technical aspects of such proofs are outside the scope of the present book and then are omitted; an exhaustive discussion is given in [111], while a shorter account can be found in a paper by Jones [97]. Denote by $\mathbf{x}_S$ any point belonging to the surface S. Addition of superscripts $+$ or $-$ to the argument $\mathbf{x}_S$ means the value of the limit of the function, evaluted for $\mathbf{x} \to \mathbf{x}_S$, from outside or inside, respectively. A superposed $\star$ indicates evaluation of the Cauchy principal value of the expression involved.

The first result is that the single-layer potential $\mathcal{S}\mathbf{w}$ is a continuous function of $\mathbf{x}$. In particular

$$\mathcal{S}\mathbf{w}(\mathbf{x}_S^\pm) = \mathcal{S}\mathbf{w}(\mathbf{x}_S). \tag{8.5}$$

Second, the double-layer potential $\mathcal{D}\mathbf{w}$ tends to a finite limit when $\mathbf{x} \to \mathbf{x}_S$, both from inside and outside, and the limits are given by

$$\mathcal{D}\mathbf{w}(\mathbf{x}_S^\pm) = \pm\tfrac{1}{2}\mathbf{w}(\mathbf{x}_S) + \mathcal{D}^\star\mathbf{w}(\mathbf{x}_S). \tag{8.6}$$

Through the stress operator $\mathbf{T}[\mathbf{U}]$ we can define the traction $\mathbf{t}[\mathbf{U}] = \mathbf{T}[\mathbf{U}]\,\mathbf{n}$, $\mathbf{U}$ being any displacement field. Then we find that

$$\mathbf{t}[\mathcal{S}\mathbf{w}](\mathbf{x}_S^\pm) = \mp\tfrac{1}{2}\mathbf{w}(\mathbf{x}_S) + \mathbf{t}^\star[\mathcal{S}\mathbf{w}](\mathbf{x}_S), \tag{8.7}$$

while the action of the traction operator on a double-layer potential is continuous across S, that is,

$$\mathbf{t}[\mathcal{D}\mathbf{w}](\mathbf{x}_S^{\pm}) = \mathbf{t}^{\star}[\mathcal{D}\mathbf{w}](\mathbf{x}_S). \tag{8.8}$$

To be more precise, if one of the limit values for $\mathbf{t}[\mathcal{D}\mathbf{w}]$ from inside (or outside) exists and is regular, then the other one also exists and the two limits coincide. Sufficient conditions for the existence are given in [111]. It goes without saying that in the right sides of (8.7) and (8.8) the operator $\mathbf{t}$ acts on the current position $\mathbf{x}$.

We are now in a position to consider the (singular) integral equations on the boundary S yielding the required information on the scattered field. First suppose that the obstacle is perfectly rigid, so that at S the total displacement field $\mathbf{U}^+$ vanishes, namely

$$\mathbf{U}^+ = \mathbf{U}^i + \mathbf{U} = 0,$$

where $\mathbf{U}^i$ is the incident field and $\mathbf{U}$ is the scattered field. We already know that $\mathbf{U}$ is completely determined by its value at S. However, before using the representation (8.3), we have to find the unknown traction at the boundary from the given datum $\mathbf{U} = -\mathbf{U}^i$ at S. Taking the limit of (8.3) as $\mathbf{x} \to \mathbf{x}_S$ from the outside, using (8.5) and (8.6), and substituting the data at S we are left with the integral equation

$$\mathcal{S}\mathbf{t}(\mathbf{x}_S) = \tfrac{1}{2}\mathbf{U}^i - \mathcal{D}^{\star}\mathbf{U}^i(\mathbf{x}_S), \tag{8.9}$$

in the unknown surface traction $\mathbf{t}(\mathbf{x}_S)$.

The other limit case considered here is that of an empty inclusion. This corresponds to a vanishing total traction at S, viz

$$\mathbf{t} = -\mathbf{t}^i.$$

Suppose that the unknown displacement has been represented in the form of a single-layer potential, namely

$$\mathbf{U}(\mathbf{x}) = \mathcal{S}\mathbf{w}(\mathbf{x}),$$

where $\mathbf{w}$ is the density defined over S, still to be determined. The action of the traction operator on both sides of this equation yields, after comparison with (8.7),

$$\mathbf{t}[\mathbf{U}](\mathbf{x}_S) = -\tfrac{1}{2}\mathbf{w}(\mathbf{x}_S) + \mathbf{t}^{\star}[\mathcal{S}\mathbf{w}](\mathbf{x}_S);$$

upon substitution of the boundary condition we obtain

$$\mathbf{t}^i(\mathbf{x}_S) = \tfrac{1}{2}\mathbf{w}(\mathbf{x}_S) - \mathbf{t}^{\star}[\mathcal{S}\mathbf{w}](\mathbf{x}_S),$$

which is to be regarded as an integral equation for the single-layer density $\mathbf{w}$.

Further alternative formulations of these problems may be discussed following the lines described in [52, 104] within the framework of the scalar theory of scattering, which is modelled through the Helmholtz equation.

Appendix. Asymptotic behaviour via the method of stationary phase

In connection with the analysis of the far field of §7.7, here we gather the essentials of the asymptotic behaviour of integrals of the form

$$I(\lambda) = \int_{\Omega} f(x) \exp[i\lambda\varphi(x)] dx$$

where φ is a real-valued function, f is possibly complex-valued, $\varphi \in C^{\infty}(\Omega)$, while Ω is a region in $\mathbb{R}^n$. Definite results follow when we estimate $I(\lambda)$ as $\lambda \to \infty$ through asymptotic expansions. This can be performed through the method of stationary phase which is briefly reviewed here.

To begin with we recall that a sequence of functions $\{\varphi_n\}$, defined in some neighbourhoods of the point $x_0 \in \Omega$ is said to be asymptotic as $x \to x_0$ if for all n and $x \to x_0$ we have $|\varphi_{n+1}(x)| = o(|\varphi_n(x)|)$. A formal series $\sum_{n=0}^{\infty} \varphi_n(x)$, $x \to \Omega$, is called an asymptotic expansion of the function $f(x)$, as $\Omega \ni x \to x_0 \in \Omega$, if the sequence of functions $\{\varphi_n\}$ is asymptotic and for any $N < \infty$ there is a neighbourhood Ω_0 of x_0 and a constant γ_N such that

$$\Big|f(x) - \sum_{n=0}^{N} \varphi_n(x)\Big| < \gamma_N |\varphi_{N+1}(x)|, \qquad x \in \Omega_0 \setminus x_0.$$

Moreover, we denote by $O(\lambda^{-\infty})$, as $\lambda \to \infty$, functions $\psi = \psi(\lambda)$ such that $\psi(\lambda) = O(\lambda^{-N})$, as $\lambda \to \infty$, for any $N < \infty$.

For formal simplicity we confine first to the one-dimensional case. Let $\Omega = [a, b]$ and denote by a prime the derivative with respect to x. The next theorem shows the asymptotic behaviour of $I(\lambda)$ and its derivatives.

Lemma A.1. *If $f \in C_0^{\infty}([a, b])$, $\varphi \in C^{\infty}([a, b])$, and $\varphi'(x)$ does not vanish in $[a, b]$ then*

$$dI(\lambda)/d\lambda = O(\lambda^{-\infty}), \qquad \text{as} \quad \lambda \to \infty.$$

Proof. Since $\varphi'(x) \neq 0$ we can write

$$I(\lambda) = \frac{1}{\lambda} \int_a^b f_0(x) i\lambda\varphi'(x) \exp[i\lambda\varphi(x)]\, dx,$$

where $f_0 = f/i\varphi'$. Integration by parts and the observation that $f_0 \in C_0^{\infty}([a, b])$ provide

$$I(\lambda) = -\frac{1}{\lambda} \int_a^b f_1(x) \exp[i\lambda\varphi(x)]\, dx$$

where $f_1 = f_0'$. Successive integrations by parts yield

$$I(\lambda) = \left(-\frac{1}{\lambda}\right)^N \int_a^b f_N(x) \exp[i\lambda\varphi(x)]\, dx$$

$f_N \in C_0^\infty([a,b])$ being the N-th derivative of f_0 with respect to x. Hence $|I(\lambda)| \leq \gamma_N |\lambda|^{-N}$ for any N, namely $I(\lambda) = O(\lambda^{-\infty})$ as $\lambda \to \infty$. □

By the same token we prove that $d^j I(\lambda)/d\lambda^j = O(\lambda^{-\infty})$, for any positive integer j, by starting from

$$\frac{d^j I(\lambda)}{d\lambda^j} = \frac{1}{\lambda} \int_a^b \frac{f(x)[i\varphi(x)]^j}{i\varphi'(x)}\, i\lambda\varphi'(x) \exp[i\lambda\varphi(x)]\, dx$$

and observing that $f\varphi^j/\varphi' \in C_0^\infty([a,b])$.

A point $x_0 \in \Omega$ such that $\varphi'(x_0) = 0$ is called a point of stationary phase. If $\varphi'(x_0) = 0$ and $\varphi''(x_0) \neq 0$ then x_0 is called a non-degenerate point of stationary phase.

A neutralizer at a point x_0 is a C^∞ function equal to unity in some neighbourhood of x_0 and zero outside some larger neighbourhood.

Lemma A.2. *Let $f \in C_0^\infty([a,b])$, $\varphi \in C^\infty([a,b])$, and $x_0 \in (a,b)$ be a point of stationary phase, and the only one. Then, for any neutralizer h at x_0,*

$$\mathcal{I}(\lambda) = \int_a^b [1 - h(x)] f(x) \exp[i\lambda\varphi(x)]\, dx = O(\lambda^{-\infty}), \qquad \text{as} \quad \lambda \to \infty.$$

Proof. Let $[x_0 - \nu, x_0 + \nu] \subset (a,b)$ be the interval where $h = 1$. Then we have

$$\mathcal{I}(\lambda) = \mathcal{I}_-(\lambda) + \mathcal{I}_+(\lambda)$$

where $\mathcal{I}_-$, $\mathcal{I}_+$ are the restrictions of $\mathcal{I}$ to $[a, x_0 - \nu]$, $[x_0 + \nu, b]$, respectively. Further, $(1-h)f \in C_0^\infty([a, x_0 - \nu]) \cup C_0^\infty([x_0 + \nu, b])$. Then, by Lemma A.1,

$$\mathcal{I}_-(\lambda) = O(\lambda^{-\infty}), \qquad \mathcal{I}_+(\lambda) = O(\lambda^{-\infty})$$

whence the desired result. □

Theorem A.1. *If $f \in C_0^\infty([a,b])$ and φ has only one stationary point $x_0 \in [a,b]$, and x_0 is non-degenerate, then*

$$I(\lambda) \sim \exp[i\lambda\varphi(x_0)] \sum_{k=0}^{\infty} a_k\, \lambda^{-(k+1/2)}, \qquad \text{as} \quad \lambda \to \infty, \tag{A.1}$$

where

$$a_0 = f(x_0)\sqrt{\frac{2\pi}{|\varphi''(x_0)|}} \exp[i\tfrac{\pi}{4}\operatorname{sgn}\varphi''(x_0)]$$

and the coefficients a_k, $k \geq 1$, are determined by the values, at $x = x_0$, of f, φ and their derivatives of order not greater than $2k$.

Proof. To fix ideas, let $\varphi''(x_0) > 0$. Since x_0 is a non-degenerate point of stationary phase, letting $\psi(x) = \varphi(x) - \varphi(x_0)$ we have

$$\psi(x) = \tfrac{1}{2}\varphi''(x_0)(x - x_0)^2 + O(|x - x_0|^3) \qquad \text{as } x \to x_0.$$

Then $g(x) := \psi(x)/(x - x_0)^2 \in C^\infty$ and $g(x_0) = \frac{1}{2}\varphi''(x_0) > 0$. Hence, in a neighbourhood of x_0, we can define the change of variables

$$x \to t = \frac{x - x_0}{\sqrt{g(x)}}. \tag{A.2}$$

We have

$$t'(x) = \frac{1}{\sqrt{g(x)}} - \frac{1}{2}\frac{(x - x_0)g'(x)}{g^{3/2}(x)}$$

and

$$g'(x) = \frac{\varphi'(x)}{(x - x_0)^2} - 2\frac{\varphi(x) - \varphi(x_0)}{(x - x_0)^3}.$$

Because $\varphi'(x_0) = 0$ we obtain $t'(x_0) = \sqrt{\varphi''(x_0)/2} > 0$. Accordingly, choose a constant $\delta > 0$ such that $t' > 0$ as $|x - x_0| \leq \delta$. Now take any function $h \in C_0^\infty([a, b])$ such that $h(x) = 1$ as $|x - x_0| < \delta/2$ and $h(x) = 0$ as $|x - x_0| > \delta$. By Lemma A.2 the asymptotic behaviour of $I(\lambda)$ coincides with that of

$$I_1(\lambda) = \int_a^b h(x)f(x)\exp[i\lambda\varphi(x)]\,dx = \exp[i\lambda\varphi(x_0)]\int_a^b h(x)f(x)\exp[i\lambda\psi(x)]\,dx.$$

By the change of variables (A.2) we can write

$$I_2(\lambda) := \int_a^b h(x)f(x)\exp[i\lambda\psi(x)]\,dx = \int_{-\infty}^{\infty} h(x(t))\,f(x(t))\,J(t)\,\exp(i\lambda t^2)\,dt$$

where $J = dx/dt \in C^\infty$, $J(0) = \sqrt{2/\varphi''(x_0)}$. Hence

$$I_2(\lambda) = \int_0^\infty q(t)\exp(i\lambda t^2)\,dt \tag{A.3}$$

where $q(t) = h(x(t))\,f(x(t))\,J(t) + h(x(-t))\,f(x(-t))\,D(-t)$. Let

$$f(x(t))\,J(t) \sim \sum_{k=0}^{\infty} c_k t^k, \qquad \text{as} \quad t \to 0,$$

be the Taylor series of fJ. Since $h = 1$ in a neighbourhood of $x = x_0$ we have

$$q(t) \sim 2\sum_{k=0}^{\infty} c_{2k}\,t^{2k}, \qquad \text{as} \quad t \to 0.$$

To obtain the asymptotic behaviour of I_2 it is convenient to consider integrals of $\exp(i\lambda t^2)$ as follows. Look at the contour Γ_t in the complex z-plane such that, starting from $z = t \in \mathbb{R}^{++}$, we run the segment $[t, t+R]$, then the arc $|z| = R$ up to $z = R\exp(i\pi/4)$ and finally the segment l_t such that $z = t + \rho\exp(i\pi/4)$ as ρ varies from R to zero. By the Cauchy theorem, the integral of $\exp(i\lambda z^2)(t-z)^{j-1}/(j-1)!$ along Γ_t vanishes for any R and then the limit as $R \to \infty$ yields

$$-\int_t^\infty \frac{(t-\tau)^{j-1}}{(j-1)!}\exp(i\lambda\tau^2)\,d\tau = \int_{l_t} \frac{(t-z)^{j-1}}{(j-1)!}\exp(i\lambda z^2)\,dz, \qquad t \in \mathbb{R}^+.$$

This allows us to regard

$$E_j(\lambda, t) = \int_{l_t} \frac{(t-z)^{j-1}}{(j-1)!}\exp(i\lambda z^2)\,dz$$

as an integral, of order j, of $\exp(i\lambda t^2)$. The choice of this form of integral, against the obvious one of the integral over the real interval $[0, t]$, is due to the behaviour as $\lambda \to \infty$. Indeed, substitution for z along l_t gives

$$E_j(\lambda, t) = \frac{(-1)^{j-1}}{(j-1)!}\exp(i\tfrac{\pi}{4}j)\int_\infty^0 \rho^{j-1}\exp[i\lambda(t^2 + 2t\rho\exp(i\tfrac{\pi}{4}) + i\rho^2)]\,d\rho \tag{A.4}$$

whence

$$|E_j(\lambda, t)| \le \frac{1}{(j-1)!}\int_0^\infty \rho^{j-1}\exp[-\lambda(2t\rho + \rho^2)]\,d\rho \le \frac{1}{(j-1)!}\int_0^\infty \rho^{j-1}\exp(-\lambda\rho^2)\,d\rho.$$

Hence, letting $\xi = \sqrt{\lambda}\rho$, we obtain

$$|E_j(\lambda, t)| \le C_j\lambda^{-j/2}, \qquad t \in \mathbb{R}^+,$$

where

$$C_j = \frac{1}{(j-1)!}\int_0^\infty \xi^{j-1}\exp(-\xi^2)\,d\xi.$$

Integration by parts, of (A.3), $2N+2$ times yields

$$I_2(\lambda) = \sum_{j=1}^{2N+2}(-1)^{j-1}\left[q^{j-1}(t)\,E_j(\lambda, t)\right]_0^\infty + \int_0^\infty q^{2N+2}(t)\,E_{2N+2}(\lambda, t)\,dt$$

and the integral is $O(\lambda^{-(N+1)})$. Meanwhile, the behaviour of $q(t)$ as $t \to 0$ shows that

$$q^h(0) = \begin{cases} 2\,h!\,c_h, & h \text{ even} \\ 0, & h \text{ odd} \end{cases}$$

Then

$$I_2(\lambda) = -2c_0 E_1(\lambda,0) - \sum_{k=1}^{N} 2\,(2k)!c_{2k}E_{2k+1}(\lambda,0) + O(\lambda^{-(N+1)}).$$

Now, by (A.4) we have

$$E_{2k+1}(\lambda,0) = -\frac{1}{(2k)!}\exp[i\tfrac{\pi}{4}(2k+1)]\int_0^\infty \rho^{2k}\exp(-\lambda\rho^2)\,d\rho$$

Hence, because

$$\int_0^\infty \rho^{2k}\exp(-\lambda\rho^2)\,d\rho = \frac{(2k-1)!!\sqrt{\pi}}{2^{k+1}}\lambda^{-(k+1/2)},$$

we have

$$E_1(\lambda,0) = \exp(i\tfrac{\pi}{4})\frac{\sqrt{\pi}}{2}\lambda^{-1/2}, \qquad E_{2k+1}(\lambda,0) = -\frac{1}{2}\Gamma(\tfrac{2k+1}{2})\frac{1}{(2k)!}\exp[i\tfrac{\pi}{4}(2k+1)]\lambda^{-(k+1/2)}.$$

Accordingly,

$$I_2(\lambda) \sim f(x_0)\sqrt{\frac{2\pi}{\varphi''(x_0)}}\exp(i\tfrac{\pi}{4})\lambda^{-1/2} + \sum_{k=1}^{\infty}\Gamma(\tfrac{2k+1}{2})\exp[i\tfrac{\pi}{4}(2k+1)]c_{2k}\lambda^{-(k+1/2)}$$

as $\lambda \to \infty$. This leads to the formula (A.1) for $I_1(\lambda)$. The coefficients a_k are then defined as

$$a_k = \Gamma(\tfrac{2k+1}{2})\exp[i\tfrac{\pi}{4}(2k+1)]\,c_{2k}.$$

By the definition of c_{2k}, it follows that they involve the derivatives of f, and then of φ, with respect to t up to order $2k$.

If, instead, $\varphi''(x_0) < 0$, then we consider the complex conjugate functional

$$I^*(\lambda) = \int_a^b f^*(x)\exp[-i\lambda\varphi(x)]\,dx.$$

Then we can repeat step by step the previous procedure to get

$$I^*(\lambda) = \exp[-i\lambda\varphi(x_0)][f^*(x_0)\sqrt{2\pi/[-\varphi''(x_0)]}\exp(i\tfrac{\pi}{4}) + \sum_{k=1}^{\infty}\bar{a}_k\lambda^{-(k+1/2)}$$

where the $\bar{a}_k$'s are now defined through the transformation $x(t)$ induced by

$$\psi(x) = -\tfrac{1}{2}\varphi''(x_0)(x-x_0)^2 + O(|x-x_0|^3).$$

Taking the complex conjugate yields

$$I(\lambda) = \exp[i\lambda\varphi(x_0)][f(x_0)\sqrt{2\pi/[-\varphi''(x_0)]}\exp(-i\tfrac{\pi}{4}) + \sum_{k=1}^{\infty}\bar{a}_k^*\lambda^{-(k+1/2)},$$

which completes the proof. □

Theorem A.1 is now carried over to the *multidimensional case*. Let $x \in \mathbb{R}^n$ and still denote by a prime the (gradient) derivative with respect to x. Consider

$$I(\lambda) = \int_{\mathbb{R}^n} f(x) \exp[i\lambda\varphi(x)]\, dx, \qquad f, \varphi \in C^\infty(\mathbb{R}^n),$$

where f is complex-valued and φ is real-valued. Moreover, assume that there is a domain $D \subset \mathbb{R}^n$ such that $f(x) = 0$ as $x \in \mathbb{R}^n \setminus D$. Preliminarily we prove the following property of $I(\lambda)$.

Lemma A.3. *If $\varphi' \neq 0$ in $\bar{D}$ then*

$$I(\lambda) = O(\lambda^{-\infty}) \qquad as \quad \lambda \to \infty.$$

Proof. Consider the first-order scalar differential operator

$$L = \sum_{j=1}^{n} \left(\frac{1}{|\varphi'|^2} \frac{\partial \varphi}{\partial x_j} \frac{\partial}{\partial x_j} \right) .$$

The formally adjoint operator $\tilde{L}$ is given by

$$\tilde{L}u = -\sum_{j=1}^{n} \frac{\partial}{\partial x_j} \left(\frac{u}{|\varphi'|^2} \frac{\partial \varphi}{\partial x_j} \right) .$$

Since $L\ \exp(i\lambda\varphi) = i\lambda \exp(i\lambda\varphi)$, we have

$$I = \frac{1}{i\lambda} \int_{\mathbb{R}^n} f\ L \exp(i\lambda\varphi)\, dx = \frac{1}{i\lambda} \int_{\mathbb{R}^n} (\tilde{L}\ f) \exp(i\lambda\varphi)\, dx.$$

Repeating the procedure N times and estimating the resulting integral in terms of the maximum modulus of the integrand proves that $I(\lambda) = O(\lambda^{-N})$. The arbitrariness of N provides the desired result for the function $I(\lambda)$. □

By first differentiating with respect to λ and then repeating the procedure we obtain the same result for the derivatives of I.

A point $x_0 \in \mathbb{R}^n$ is said to be a point of stationary phase if $\varphi'(x_0) = 0$. A point of stationary phase x_0 is said to be non-degenerate if det $\varphi''(x_0) \neq 0$, where φ'' denotes the Hessian matrix.

Theorem A.2. *Let x_0 be one point, and the only one, of stationary phase in D. Moreover let x_0 be non-degenerate and lie strictly inside D. Then*

$$I(\lambda) \simeq \exp[i\lambda\varphi(x_0)] \sum_{k=0}^{\infty} a_k \lambda^{-(k+n/2)}, \qquad as \quad \lambda \to \infty, \tag{A.5}$$

where

$$a_0 = \frac{f(x_0)\,(2\pi)^{n/2}}{\sqrt{|\det\varphi''(x_0)|}} \exp[i\tfrac{\pi}{4}\mathrm{sgn}\varphi''(x_0)].$$

Proof. Since x_0 is nondegenerate we can write

$$\varphi(x) - \varphi(x_0) = \tfrac{1}{2}(x - x_0, \varphi''(x_0)(x - x_0)) + O(|x - x_0|^3)$$

where $(\cdot,\cdot)$ denotes the usual inner product in $\mathbb{R}^n$. Let μ_i be the i-th eigenvalue of the symmetric matrix $\varphi''(x_0)$. Then by Morse Lemma (cf. [127]) there exists a neighbourhood U_δ of x_0, $U_\delta = \{x : |x - x_0| \leq \delta\}$, such that a change of variables

$$x \to y = y(x), \quad y(x_0) = 0,$$

exists with $y \in C^\infty(U_\delta)$, $y'(x_0) = \mathbf{1}$, and that in the new variables the function φ becomes

$$\varphi(y) - \varphi(0) = \tfrac{1}{2}\sum_{j=1}^{n} \mu_j y_j^2.$$

Further, letting $\mu_1, .., \mu_r$ be the positive eigenvalues and $\mu_{r+1}, .., \mu_n$ the negative ones, the change of variables $z_j = \sqrt{|\mu_j|}y_j$ allows us to write

$$\varphi(z) - \varphi(0) = \frac{1}{2}\Big(\sum_{j=1}^{r} z_j^2 - \sum_{j=r+1}^{n} z_j^2\Big) =: \gamma(z).$$

So we let $n - r$ be the index of $\varphi''(x_0)$. Consider a neutralizer $h(x)$ such that $h(x) = 1$ as $|x - x_0| < \delta/2$ and $h(x) = 0$ as $|x - x_0| > \delta$. Then, by Lemma A.3,

$$\int_{\mathbb{R}^n} (1 - h) f \exp(i\lambda\varphi)\, dx = O(\lambda^{-\infty}) \qquad \text{as} \quad \lambda \to \infty.$$

Accordingly it is enough to prove the theorem for

$$I_1(\lambda) = \int_{\mathbb{R}^n} h(x)\, f(x)\, \exp[i\lambda\varphi(x)]\, dx = \exp[i\lambda\varphi(x_0)] \int_{\mathbb{R}^n} (hfJ)(x(z)) \exp[i\lambda\gamma(z)] dz$$

where $w = (z_1, .., z_r, -z_{r+1}, .., -z_n)$ and J is now the Jacobian of the change of variables $z \to x$. Since $\det y'(x_0) = 1$ then, at $z = 0$,

$$J = |\prod_{j=1}^{n} \mu_j|^{(-1/2)} = |\det \varphi''(x_0)|^{(-1/2)}.$$

Now, following [15], let $F(z) = (fJ)(z)$ and consider the identity

$$F(z) = F(0) + (w, H)$$

where H is the n-tuple defined by

$$H_1 = \frac{1}{z_1}[F(z_1,..,z_n) - F(0,z_1,..,z_n)],$$

$$H_2 = \frac{1}{z_2}[F(0,z_2,..,z_n) - F(0,0,z_3,..,z_n)],$$

$$\vdots$$

$$H_r = \frac{1}{z_r}[F(0,..,z_r,z_{r+1},..,z_n) - F(0,..,z_{r+1},..,z_n)],$$

$$H_{r+1} = -\frac{1}{z_{r+1}}[F(0,..,0,z_{r+1},..,z_n) - F(0,..,0,z_{r+2},..,z_n)],$$

$$\vdots$$

$$H_n = -\frac{1}{z_n}[F(0,..,0,z_n) - F(0,..,0,0)].$$

Then, letting

$$\mathcal{I}_0(\lambda) = F(0)\int_{\mathbb{R}^n} h(z)\exp[i\lambda\gamma(z)]\,dz, \qquad \mathcal{I}_1(\lambda) = \int_{\mathbb{R}^n}(w,H)\,h(z)\exp[i\lambda\gamma(z)]\,dz,$$

we can write

$$I_1(\lambda) = \exp[i\lambda\varphi(x_0)][\mathcal{I}_0(\lambda) + \mathcal{I}_1(\lambda)].$$

As regards I_1, observe that $w\exp[i\lambda\gamma(z)] = (i\lambda)^{-1}\partial\exp[i\lambda\gamma(z)]/\partial z$ and then

$$(w,H)\,h(z)\,\exp[i\lambda\gamma(z)] = (i\lambda)^{-1}\left\{\mathrm{div}\big[H\,h(z)\exp[i\lambda\gamma(z)]\big] - \exp[i\lambda\gamma(z)]\mathrm{div}\,[H\,h(z)]\right\}$$

where div denotes the divergence with respect to z. The divergence theorem and the vanishing of h outside a compact support make the contribution of the first term vanish. Hence

$$\mathcal{I}_1(\lambda) = -\frac{1}{i\lambda}\int_{\mathbb{R}^n}\big[h\,\mathrm{div}H + \big(H,\frac{\partial h}{\partial z}\big)\big]\exp[i\lambda\gamma(z)]\,dz.$$

Owing to the factor λ^{-1}, $\mathcal{I}_1(\lambda)$ is of lower order asymptotic than $I_1(\lambda)$ and then $\mathcal{I}_0(\lambda)$ must contain the leading term. Let $h_j(z_j), j = 1,..,n$, be one-dimensional neutralizers and write

$$\mathcal{I}_0(\lambda) = \int_{\mathbb{R}^n}\prod_{j=1}^{n} h_j(z_j)\exp[i\lambda\gamma(z)]\,dz + \int_{\mathbb{R}^n}[h(z) - \prod_{j=1}^{n} h_j(z_j)]\exp[i\lambda\gamma(z)]\,dz.$$

The second integral vanishes identically at a neighbourhood of the point of stationary phase and then it is $O(\lambda^{-\infty})$. The evaluation of the first integral is immediate by observing that

$$\int_{\mathbb{R}^n}\prod_{j=1}^{n} h_j(z_j)\exp[i\lambda\gamma(z)]\,dz = \prod_{j=1}^{n}\int_{-\infty}^{\infty} h_j(z_j)\exp[i\lambda w_j z_j/2]\,dz_j.$$

For each value of j we can thus apply the procedure for the evaluation of $I_2(\lambda)$ in Theorem A.2 to get

$$\mathcal{I}_0(\lambda) = F(x_0)\left(\frac{2\pi}{\lambda}\right)^{n/2} \exp[i\tfrac{\pi}{4}\operatorname{sgn}\varphi''(x_0)] + O(\lambda^{-(1+n/2)})$$

where $O(\lambda^{-(1+n/2)})$ stands for a power series in $\lambda^{-(k+n/2)}$, $k = 1, 2, \dots$ Then we evaluate

$$I_1(\lambda) \simeq \exp[i\lambda\varphi(x_0)]\mathcal{I}_0(\lambda)$$

whence the sought result (A.5) for $I(\lambda)$. □

In applications we have sometimes to deal with integrals of the form

$$\tilde{I}(\lambda) = \int_{\mathbb{R}^n} f(x,\lambda)\exp[i\lambda\varphi(x)]\,dx, \qquad \varphi \in C^\infty(\mathbb{R}^n),$$

where f is viewed as a function of x parameterized by λ. Further, for the cases we are interested in, f admits the asymptotic limit

$$f_\infty(x) = \lim_{\lambda\to\infty} f(x,\lambda) \in C^\infty(\mathbb{R}^n). \tag{A.6}$$

Then the leading term of $\tilde{I}(\lambda)$, as $\lambda \to \infty$, corresponds to $f_\infty(x)$ for which Theorem A.2 applies as far as a_0 is concerned.

Corollary A.1. *Let x_0 be a point, and the only one, of stationary phase in D. Moreover let x_0 be non-degenerate and lie strictly inside D. If $f(x,\lambda)$ meets the requirement (A.6) then*

$$\tilde{I}(\lambda) \simeq \exp[i\lambda\varphi(x_0)]\frac{f_\infty(x_0)(2\pi)^{n/2}}{\sqrt{|\det\varphi''(x_0)|}}\exp[i\tfrac{\pi}{4}\operatorname{sgn}\varphi''(x_0)]\,\lambda^{-n/2} + O(\lambda^{-(1+n/2)})$$

as $\lambda \to \infty$.

8 PERTURBATION METHODS IN HETEROGENEOUS MEDIA

The model of homogeneous body is obviously a limit case in that the material properties of real bodies are usually dependent on the position. There are cases where the variation of the material properties over the region affected by a propagating wave is small enough to allow the approximation of homogeneous body with admissible errors. Against these errors is the fact that the mathematical apparatus is much simpler and often solutions can be obtained in closed form. There are situations, though, where the material properties vary rapidly with distance or the region interested by the propagation of the wave is very large, as it happens, e.g., in underwater acoustic exploration. Of course in such cases the heterogeneity of the body has to be incorporated and appropriate mathematical procedures are required to obtain definite results. Needless to say, inhomogeneous wave solutions as such are ruled out by the heterogeneity as well as plane waves that are not allowed in heterogeneous elastic bodies.

Apart from a limited number of cases, wave propagation in heterogeneous bodies is investigated by means of approximation methods. This chapter exhibits some of these methods. The first one may be viewed as the Born approximation; the heterogeneity is ultimately modelled as an equivalent body force and the mathematical problem is treated through the use of Green's functions for the background homogeneous body. The second one is the WKB method. Again at the expense of a preliminary approximation, the method yields closed-form solutions. In this framework, turning points and successive approximations are topics and methods associated with suggestive interpretations about wave behaviour. The ray method involves another procedure for the study of wave propagation in heterogeneous media. Owing to the peculiar features connected with dissipation, it is investigated at length in the next chapter.

8.1 The Born approximation

In this section we consider a time-harmonic wave travelling within a heterogeneous isotropic viscoelastic solid. We have in mind, however, that the heterogeneity is in fact a small perturbation of the equilibrium, homogeneous state. Our problem here is to establish an approximate solution for wave propagation in the perturbed material. This problem is of wide interest in seismology or non-destructive testing of materials [77]. Indeed, it is usual to regard the earth as a homogeneous material affected by small-scale heterogeneities; hence any background signal, say a plane wave, propagating within the earth is distorted by the heterogeneities and the scattered field is then identified with the first-order perturbation.

For formal convenience we denote by a superposed bar the actual values of the material parameters $\rho, \mu_0, \lambda_0, \mu', \lambda'$ which are allowed to depend on the position $\mathbf{x}$. Accordingly, we write the equations of motion as

$$\bar{\rho}\ddot{\mathbf{u}}(\mathbf{x},t) = \nabla \cdot \mathbf{T},$$

where the Cauchy stress $\mathbf{T}(\mathbf{x},t)$ is given by

$$\mathbf{T} = 2\bar{\mu}_0 \mathbf{E} + \bar{\lambda}_0 \,\mathrm{tr}\mathbf{E}\,\mathbf{1} + \int_0^\infty \left[2\bar{\mu}'(s)\mathbf{E}^t + \bar{\lambda}'(s)\,\mathrm{tr}\mathbf{E}^t\,\mathbf{1}\right] ds.$$

The dependence of $\bar{\mu}_0$, $\bar{\lambda}_0$, $\bar{\mu}'(s)$, $\bar{\lambda}'(s)$ on the position $\mathbf{x}$ is understood. Because of the thermodynamic restrictions (2.4.7) and the constitutive property $\mathbf{G}_\infty > 0$, these quantities are required to satisfy (cf. [70])

$$\bar{\mu}_0 > \bar{\mu}_\infty > 0, \qquad 3\bar{\lambda}_0 + 2\bar{\mu}_0 > 3\bar{\lambda}_\infty + 2\bar{\mu}_\infty > 0, \tag{1.1}$$

and

$$\bar{\mu}'_s(\omega) < 0, \qquad 3\bar{\lambda}'_s(\omega) + 2\bar{\mu}'_s(\omega) < 0, \quad \omega \in \mathbb{R}^{++}. \tag{1.2}$$

We look for solutions to the equation of motion in the form of time-harmonic waves, namely

$$\mathbf{u}(\mathbf{x},t) = \mathbf{U}(\mathbf{x};\omega)\exp(-i\omega t);$$

owing to the heterogeneity of the medium we cannot expect that the dependence of $\mathbf{U}$ on $\mathbf{x}$ is in the form of plane waves. Upon evaluation of the Cauchy stress, the equation of motion takes the form

$$\mathcal{L}\mathbf{U} = 0 \tag{1.3}$$

where the linear operator $\mathcal{L}$ is defined by

$$\begin{aligned}\mathcal{L}\mathbf{U} &= \bar{\rho}\omega^2\mathbf{U} + \nabla \cdot \left[\bar{\mu}(\nabla\mathbf{U} + \nabla\mathbf{U}^\dagger) + \bar{\lambda}(\nabla\cdot\mathbf{U})\,\mathbf{1}\right] \\ &= \bar{\rho}\omega^2\mathbf{U} + \bar{\mu}\Delta\mathbf{U} + (\bar{\mu}+\bar{\lambda})\nabla(\nabla\cdot\mathbf{U}) + \nabla\bar{\mu}(\nabla\mathbf{U} + \nabla\mathbf{U}^\dagger) + \nabla\bar{\lambda}(\nabla\cdot\mathbf{U}),\end{aligned}$$

and the material parameters $\bar{\mu}$, $\bar{\lambda}$ are defined as

$$\begin{aligned}\bar{\lambda}(\mathbf{x};\omega) &= \bar{\lambda}_0(\mathbf{x}) + \int_0^\infty \bar{\lambda}'(\mathbf{x},s)\exp(i\omega s)ds, \\ \bar{\mu}(\mathbf{x};\omega) &= \bar{\mu}_0(\mathbf{x}) + \int_0^\infty \bar{\mu}'(\mathbf{x},s)\exp(i\omega s)ds.\end{aligned}$$

We let the state of the material be obtained by a weak perturbation of a given background state and regard as known a solution to the equations of motion in the unperturbed state. Accordingly we represent the complex material parameters as

$$\begin{cases} \bar{\rho} = \rho_0 + \varepsilon\rho_1, \\ \bar{\lambda} = l_0 + \varepsilon l_1, \\ \bar{\mu} = m_0 + \varepsilon m_1, \end{cases}$$

where ρ_0, l_0, m_0 are the background values, while ρ_1, l_1 and m_1 are the corresponding perturbations, with common compact support Ω; the real parameter ε is introduced to account for the smallness of the perturbation. The operator $\mathcal{L}$ can be written in the form

$$\mathcal{L} = \mathcal{L}_0 + \varepsilon\mathcal{L}_1, \tag{1.4}$$

where

$$\mathcal{L}_0\mathbf{U} = \rho_0\omega^2\mathbf{U} + \nabla\cdot\left[m_0(\nabla\mathbf{U} + \nabla\mathbf{U}^\dagger) + l_0(\nabla\cdot\mathbf{U})\mathbf{1}\right],$$

$$\mathcal{L}_1\mathbf{U} = \rho_1\omega^2\mathbf{U} + \nabla\cdot\left[m_1(\nabla\mathbf{U} + \nabla\mathbf{U}^\dagger) + l_1(\nabla\cdot\mathbf{U})\mathbf{1}\right].$$

Consider the equation (1.3) and represent the displacement field $\mathbf{U}$ as

$$\mathbf{U} = \mathbf{U}_0 + \varepsilon\mathbf{U}_1, \tag{1.5}$$

where $\mathbf{U}_0$ solves the equation of motion in the background medium, that is

$$\mathcal{L}_0\mathbf{U}_0 = 0.$$

Then, upon substitution of (1.4) in (1.3), we have

$$\mathcal{L}_0\mathbf{U}_1 = -\mathcal{L}_1\mathbf{U}.$$

Suppose that the background operator $\mathcal{L}_0$ may be inverted, and denote by $\mathcal{L}_0^{-1}$ the inverse operator. Indeed, $\mathcal{L}_0^{-1}$ may be thought of in the form of a functional involving the Green's function of the background medium and depending on the boundary conditions. Formal inversion gives the relation

$$\mathbf{U}_1 = -\mathcal{L}_0^{-1}\mathcal{L}_1\mathbf{U},$$

which, by (1.5), may be regarded as an integral equation for $\mathbf{U}_1$. Under the assumption that $\varepsilon\mathbf{U}_1$ may be disregarded in comparison with $\mathbf{U}_0$, the field $\mathbf{U}$ in the right side of the expression of $\mathbf{U}_1$ may be replaced simply by $\mathbf{U}_0$, to give the representation

$$\mathbf{U}_1 = -\mathcal{L}_0^{-1}\mathcal{L}_1\mathbf{U}_0.$$

This approximation is usually named after Born, because Born used it in atomic theory, although it had already been introduced by Rayleigh and Gans. A discussion of the accuracy of this approximation in the framework of seismic scattering from small-scale heterogeneities can be found in [93, 175].

Alternatively [156, 60], we may well view the Born approximation as the first step in the search for solutions to (1.3) that are expressed in the form of a perturbation series as

$$\mathbf{U} = \mathbf{U}_0 + \varepsilon \mathbf{U}_1 + O(\varepsilon^2). \tag{1.6}$$

By means of (1.4) we write (1.3) as

$$(\mathcal{L}_0 + \varepsilon \mathcal{L}_1)\mathbf{U} = 0. \tag{1.7}$$

The condition $\mathcal{L}_0 \mathbf{U}_0 = 0$ results in the vanishing of the zeroth-order term in ε. Then, by (1.6) and (1.7), the vanishing of the first-order term in ε yields

$$\mathcal{L}_0 \mathbf{U}_1 = -\mathcal{L}_1 \mathbf{U}_0, \tag{1.8}$$

which is the differential counterpart for the representation of $\mathbf{U}_1$ given by the Born approximation. Here the field $-\mathcal{L}_1 \mathbf{U}_0$ plays the role of a source for the unknown first-order perturbation field $\mathbf{U}_1$.

In general, finding a solution to (1.8) with a generic heterogeneity is a difficult task. An interesting procedure is developed in [12], where the representation of an elastic scattered field $\mathbf{U}_1$ is evaluated and then employed as the starting point for an algorithm devised to determine the solution of an inverse scattering problem in a medium with jump discontinuities in the material parameters. In the next section we examine the case when the background medium is homogeneous and the heterogeneities are confined to a finite region.

8.2 Perturbation field generated by small heterogeneities

For definiteness we regard the material as a viscoelastic solid. On applying the Born approximation we investigate the scattering of inhomogeneous waves by small heterogeneities confined to a finite region. To this end we essentially follow the method of [37]. An alternative procedure is exhibited in [175], through the equivalent-source method, in connection with elastic bodies.

It is reasonable, in many circumstances, to regard the memory effects as independent of the position in the sense that the dependences of $\bar{\lambda}'$ and $\bar{\mu}'$ on $\mathbf{x}$ and s can be factorized as

$$\bar{\lambda}'(\mathbf{x}, s) = -\bar{\lambda}_0(\mathbf{x})\, \tau(s), \qquad \bar{\mu}'(\mathbf{x}, s) = -\bar{\mu}_0(\mathbf{x})\, \tau(s);$$

although inessential, the function τ on $\mathbb{R}^+$ is often regarded as monotone decreasing. This implies that

$$\bar{\lambda}(\mathbf{x}) = \bar{\lambda}_0(\mathbf{x})\,(1 - \hat{\tau}), \qquad \bar{\mu}(\mathbf{x}) = \bar{\mu}_0(\mathbf{x})\,(1 - \hat{\tau}),$$

where $\hat{\tau}$ is the complex constant parameter

$$\hat{\tau} = \int_0^\infty \tau(s)\exp(i\omega s)ds.$$

The thermodynamic restrictions (1.1)-(1.2) demand that the non-zero function τ, on $\mathbb{R}^+$, meets the condition

$$\operatorname{Im}\hat{\tau}(\omega) = \int_0^\infty \tau(s)\sin\omega s\, ds > 0, \qquad \omega \in \mathbb{R}^{++}.$$

Of course, letting $\tau = 0$ means that the solid is elastic.

Letting the background state be homogeneous we can take

$$\bar{\lambda}_0(\mathbf{x}) = \lambda_0 + \varepsilon\, \lambda_1(\mathbf{x}),$$
$$\bar{\mu}_0(\mathbf{x}) = \mu_0 + \varepsilon\, \mu_1(\mathbf{x}),$$

where λ_0 and μ_0 are real constants, while the real functions λ_1 and μ_1 of $\mathbf{x}$ have the region Ω as their common support. The assumed properties are then summarized as

$$\begin{cases} \bar{\lambda}(\mathbf{x}) = [\lambda_0 + \varepsilon\, \lambda_1(\mathbf{x})]\,(1-\hat{\tau}), \\ \bar{\mu}(\mathbf{x}) = [\mu_0 + \varepsilon\, \mu_1(\mathbf{x})]\,(1-\hat{\tau}), \\ \bar{\rho}(\mathbf{x}) = \rho_0 + \varepsilon\, \rho_1(\mathbf{x}), \end{cases} \tag{2.1}$$

whence we see that the complex parameters l_0, m_0, l_1, m_1 are expressed as

$$l_0 = \lambda_0(1-\hat{\tau}), \qquad m_0 = \mu_0(1-\hat{\tau}),$$

$$l_1 = \lambda_1(1-\hat{\tau}), \qquad m_1 = \mu_1(1-\hat{\tau}).$$

We now investigate some properties of the solution $\mathbf{U}_1$ to the differential equation (1.6). We can write this equation as

$$\rho_0\omega^2\mathbf{U}_1 + \nabla\cdot\{m_0[\nabla\mathbf{U}_1 + (\nabla\mathbf{U}_1)^\dagger] + l_0(\nabla\cdot\mathbf{U}_1)\mathbf{1}\} = -\mathbf{f} \tag{2.2}$$

where

$$\mathbf{f} = \rho_1\omega^2\mathbf{U}_0 + (1-\hat{\tau})\,\nabla\cdot\{\mu_1[\nabla\mathbf{U}_0 + (\nabla\mathbf{U}_0)^\dagger] + \lambda_1(\nabla\cdot\mathbf{U}_0)\mathbf{1}\}. \tag{2.3}$$

Apart from trivial changes, equation (2.2) is just of the form (7.2.5). This means that the behaviour of the perturbation field $\mathbf{U}_1$ is governed by the same differential equations as is the background field, with a source term that depends on the incident (background) field and the spatial distribution of heterogeneities. Accordingly the integral representation theorem (7.2.16) applies to solutions of (2.2) to give

$$\mathbf{U}^1(\mathbf{x}) = \int_{\mathcal{V}} \boldsymbol{\mathcal{G}}\,\mathbf{f}\, dy + \int_{\partial\mathcal{V}} \{\boldsymbol{\mathcal{G}}\,\mathbf{t}_1 - l_0(\mathbf{U}_1\cdot\mathbf{n})\,\nabla\cdot\boldsymbol{\mathcal{G}} - 2m_0\,[\operatorname{sym}(\mathbf{U}_1\otimes\mathbf{n})\nabla]\boldsymbol{\mathcal{G}}\}\, da_y, \tag{2.4}$$

where $\mathcal{V}$ is any regular region in space containing the support Ω of the perturbing terms ρ_1, λ_1, μ_1 and $\partial\mathcal{V}$ denotes the boundary of $\mathcal{V}$. Since $\mathbf{x}$ is now the position where we determine the unknown $\mathbf{U}_1$, we denote by $\mathbf{y}$ the current position in $\mathcal{V}$, and by dy, da_y the volume element and the surface element. Here $\mathcal{G}(\mathbf{y},\mathbf{x})$ denotes the Green's tensor function of the background state; letting $r = |\mathbf{y} - \mathbf{x}|$, by (7.2.9) and (7.2.10) we can write

$$\mathcal{G}(\mathbf{y},\mathbf{x}) = \frac{1}{4\pi\rho_0\omega^2}[k_T^2\, g_T\, \mathbf{1} - \nabla \otimes \nabla(g_L - g_T)],$$

where

$$g_{L,T} = \frac{\exp(ik_{L,T}r)}{r},$$

and

$$k_T^2 = \frac{\rho_0\omega^2}{m_0}, \qquad k_L^2 = \frac{\rho_0\omega^2}{2m_0 + l_0}.$$

Incidentally, $\mathrm{Re}\, k_{L,T} > 0$.

Although (2.4) provides the expression of $\mathbf{U}_1$ in terms of $\mathbf{U}_0$, the heterogeneities, and the values assumed by $\mathbf{U}_1$ and $\mathbf{t}_1$ at the boundary, the determination of the last set of data is a rather formidable task. A remarkable simplification occurs if, motivated by the localization of the sources for the field $\mathbf{U}_1$, we allow $\mathbf{U}_1$ to satisfy the radiation condition. Then the volume $\mathcal{V}$ may be identified with, say, a sphere centred at $\mathbf{x}$ with radius approaching infinity, and the radiation condition guarantees that the contribution coming from the surface integral approaches zero. In this context we have

$$\mathbf{U}_1(\mathbf{x}) = \int_{\mathcal{V}} \mathcal{G}(\mathbf{y},\mathbf{x})\, \mathbf{f}(\mathbf{y})\, dy, \tag{2.5}$$

where $\mathcal{V}$ may be identified with the whole three-dimensional space. Really, the vector function $\mathbf{f}$ has a compact support which is determined by the support of the perturbation functions ρ_1, λ_1, and μ_1. Indeed, consistent with the assumption that the radiation condition holds, we restrict attention to those situations when the diameter of the support Ω is small relative to the distance between $\mathbf{x}$ and Ω itself.

We now investigate the far-field limit of the scattered field $\mathbf{U}_1(\mathbf{x})$ as given by (2.5). In a sense, this point may be regarded as a natural complement to the analysis of the asymptotic behaviour of the scattered field that is performed in the previous chapter. There we start from the integral representation of the scattered field and examine the asymptotic form, but in the absence of the volume integral of (volume) sources. Here instead, we are considering the effect of volume sources, while surface sources related to discontinuities in the material parameters are disregarded.

Describe the position vector relative to an origin $O \in \Omega$. By following the notation of §7.4 we set

$$R = |\mathbf{x}|, \qquad \hat{\mathbf{x}} = \mathbf{x}/R, \qquad r = |\mathbf{y} - \mathbf{x}|.$$

We have

$$g_{L,T} = \frac{\exp(ik_{L,T}r)}{r} = \frac{\exp(ik_{L,T}R)}{R}\exp(-ik_{L,T}\,\hat{\mathbf{x}}\cdot\mathbf{y})\big[1+O(R^{-1})\big],$$

$$\nabla\otimes\nabla g_{L,T} = -g_{L,T}\big[k_{L,T}^2\hat{\mathbf{x}}\otimes\hat{\mathbf{x}}+O(R^{-1})\big].$$

Substitution into the expression of $\mathcal{G}$ yields

$$\begin{aligned}\mathcal{G}(\mathbf{y},\mathbf{x}) = \frac{1}{4\pi\rho_0\omega^2}\Big\{&k_T^2\,\frac{\exp(ik_T R)}{R}\exp(-ik_T\,\hat{\mathbf{x}}\cdot\mathbf{y})\,[\mathbf{1}-\hat{\mathbf{x}}\otimes\hat{\mathbf{x}}+O(R^{-1})]\\ &+k_L^2\,\frac{\exp(ik_L R)}{R}\exp(-ik_L\,\hat{\mathbf{x}}\cdot\mathbf{y})\,[\hat{\mathbf{x}}\otimes\hat{\mathbf{x}}+O(R^{-1})]\Big\},\end{aligned}$$

whence it follows that

$$\begin{aligned}\mathcal{G}(\mathbf{y},\mathbf{x})\mathbf{f} = \frac{1}{4\pi\rho_0\omega^2}\Big\{&k_T^2\,\frac{\exp(ik_T R)}{R}\exp(-ik_T\,\hat{\mathbf{x}}\cdot\mathbf{y})\,[\mathbf{f}-(\mathbf{f}\cdot\hat{\mathbf{x}})\,\hat{\mathbf{x}}+O(R^{-1})]\\ &+k_L^2\,\frac{\exp(ik_L R)}{R}\exp(-ik_L\,\hat{\mathbf{x}}\cdot\mathbf{y})\,[(\mathbf{f}\cdot\hat{\mathbf{x}})\,\hat{\mathbf{x}}+O(R^{-1})]\Big\}.\end{aligned}$$

Comparison with (2.5) shows that

$$\mathbf{U}_1(\mathbf{x}) = \frac{\exp(ik_T R)}{R}\big[\mathbf{U}_{1\perp}+O\left(R^{-1}\right)\big]+\frac{\exp(ik_L R)}{R}\big[\mathbf{U}_{1\parallel}+O\left(R^{-1}\right)\big], \tag{2.6}$$

where

$$\mathbf{U}_{1\perp} = \frac{k_T^2}{4\pi\rho_0\omega^2}\int_{\mathcal{V}}\exp(-ik_T\,\hat{\mathbf{x}}\cdot\mathbf{y})\,\big[\mathbf{f}-(\hat{\mathbf{x}}\cdot\mathbf{f})\hat{\mathbf{x}}\big]\,dy, \tag{2.7}$$

$$\mathbf{U}_{1\parallel} = \frac{k_L^2}{4\pi\rho_0\omega^2}\left[\int_{\mathcal{V}}(\hat{\mathbf{x}}\cdot\mathbf{f})\,\exp(-ik_L\,\hat{\mathbf{x}}\cdot\mathbf{y})\,dy\right]\hat{\mathbf{x}}. \tag{2.8}$$

The representation (2.6) for the perturbation field is the analogue of (7.4.8) that yields the field scattered by an inclusion inserted into a homogeneous unbounded medium. The asymptotic expression of $\mathbf{U}_1$ results from superposition of a radial and a transverse field, propagating with the speeds of longitudinal and transverse waves, respectively. Indeed, $\mathbf{U}_{1\perp}$ and $\mathbf{U}_{1\parallel}$ depend only on the direction identified by $\hat{\mathbf{x}}$, while the dependence on the distance R has been factorized in the "scalar-wave type" contributions $\exp(ik_{L,T}R)/R$. Moreover, consistent with the notation, $\mathbf{U}_{1\perp}$ is perpendicular and $\mathbf{U}_{1\parallel}$ is parallel to $\hat{\mathbf{x}}$.

For the sake of definiteness assume that the incident wave is longitudinal and then let

$$\mathbf{U}_0(\mathbf{y}) = a\mathbf{k}_L\exp(i\mathbf{k}_L\cdot\mathbf{y}),$$

a being complex-valued. Hence, by the definition of $\mathbf{f}$,

$$\mathbf{f}(\mathbf{y}) = a\,(\alpha\,\mathbf{k}_L+\beta\,\nabla\lambda_1)\,\exp(i\mathbf{k}_L\cdot\mathbf{y}),$$

where

$$\alpha = \omega^2 \rho_1 + (1 - \hat{\tau})[-k_L^2(2\mu_1 + \lambda_1) + 2i(\nabla \mu_1 \cdot \mathbf{k}_L)]$$

and

$$\beta = ik_L^2(1 - \hat{\tau}).$$

Substitution into (2.7) and (2.8) leads to the determination of the far-field limit of $\mathbf{U}_1$. Specifically, we find that

$$\mathbf{U}_{1\perp} = \frac{ak_T^2}{4\pi\rho_0\omega^2}\left\{\chi_T\left[\mathbf{k}_L - (\mathbf{k}_L \cdot \hat{\mathbf{x}})\,\hat{\mathbf{x}}\right] + \left[\boldsymbol{\sigma}_T - (\boldsymbol{\sigma}_T \cdot \hat{\mathbf{x}})\,\hat{\mathbf{x}}\right]\right\}, \tag{2.9}$$

where

$$\chi_T = \int_{\mathcal{V}} \alpha \, \exp[i(\mathbf{k}_L - k_T\hat{\mathbf{x}}) \cdot \mathbf{y}]\, dy$$

$$\boldsymbol{\sigma}_T = \int_{\mathcal{V}} \beta\, \nabla\lambda_1 \, \exp[i(\mathbf{k}_L - k_T\hat{\mathbf{x}}) \cdot \mathbf{y}]\, dy.$$

In connection with $\mathbf{U}_{1\parallel}$ we examine (2.8) to obtain

$$\mathbf{U}_{1\parallel} = \frac{ak_L^2}{4\pi\rho_0\omega^2}(\chi_L\, \mathbf{k}_L \cdot \hat{\mathbf{x}} + \boldsymbol{\sigma}_L \cdot \hat{\mathbf{x}})\,\hat{\mathbf{x}}, \tag{2.10}$$

where

$$\chi_L = \int_{\mathcal{V}} \alpha \, \exp[i(\mathbf{k}_L - k_L\hat{\mathbf{x}}) \cdot \mathbf{y}]\, dy$$

$$\boldsymbol{\sigma}_L = \int_{\mathcal{V}} \beta\, \nabla\lambda_1 \, \exp[i(\mathbf{k}_L - k_L\hat{\mathbf{x}}) \cdot \mathbf{y}]\, dy.$$

By (2.9), a direct calculation shows that the transverse part $\mathbf{U}_{1\perp}$ of $\mathbf{U}_1$ satisfies

$$\Delta\mathbf{U}_{1\perp} + k_T^2\, \mathbf{U}_{1\perp} = O\left(1/R^3\right).$$

Then, in the far-field approximation, $\mathbf{U}_{1\perp}$ is a solution to the Helmholtz equation with wavenumber equal to that of plane, transverse waves. Similarly, by (2.10) we find that

$$\Delta\mathbf{U}_{1\parallel} + k_L^2\, \mathbf{U}_{1\parallel} = O\left(1/R^3\right).$$

This means that, in the far-field approximation, $\mathbf{U}_{1\parallel}$ is a solution to the Helmholtz equation with wavenumber equal to that of plane, longitudinal waves. This agrees with the previous interpretation that, in the far-field approximation, the perturbation field $\mathbf{U}_1$ may be viewed as the resultant of a transverse wave $\mathbf{U}_{1\perp}$ and a longitudinal wave $\mathbf{U}_{1\parallel}$.

In view of (2.6), (2.9), and (2.10) it follows that, because of the heterogeneity of the body, the far-field is affected by the heterogeneities through the integrals $\chi_{L,T}$, $\boldsymbol{\sigma}_{L,T}$. The gradients $\nabla\mu_1$ and $\nabla\lambda_1$ enter linearly the integrands of $\chi_{L,T}$ and $\boldsymbol{\sigma}_{L,T}$, respectively. As it must be, their effects vanish when the body is homogeneous ($\chi_{L,T} = 0$, $\boldsymbol{\sigma}_{L,T} = 0$).

8.3 The WKB method

The WKB method can be traced back to the Italian astronomer Carlini [31] who, in 1817, wrote a paper about the calculation of planetary orbits. The accurate asymptotic approximations derived there were supported by an analysis which has been reinvestigated by Schlissel. Apart from Carlini's work, the asymptotic theory of differential equations begins in 1837 when Liouville and G. Green published papers about differential equations of the form

$$\epsilon^2 u'' + \varphi u = 0$$

where ϵ is taken to assume small values and φ is a smooth non-vanishing function of the independent variable z; a prime denotes differentiation with respect to z or the pertinent independent variable. The first investigation where φ is allowed to vanish was performed by the theoretical physicist Gans in connection with Maxwell's equations in heterogeneous media.

A remarkable progress in asymptotic methods was favoured by the appearance of Schrödinger's equation, namely

$$\psi'' + \frac{1}{h^2} f\psi = 0, \tag{3.1}$$

in the unknown function ψ of z with some given, positive-valued (potential) function f of z. Here h is Planck's constant and, mathematically, plays the role of the (small) parameter ϵ. In 1926 the physicists Wentzel, Kramers, and Brillouin, unaware of Gans' results, rediscovered them through different procedures and added new results about the pertinent eigenvalue problem. In particular, the term "turning points" for the zeros of f (or φ) came into use at that time probably because of Kramers. The three papers of 1926 had such an impact that the abbreviation "WKB method", after the initials of the authors, has remained in common use in connection with asymptotic solutions to the differential equation (3.1) or more general ones. Soon after 1926 it was remarked that Jeffreys had independently rediscovered Gans' method in 1924; that is why quite often the procedure is referred to as WKBJ method. A historical survey of the subject can be found in [90].

Before entering the necessary technical details, observe that the equation (3.1) is not so special as may seem. For, consider the differential equation

$$w'' + aw' + bw = 0$$

in the unknown function $w(z)$ while a, b are given functions of z. Upon the transformation

$$\psi = w \exp\left[\tfrac{1}{2}\int_{z_0}^{z} a(s)\, ds\right]$$

we obtain

$$\psi'' + (b - \tfrac{1}{4}a^2)\psi = 0.$$

Then the term in the first-order derivative can always be removed without any loss of generality.

Back to (3.1), we begin from the basic idea of the WKB method. Consider the unknown function $\psi(z)$ in the form $A(z)\exp[iS(z)/h]$ where A, S are to be determined. Substitution for ψ in (3.1) yields

$$h^2 A'' - A(S')^2 + fA = 0, \tag{3.2}$$

$$2A'S' + AS'' = 0. \tag{3.3}$$

By (3.3) we have

$$A^2 S' = \text{const.} \tag{3.4}$$

Now assume that A is a slowly varying function of z so that $h^2 A''/A$ is negligible. Then, by (3.2),

$$S(z) = \pm \int_{z_0}^{z} f^{1/2}(s)\, ds.$$

Hence, by (3.4), we have the WKB solution

$$\psi(z) = \text{const.}\, f^{-1/4}(z) \exp\Big[\pm \frac{i}{h} \int_{z_0}^{z} f^{1/2}(s)\, ds\Big]. \tag{3.5}$$

The solution (3.5) may be obtained also in a way that emphasizes the asymptotic behaviour as $h \to 0$. Consider the function

$$\exp\Big[\frac{1}{h} \int_{z_0}^{z} (\phi_0 + h\phi_1 + h^2\phi_2 + \dots)ds\Big].$$

Substitution into (3.1) provides an expansion in powers of h which must vanish. Letting the coefficient of any power vanish we obtain

$$\phi_0^2 + f = 0,$$

$$2\phi_0\phi_1 + \phi_0' = 0,$$

$$2\phi_0\phi_2 + \phi_1^2 + \phi_1' = 0,$$

$$\dots\dots\dots\dots ;$$

the first equation may be viewed as the definition of ϕ_0, the second one the definition of ϕ_1, and so on. The restriction to the first two equations provides just the solution (3.5).

To get an idea of the accuracy of the solution, observe that (3.5) is an exact solution of

$$\psi_1'' + \psi_1\Big\{\frac{1}{h^2} f + \Big[\frac{f''}{4f} - \frac{5(f')^2}{16f^2}\Big]\Big\} = 0. \tag{3.6}$$

The differential equation (3.6) becomes closer and closer to (3.1), for $f \neq 0$, as $h \to 0$.

Before examining some applications, it is worth having a look at a wave feature whereby the WKB approximation is viewed as the first term of a geometrical optical series. Consider a heterogeneous, elastic half-space $z > 0$ which consists of a set of homogeneous layers $(0, z_1), (z_1, z_2), \dots$. The material properties are supposed to depend only on the z-coordinate. A plane wave $\exp[i(k_0 z - \omega t)]$ arrives from the homogeneous, elastic half-space $z < 0$ and travels in the direction of increasing z. As usual we omit writing the factor $\exp(-i\omega t)$. Here we refer to both transverse waves and longitudinal waves; accordingly we let γ stand for the coefficients μ or $2\mu + \lambda$ as appropriate. At the boundary $z = 0$ of the first layer the incident wave splits into a transmitted wave T_1 and a reflected wave R_1; they are expressed by

$$\mathcal{R}_0 \exp(-ik_0 z), \qquad \mathcal{T}_0 \exp(ik_1 z).$$

The coefficients $\mathcal{R}_0$ and $\mathcal{T}_0$ are given by the relations (4.5.22), in the special case of elastic-elastic interface and normal incidence. For any case of (4.5.22) we can write the reflection coefficient $\mathcal{R}$ and the refraction coefficient $\mathcal{T}$ as

$$\mathcal{R} = \frac{\breve{\rho} k - \rho \breve{k}}{\breve{\rho} k + \rho \breve{k}}, \qquad \mathcal{T} = \frac{2\breve{\rho} k}{\breve{\rho} k + \rho \breve{k}}.$$

Let γ_0 be the value of γ for $z < 0$ and $\gamma_1, \gamma_2, \dots$ the values in the layers $(0, z_1), (z_1, z_2), \dots$. Then, letting $\eta = 1/\gamma k = k/\omega^2 \rho$ we have

$$\mathcal{R}_0 = \frac{\eta_0 - \eta_1}{\eta_0 + \eta_1}, \qquad \mathcal{T}_0 = \frac{2\eta_0}{\eta_0 + \eta_1}.$$

The wave T_1 splits, at the boundary $z = z_1$, into a transmitted wave T_2 travelling in the second layer and a reflected wave R_2 travelling downwards in the first layer. At the boundary $z = z_j$ the incident wave splits into a transmitted wave T_{j+1} and a reflected wave R_{j+1}. The quantity

$$\mathcal{T}_j = \frac{2\eta_j}{\eta_j + \eta_{j+1}}$$

is the ratio between the amplitude of T_{j+1} and that of T_j. Consider the sequence of waves $T_1, T_2, \dots, T_{n-1}$ and let $\Delta z_j = z_j - z_{j-1}$, $\Delta \eta_j = \eta_j - \eta_{j-1}$, $j = 1, 2, \dots, n$. Letting $u(z,t) = U(z)\exp(-i\omega t)$, we can express the value of the field U just before the surface $z = z_n$ as

$$\begin{aligned} U(z_n^-) &= \prod_{j=1}^{n} \frac{2\eta_{j-1}}{\eta_{j-1} + \eta_j} \exp(ik_j \Delta z_j) \\ &= \exp\Big[-\sum_{j=1}^{n} \ln\Big(1 + \frac{\Delta\eta_j}{2\eta_{j-1}}\Big)\Big] \exp\Big[i \sum_{j=1}^{n} k_j\, \Delta z_j \Big]. \end{aligned}$$

Now we pass to the limit of a continuous body. For small values of $\Delta\eta_j$ we can write

$$\ln\Big(1 + \frac{\Delta\eta_j}{2\eta_{j-1}}\Big) \simeq \frac{\Delta\eta_j}{2\eta_{j-1}}.$$

Then the continuum limit provides

$$U(z) = \exp\Big(-\tfrac{1}{2} \int_{\eta_0}^{\eta(z)} \frac{d\eta}{\eta} \Big) \exp\Big[i \int_0^z k(s)\, ds \Big]$$

whence, by the definition of η,

$$U(z) = \Big(\frac{k_0 \rho(z)}{\rho_0 k(z)} \Big)^{1/2} \exp\Big[i \int_0^z k(s)\, ds \Big]. \tag{3.7}$$

Formally the procedure developed in [24] corresponds to letting ρ be independent of z. In such a case the field (3.7) takes the form (3.5) with $f = k^2$ (and $h = 1$). On the basis of this result, Bremmer gave the WKB approximation the interpretation that it represents the wave originating by refractions, directly from the primary wave which arrives from the homogeneous space. As we show in a moment, though, this interpretation deserves some remarks.

Still we have in mind transverse and longitudinal waves for one-dimensional motions along the z-axis. Following §6.5 we can write the wave equation as

$$(\gamma U')' + \rho \omega^2 U = 0. \tag{3.8}$$

The introduction of $\psi = \sqrt{\gamma}\, U$ allows (3.8) to be written as

$$\psi'' + g \omega^2 \psi = 0 \tag{3.9}$$

where

$$g = \frac{\rho}{\gamma} + \Big[\frac{(\gamma')^2}{4\gamma^2} - \frac{\gamma''}{2\gamma} \Big] \frac{1}{\omega^2};$$

throughout this section we regard g as a strictly-positive valued function. In view of (3.5) we can write at once the WKB solution ψ to (3.9) whence

$$U(z) = \text{const.}\, [\gamma^2(z) g(z)]^{-1/4} \exp\Big[\pm i\omega \int_{z_0}^z g^{1/2}(s)\, ds \Big]. \tag{3.10}$$

As $\omega \to \infty$, we have $g \simeq \rho/\gamma$ and then the amplitude of U is proportional to $(\gamma\rho)^{-1/4}$ or $(k/\rho)^{1/2}$. According to (3.7), instead, the amplitude of U is proportional to $(\rho/k)^{1/2}$. Incidentally, the result (3.7) follows also by replacing (3.8) with $U'' + (\rho\omega^2/\gamma)U = 0$. This shows the drastic effect of disregarding the term $\gamma' U'$ in the differential equation.

By analogy with (3.6) we can say that the function (3.10) is the exact solution to (3.9) if and only if

$$5(g')^2 - 4gg'' = 0.$$

This occurs if $g = \text{const.}$ or $g = \text{const.}\,(a + z)^{-4}$, a being a constant. We do not discuss here specific examples for γ and ρ that meet these conditions. The interested reader is referred to [137].

Insights into the behaviour of the solution to the differential equation (3.9) may be gained through an analysis which is based on a suitable transformation of the unknown function and independent variable. Let

$$\xi(z) = \omega \int_0^z g^{1/2}(s)\, ds,$$

$$\phi = \omega^{1/2} g^{1/4} \psi.$$

Since $\omega g^{1/2}$ may be viewed as the local wavenumber then $d\xi = \omega g^{1/2} dz$ is 2π times the variation of the coordinate, dz, relative to the wavelength $2\pi/\omega g^{1/2}$. Now, upon substitution and straightforward calculations we can write (3.9) in the form

$$\frac{d^2\phi}{d\xi^2} + \left[1 + \frac{1}{4g^2}\left(\frac{dg}{d\xi}\right)^2 - \frac{1}{2g}\frac{d^2 g}{d\xi^2}\right]\phi = 0. \tag{3.11}$$

Equation (3.11) is convenient in describing the behaviour of the field under consideration when the function g varies slowly. Let

$$\nu = -\tfrac{1}{2}\frac{d}{d\xi}\ln g^{1/2}.$$

Then (3.11) can be written as

$$\frac{d^2\phi}{d\xi^2} + \left(1 + \frac{d\nu}{d\xi} - \nu^2\right)\phi = 0. \tag{3.12}$$

A solution to (3.12) is found in the particular case when g has the form

$$g^{1/2}(z) = \begin{cases} g_0^{1/2} L/z, & z > L, \\ g_0^{1/2} & z < L; \end{cases}$$

this means that ρ and γ are constant as $z < L$ and then, for example, $\rho^{1/2}$ decreases as $1/z$ as $z > L$. Choose the origin of ξ at $z = 0$ and then we have

$$\xi = \begin{cases} \omega g_0^{1/2} L \ln(z/L), & z > L, \\ \omega g_0^{1/2}(z - L), & z < L. \end{cases}$$

In terms of ξ, the functions g and ν are given by

$$g^{1/2}(\xi) = \begin{cases} g_0^{1/2} \exp(-\xi/\omega g_0^{1/2} L), & \xi > 0, \\ g_0^{1/2}, & \xi < 0, \end{cases}$$

and

$$\nu(\xi) = \begin{cases} 1/2\omega g_0^{1/2} L, & \xi > 0, \\ 0, & \xi < 0. \end{cases}$$

Then, letting $\epsilon = 1/2\omega g_0^{1/2} L$ we have

$$\frac{d^2\phi}{d\xi^2} + (1-\epsilon^2)\phi = 0, \ \xi > 0, \tag{3.13}$$

$$\frac{d^2\phi}{d\xi^2} + \phi = 0, \ \xi < 0. \tag{3.14}$$

The solution of (3.13)-(3.14) is subject to continuity requirements at $\xi = 0$. Having in mind the mechanical context, we require that the quantities U and $\gamma U'$ be continuous in that they represent the displacement and the stress. This in turn implies that, at $\xi = 0$,

$$\phi(0^-) = \phi(0^+),$$

$$\frac{d\phi}{d\xi}(0^+) - \frac{d\phi}{d\xi}(0^-) = -\frac{\gamma^{1/2}}{\omega g^{1/4}}(\gamma^{-1/2}g^{-1/4})'\phi(0^+) + \frac{\gamma^{1/2}}{\omega g^{1/4}}(\gamma^{-1/2}g^{-1/4})'\phi(0^-).$$

In particular, if γ is constant then the jump condition

$$\left[\frac{d\phi}{d\xi}\right] = \frac{1}{2\omega g}[(g^{1/2})]\phi \tag{3.15}$$

follows.

Since (3.13)-(3.14) allow for plane wave solutions, suppose that an incident wave propagates in $\xi < 0$, in the ξ-direction and that a reflected and a transmitted wave arise at $\xi = 0$. Then we have

$$\phi = \begin{cases} \exp(i\xi) + \mathcal{R}\exp(-i\xi), & \xi < 0, \\ \mathcal{T}\exp(i\sqrt{1-\epsilon^2}\,\xi), & \xi > 0, \end{cases}$$

$\mathcal{R}$ and $\mathcal{T}$ denoting the reflection and transmission coefficients. The continuity of ϕ and (3.15) yield

$$1 + \mathcal{R} = \mathcal{T},$$

$$i\sqrt{1-\epsilon^2}\,\mathcal{T} - i(1-\mathcal{R}) = -\epsilon\mathcal{T},$$

whence

$$\mathcal{R} = \frac{i}{\epsilon}(1-\sqrt{1-\epsilon^2}), \qquad \mathcal{T} = \frac{i}{\epsilon}(1-\sqrt{1-\epsilon^2} - i\epsilon). \tag{3.16}$$

Because

$$\sqrt{1-y} = 1 + \sum_{n=1}^{\infty}(-1)^n \binom{1/2}{n} y^n, \quad y < 1,$$

we can write

$$\mathcal{R} = i\Big[\frac{\epsilon}{2} + \frac{\epsilon^3}{8} + \frac{\epsilon^5}{16} + ...\Big] = i\Big[\frac{1}{4\omega g_0^{1/2}L} + \frac{1}{64(\omega g_0^{1/2}L)^3} + \frac{1}{512(\omega g_0^{1/2}L)^5} + ...\Big]. \tag{3.17}$$

Analogously for $\mathcal{T} = 1 + \mathcal{R}$. The result (3.17) and the analogue for $\mathcal{T}$ are often viewed as due to a number of reflections in the heterogeneous region.

On account of the expression (3.16) for $\mathcal{T}$, using the expansion

$$\exp(i\sqrt{1-\epsilon^2}\xi) = \exp(i\xi) - \frac{i}{2}\epsilon^2 \xi \exp(i\xi) + O(\epsilon^4)$$

provides the transmitted wave (in $\xi > 0$) in the form

$$\phi(\xi) = \exp(i\xi) + \frac{i\epsilon}{2}\exp(i\xi) - \frac{i\epsilon^2}{2}\xi \exp(i\xi) + \epsilon^3\big(\frac{i}{8} + \frac{\xi}{4}\big)\exp(i\xi)$$

or, in the original variables,

$$U(z) = [\omega^2\gamma^2 g_0]^{-1/4} \exp\left[i\omega \int_L^z g^{1/2}(s)ds\right]\left[1 + \frac{i}{4\omega g_0^{1/2} L} + ...\right]$$

whence

$$U(z) = [\omega^2\gamma^2 g_0]^{-1/4} \left(\frac{z}{L}\right)^{i\omega g_0^{1/2} L} \left[1 + \frac{i}{4\omega g_0^{1/2} L} + ...\right].$$

The first term is just in the form of the WKB solution (3.10) and this again may motivate saying that the WKB solution is the first term in an asymptotic series for the transmitted wave (cf. §8.5). An analogous procedure and similar results hold for the solution to Maxwell's equations in heterogeneous media [168].

It is worth mentioning that the WKB method has been applied to the case when the "refractive index", the function $f(z)$ in (3.1), is an analytic function. An investigation by Meyer [125] resolved the difficulty that, in such a case, no wave reflection occurs. He proved that, for small values of h, the mathematical problem is well posed and that the reflection coefficient is trascendentally small. He also showed that, apart from its exponential order, the reflection coefficient is essentially determined by the local behaviour of the analytic continuation of the solution in the neighbourhood of a few marked points in the complex plane where the refractive index is either singular or zero. Later Chapman and Mahony [48] proved that the reflection coefficient can be described by a few parameters specifying the dominant terms in the differential equation at such points. The interested reader is referred to [48] for the proof and technical details. In this regard, though, we observe that the minimal hypothesis in such investigations is the continuity of $f(z)$ in $\mathbb{R}$. Since, in practical models, f involves derivatives of physical functions up to second order then the continuity of f turns out to be a strong hypothesis. That is why here we do not pursue the subject.

While the WKB method, as it stands, works for one-dimensional problems, generalizations have been set up and applied, e.g., to acoustic propagation in horizontally stratified oceans [170, 171]. A generalization of the WKB method consists in allowing for oblique incidence; in this chapter oblique incidence is treated in §§8.4, 8.5 through solutions with separable variables.

8.4 Turning points and extended WKB method

Consider the Helmholtz equation in the form

$$u'' + \omega^2 q^2 u = 0$$

where q represents the slowness or the refractive index. When this equation describes the Z−term in an oblique incidence problem (cf. §6.4) q^2 may vanish or even be negative. Regard as turning points (or transition points) the values of z where q vanishes. Our attention in this section is confined to the behaviour of u and the overall behaviour of an Epstein layer when turning points occur. Needless to say, turning points occur depending on both the constitutive properties of the medium and the direction of the incident wave.

As shown in the previous section, the phase S in the WKB method is derived on the assumption that A''/A is negligible, which is reasonable when q is relatively large. Around turning points this condition fails. That is why the behaviour around turning points deserve a specific description. We essentially review Langer's method (cf. [100], §6.24; [168], §4.7).

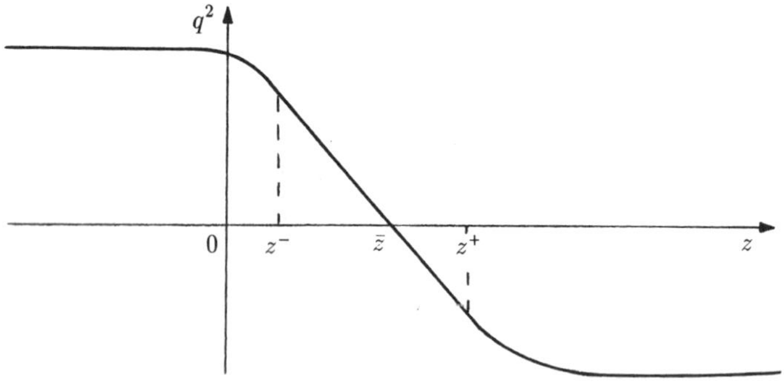

Fig. 8.1 Sketch of $q^2(z)$.

Let q^2 depend on z as in Fig. 8.1. In the region $z < z^-$, q^2 is sufficiently slowly varying and the WKB solution is valid. However, by analogy with the WKB solution take u in the form

$$u(z) = \frac{P}{\sqrt{q(z)}}\Big[\exp\Big(i\omega\int_0^z q(s)\,ds\Big) + R\exp\Big(-i\omega\int_0^z q(s)\,ds\Big)\Big] \tag{4.1}$$

where P and R are arbitrary constants. Let c_0 be the phase speed as $z \to -\infty$. Then, as $z \to -\infty$, we have $\int_0^z q(s)\,ds \simeq z/c_0$ and

$$u(z) \simeq Pc_0[\exp(ik_0 z) + R\exp(-ik_0 z)] \tag{4.2}$$

where $k_0 = \omega/c_0$. We may comment on (4.2) by saying that, as $z \to -\infty$, the solution u represents the superposition of an incident wave $\exp(ik_0 z)$ and a reflected wave $R\exp(-ik_0 z)$. Meanwhile this allows us to view R as the reflection coefficient which has yet to be determined. When $z > \bar{z}$ we have $q^2 < 0$. Then we can express u in the region $z > z^+$ as

$$u(z) \simeq \frac{PT}{(-q^2)^{1/4}} \exp\big[-\omega \int_{\bar{z}}^{z} (-q^2)^{1/2}(s)\, ds\big]. \tag{4.3}$$

The constant T can be viewed as the transmission coefficient and is also to be determined.

Around the turning point $\bar{z}$ we can approximate $q^2(z)$ as

$$q^2(z) = -\frac{1}{c_0^2}\alpha(z-\bar{z})$$

α being a positive constant. Upon substitution we have

$$u'' - k_0^2\alpha(z-\bar{z})u = 0$$

or, in terms of the new variable $\tau = (k_0^2\alpha)^{1/3}(z-\bar{z})$,

$$\frac{d^2u}{d\tau^2} - \tau u = 0. \tag{4.4}$$

Equation (4.4) is a form of the well known Airy's equation and its solution is given in terms of the Airy's functions. By requiring that the solution give rise to an exponentially damped wave for large positive values of τ, or z, we have

$$u(\tau) \simeq \frac{1}{2i(-\tau)^{1/4}}\big\{\exp\big[\tfrac{2i}{3}(-\tau)^{3/2}\big]\exp\big(\tfrac{i\pi}{4}\big) - \exp\big[-\tfrac{2i}{3}(-\tau)^{3/2}\big]\exp\big(-\tfrac{i\pi}{4}\big)\big\}$$

when τ is a large, negative number while

$$u(\tau) \simeq \frac{1}{2\tau^{1/4}}\exp\big(-\tfrac{2}{3}\tau^{3/2}\big)$$

when τ is a large positive number. In terms of the original variable z we have

$$u(z) \simeq \frac{(k_0^2\alpha)^{1/6}}{2i\sqrt{\omega q}}\big[\exp\big(-i\omega\int_{\bar{z}}^{z} q(s)\,ds\big)\exp\big(\frac{i\pi}{4}\big) - \exp\big(i\omega\int_{\bar{z}}^{z} q(s)\,ds\big)\exp\big(-\frac{i\pi}{4}\big)\big] \tag{4.5}$$

when $\bar{z} - z \gg (k_0^2\alpha)^{-1/3}$, and

$$u(z) \simeq \frac{(k_0^2\alpha)^{1/6}}{2(-q^2)^{1/4}}\exp\big[-\omega\int_{\bar{z}}^{z}(-q^2)^{1/2}(s)\,ds\big] \tag{4.6}$$

when $z - \bar{z} \gg (k_0^2\alpha)^{-1/3}$. Comparison with (4.1) shows that we can identify R as

$$R = -i\exp\big[2i\omega\int_0^{\bar{z}} q(z)\,dz\big].$$

Similarly, comparison with (4.3) yields

$$T = -\exp(-i\pi/4)\exp\big[-i\omega\int_0^{\bar{z}} q(s)\,ds\big].$$

The expressions for u, both near $\bar{z}$ and far from $\bar{z}$ are recovered by letting

$$\tau = \Big(i\frac{3}{2}\int_{\bar{z}}^{z} q(s)\,ds\Big)^{2/3}$$

where q is meant as the square root of q^2. This is the essence of the so-called "extended WKB approximation".

There are circumstances where oblique incidence leads to an equation of the form

$$u'' + \omega^2 q^2 u - \omega^2 \frac{1}{n} n' u' = 0 \tag{4.7}$$

as is the case for waves with horizontal polarization. Letting $\psi = u/\sqrt{n}$ we can write (4.7) in the form

$$\psi'' + \omega^2 Q^2 \psi = 0$$

where

$$Q^2 = q^2 + \frac{1}{2n}n'' - \frac{3}{4n^2}(n')^2.$$

The previous analysis about the behaviour at the turning point applies by replacing formally q by Q.

8.5 Separable variables and successive approximations

The purpose of this section is to compare descriptions of wave solutions in heterogeneous media, particularly in the form of Epstein layer, and to show connections between solutions obtained by different methods, viz separable variables, ray methods, successive approximations. Consider the Helmholtz equation

$$\Delta u + \omega^2 q^2 u = 0$$

and observe that solutions exist with separable variables (cf. §6.4, eq. (6.4.3)) provided that

$$Z'' + (\omega^2 q^2 - \beta - \gamma)Z = 0.$$

This means that the Z-term of the solution satisfies a one-dimensional Helmholtz equation with the reduced slowness $\sqrt{\omega^2 q^2 - \beta - \gamma}$. Accordingly, provided we take into account the reduced slowness we can confine the attention to the equivalent one-dimensional problem; for simplicity we consider

$$u'' + \omega^2 q^2 u = 0 \tag{5.1}$$

where, with abuse of notation, q is the effective slowness in one dimension. Look for solutions of (5.1) in the form

$$u(\mathbf{x};\omega) = \exp[i\omega\tau(\mathbf{x})] \sum_{j=0}^{\infty} u_j (i\omega)^{-j}$$

which is characteristic of the ray method (cf. Ch. 9). Then we have

$$(\tau')^2 = q^2, \tag{5.2}$$

$$2\tau' u_0' + u_0 \tau'' = 0, \tag{5.3}$$

$$2\tau' u_j' + u_j \tau'' = -u_{j-1}'', \qquad j = 1, 2, \ldots. \tag{5.4}$$

The integration of (5.1)-(5.3) is immediate. First,

$$\tau = \pm \int_{z_0}^{z} q(s)\, ds.$$

Then (5.3) yields

$$u_0 = \frac{d_0}{\sqrt{q}}$$

where d_0 is a constant. Integration of (5.4) in the unknown u_j yields

$$u_j = \frac{1}{\sqrt{q}} \Big[\mp \int_{z_0}^{z} \frac{1}{2\sqrt{q(s)}} u_{j-1}''(s)\, ds + d_j \Big]$$

where $j = 1, 2, \ldots$ and the d_j's are constants. To within an inessential constant factor (d_0), two solutions can be taken as the two asymptotic expansions

$$v(z) = \frac{1}{\sqrt{q(z)}} \Big\{ 1 - \sum_{j=1}^{\infty} \frac{1}{(i\omega)^j} \Big[\int_{z_0}^{z} \frac{1}{2\sqrt{q(s)}} u_{j-1}''(s)\, ds + d_j \Big] \Big\} \exp\Big(i\omega \int_{z_0}^{z} q(s)\, ds \Big) \tag{5.5}$$

and

$$w(z) = \frac{1}{\sqrt{q(z)}} \Big\{ 1 + \sum_{j=1}^{\infty} \frac{1}{(i\omega)^j} \Big[\int_{z_0}^{z} \frac{1}{2\sqrt{q(s)}} u_{j-1}''(s)\, ds + d_j \Big] \Big\} \exp\Big(-i\omega \int_{z_0}^{z} q(s)\, ds \Big). \tag{5.6}$$

We may take such functions v, w as the pair of fundamental solutions in $z \in [0, h]$. Correspondingly we may evaluate the reflection coefficient $\mathcal{R}$ and the transmission coefficient $\mathcal{T}$ through the formulae (6.4.9), (6.4.10).

Also by analogy with (5.5) and (5.6), we may look for a solution u to (5.1) as follows. Let

$$u(z) = \frac{1}{\sqrt{q(z)}}\Big[f(z)\exp\big(i\omega\int_\alpha^z q(s)\,ds\big) + g(z)\exp\big(-i\omega\int_\beta^z q(s)\,ds\big)\Big] \tag{5.7}$$

where f, g are functions to be determined and α, β are suitable, real quantities. In this regard it is convenient to introduce an auxiliary function ψ and write (5.1) as a first-order system of equations, viz

$$u' = \psi, \tag{5.8}$$

$$\psi' = -\omega^2 q^2 u. \tag{5.9}$$

In view of the freedom connected with the functions f, g we can look for ψ in the form

$$\psi(z) = i\omega\sqrt{q(z)}\Big[f(z)\exp\big(i\omega\int_\alpha^z q(s)\,ds\big) - g(z)\exp\big(-i\omega\int_\beta^z q(s)\,ds\big)\Big].$$

Substitution in (5.8)-(5.9) and use of (5.7) yield

$$\begin{aligned} f' &= \frac{q'}{2q} g\exp\big(-i\omega\int_\alpha^z q(s)\,ds\big)\exp\big(-i\omega\int_\beta^z q(s)\,ds\big),\\ g' &= \frac{q'}{2q} f\exp\big(i\omega\int_\alpha^z q(s)\,ds\big)\exp\big(i\omega\int_\beta^z q(s)\,ds\big). \end{aligned} \tag{5.10}$$

We follow the usual approximation that q'/q is uniformly small. Accordingly we let $q'/2q = \epsilon\,\delta(z)$ and regard ϵ, for example the mean value of $q'/2q$, as a small parameter. Then we solve (5.10) through the successive approximations method where the unknown functions are written as powers of ϵ. Letting

$$\lambda_\pm(z) = \delta(z)\exp\big(\pm i\omega\int_\alpha^z q(s)\,ds\big)\exp\big(\pm i\omega\int_\beta^z q(s)\,ds\big),$$

we can write (5.10) as

$$\begin{aligned} f' &= \epsilon\lambda_- g,\\ g' &= \epsilon\lambda_+ f. \end{aligned} \tag{5.11}$$

Represent f and g through the expansions

$$f = \sum_j f_j\epsilon^j, \qquad g = \sum_j g_j\epsilon^j.$$

Substituting in (5.11) and setting the coefficient of each power of ϵ separately equal to zero yields

$$\begin{aligned} f_0' &= 0, \qquad f_j' = \lambda_- g_{j-1}, \quad j = 1, 2, ...,\\ g_0' &= 0, \qquad g_j' = \lambda_+ f_{j-1}, \quad j = 1, 2, ... \end{aligned}$$

This implies that f_0 and g_0 are constant across the layer, say

$$f_0 = a, \qquad g_0 = b.$$

To determine a, b and the integration constants for f_j, g_j we need appropriate boundary conditions. We regard f as related to a wave propagating downwards and g to a wave propagating upwards. To fix ideas we consider the process produced by an incident wave coming from the half-space $z < 0$ and set

$$f(0) = 1, \qquad g(h) = 0.$$

It is convenient, if not imperative, to satisfy these conditions by requiring that

$$f_0(0) = 1, \quad g_0(h) = 0, \quad f_j(0), \ g_j(h) = 0, \ j = 1, 2, \dots .$$

Then the functions f_j, g_j, $j = 1, 2, \dots$, satisfy the recurrence relations

$$f_j(z) = \int_0^z \lambda_-(s)\, g_{j-1}(s)\, ds, \qquad g_j(z) = \int_h^z \lambda_+(s)\, f_{j-1}(s)\, ds$$

whereby f_j and g_j are eventually given by j-fold integral representations. Indeed, letting

$$\phi(z) = \int_0^z q(\sigma)\, d\sigma,$$

upon immediate simplifications, if j is even we have

$$\begin{aligned} \epsilon^j f_j(z) = & \int_0^z ds\, \frac{q'}{2q}(s) \exp[-2i\omega\phi(s)] \int_h^s dt\, \frac{q'}{2q}(t) \exp[2i\omega\phi(t)] \\ & \dots \int_0^{\cdot} dv\, \frac{q'}{2q}(v) \exp[-2i\omega\phi(v)] \int_h^v dw\, \frac{q'}{2q}(w) \exp[2i\omega\phi(w)] \end{aligned}$$

and $g_j = 0$ while, if j is odd, $f_j = 0$ and

$$\begin{aligned} \epsilon^j g_j(z) = & \exp[-i\omega\phi(\alpha)] \exp[-i\omega\phi(\beta)] \int_h^z ds\, \frac{q'}{2q}(s) \exp[2i\omega\phi(s)] \\ & \dots \int_0 dv\, \frac{q'}{2q}(v) \exp[-2i\omega\phi(v)] \int_h^v dw\, \frac{q'}{2q}(w) \exp[2i\omega\phi(w)] \\ =: & \exp\Big(i\omega \int_\alpha^0 q(\sigma)\, d\sigma\Big) \exp\Big(i\omega \int_\beta^0 q(\sigma)\, d\sigma\Big) \epsilon^j \bar{g}_j(z). \end{aligned}$$

Accordingly we can write

$$\begin{aligned} u(z) = \frac{1}{\sqrt{q(z)}} \{ & [1 + \epsilon^2 f_2(z) + \epsilon^4 f_4(z) + \dots] \exp[i\omega(\phi(z) - \phi(\alpha))] \\ & + [\epsilon g_1(z) + \epsilon^3 g_3(z) + \dots] \exp[-i\omega(\phi(z) - \phi(\beta))] \} \end{aligned}$$

and then we have

$$u(z) = \frac{\exp[-i\omega\phi(\alpha)]}{\sqrt{q(z)}}\{[1 + \epsilon^2 f_2(z) + \epsilon^4 f_4(z) + ...]\exp[i\omega\phi(z)] + [\epsilon\bar{g}_1(z) + \epsilon^3\bar{g}_3(z) + ...]\exp[-i\omega\phi(z)]\}. \tag{5.12}$$

The result (5.12) deserves some comments and interpretations. Observe that the factor

$$\exp[-i\omega\phi(\alpha)] = \exp\left[i\omega\int_\alpha^0 q(\sigma\, d\sigma\right]$$

accounts for the "phase" of the incident wave at $z = 0$. Then, at the zeroth order in ϵ, the solution

$$u(z) = \frac{\exp[-i\omega\phi(\alpha)]}{\sqrt{q(z)}}\{[1 + \epsilon^2 f_2(z) + \epsilon^4 f_4(z) + ...]\exp[i\omega\phi(z)]$$

is just the usual WKB approximation (cf. (3.5)).

To obtain an interpretation of the other terms in (5.12) it is worth reconsidering the reflection process. At an interface between two homogeneous media where the slowness is q and $q + \eta$ the reflection coefficient in the medium with slowness q is given by

$$\mathcal{R} = \frac{q - (q + \eta)}{q + (q + \eta)}.$$

If the slowness is a C^1 function, of z, then we can say that the amplitude reflected between z and $z + dz$ is

$$d\mathcal{R} = -\frac{q'}{2q}(z)\, dz.$$

So, $-q'/2q$ is the reflection coefficient, per unit length, for incident waves which propagate in the z-direction while $q'/2q$ is the reflection coefficient for propagation in the opposite direction.

With this in mind, consider (5.12) at the first order in ϵ, viz

$$u(z) = \frac{\exp[-i\omega\phi(\alpha)]}{\sqrt{q(z)}}\{\exp[i\omega\phi(z)] + \int_h^z \frac{q'}{2q}(s)\exp[2i\omega\phi(s)]\exp[-i\omega\phi(z)]\}.$$

Since $z \in [0, h]$, the new term can be more properly written as

$$\int_z^h \exp[i\omega(\phi(s) - \phi(\alpha))]\,\frac{-q'}{2q}(s)\,\exp[-i\omega(\phi(z) - \phi(s))]\, ds.$$

The first exponential propagates the wave from the height α to the height s, downwards; the fraction reflects (partially) the wave; the second exponential propagates the reflected wave up to the height z. The integral on s in $[z, h]$ allows us to view the overall contribution as follows: while running from z to h the incident wave is reflected in a continuous way; the waves produced by reflection determine the contribution at z.

The interpretation of the second-order term follows in a similar way. At the second order in ϵ we have the additional term

$$\int_0^z ds\Big[\int_s^h dv\, \exp[i\omega(\phi(v)-\phi(\alpha))\, \frac{-q'}{2q}(v)\, \exp[-i\omega(\phi(s)-\phi(v))]\Big]\, \frac{q'}{2q}(s)\, \exp[i\omega(\phi(z)-\phi(s))].$$

The expression $\exp[i\omega(\phi(v) - \phi(\alpha))]$ makes the incident wave propagate from α to v in $[s,h]$; $(-q'/2q)(v)$ describes the reflection at v; $\exp[-i\omega(\phi(s) - \phi(v))]$ makes the reflected wave propagate to s in $[0,z]$; $(q'/2q)(s)$ describes the reflection at s; $\exp[i\omega(\phi(z) - \phi(s))]$ makes the secondary reflected wave propagate downwards to z. The double integration allows us to view the result as due to the continuous reflection of the incident wave across the layer and the continuous reflection of these secondary waves in the layer $[0,z]$.

The analysis of higher-order terms shows the contribution of higher-order waves generated by a continuous reflection process in the layer.

9 RAY METHOD FOR HETEROGENEOUS, DISSIPATIVE MEDIA

Within the context of direct asymptotic methods, the so-called ray method has rightly received noticeable attention. Likewise other asymptotic methods, for time-harmonic waves the ray method involves series solutions whose nth term has a coefficient of the form $(i\omega)^{-n}$. The vanishing of the lowest-order term results in the trajectory of the ray while the vanishing of the remaining coefficients yields the evolution of the pertinent amplitudes along the ray. All this is well-known for the scalar Helmholtz equation or the problem connected to wave propagation in unstressed elastic solids.

New, interesting features arise in the application of the ray method to prestressed viscoelastic bodies. The occurrence of a prestress produces a qualitative change of the eikonal equations which again have solutions in terms of transverse waves and longitudinal waves. The prestress makes a (suitably defined) ray in solids be no longer orthogonal to the surfaces of constant phase, this circumstance being quite unusual in the literature of the ray method. The analysis of the amplitude evolution shows that, along ray tubes, the amplitude decays according to the rate of dissipation. Further, energy partition at an interface is shown to be affected by the prestress but not by the dissipative properties of the materials.

The dissipation effect is qualitatively the same in viscoelastic fluids but, there, rays are orthogonal to surfaces of constant phase. Indeed, except for the decay due to dissipation, there is a close connection between rays in viscoelastic fluids and rays in unstressed elastic solids.

Reflection and refraction of rays is investigated by determining the analogue of Snell's law and then the amplitudes pertaining to the rays emanating from the interface. The effect of prestress is shown to be remarkable also in this framework. Explicit formulae are established for reflection at a traction-free surface.

9.1 Ray method for the Helmholtz equation

The Helmholtz equation describes time-harmonic wave propagation in homogeneous elastic bodies, relative to a stress-free placement with a vanishing body force, and in homogeneous non-conducting dielectrics. Rays in elastic homogeneous bodies are thoroughly investigated in [3], Ch. 4. Upon commonly accepted approximations, the Helmholtz equation is regarded to model also propagation in the corresponding heterogeneous media. Within this simple framework and with the purpose of a useful introduction of the subject, we recall the procedure and results of the ray method for the scalar Helmholtz equation.

Assume that, in connection with a time-harmonic dependence, the pertinent field $U(\mathbf{x})$ satisfies the Helmholtz equation

$$\Delta U + \omega^2 q^2 U = 0,$$

where q is a known, real function of the position $\mathbf{x}$ which is to be viewed as the slowness. The function U will obviously depend on the parameter $\omega \in \mathbb{R}^+$ and we look for an expression of U which is meaningful for large values of ω. Then we consider a series expansion in the form

$$U(\mathbf{x};\omega) \simeq \omega^\beta \exp[i\omega\tau(\mathbf{x})] \sum_{j=0}^{\infty} U_j(\mathbf{x})(i\omega)^{-j}. \tag{1.1}$$

The exponent β is left undetermined by the present development; it is determined as soon as we match the solution U with prescribed data. Upon substitution of (1.1) in the Helmholtz equation we have

$$\sum_{j=0}^{\infty}\{(i\omega)^{2-j}U_j[(\nabla\tau)^2 - q^2] + (i\omega)^{1-j}[2\nabla\tau\cdot\nabla U_j + U_j\Delta\tau] + (i\omega)^{-j}\Delta U_j\} = 0.$$

The series solution (1.1) is determined by setting the coefficients of each power of ω equal to zero. The largest exponent is two and occurs when $j = 0$; a non-zero U_0 is allowed only if

$$(\nabla\tau)^2 = q^2, \tag{1.2}$$

that is, the phase τ is a solution of the eikonal equation. Because of (1.2) each term of the first series vanishes. Consider the remaining highest power; the vanishing of its coefficient yields the transport equation

$$2\nabla\tau\cdot\nabla U_0 + U_0\Delta\tau = 0. \tag{1.3}$$

Meanwhile the vanishing of the coefficient of the subsequent powers results in

$$2\nabla\tau\cdot\nabla U_j + U_j\Delta\tau = -\Delta U_{j-1}, \qquad j = 1,2,\dots . \tag{1.4}$$

Once U_0 is determined by (1.3) then $U_1, U_2, \dots$ are provided by (1.4). Let us see how to proceed.

Consistent with (1.2), here we regard the rays as the flow of the vector field (slowness vector)

$$\mathbf{q} = \nabla\tau. \tag{1.5}$$

Denote by σ the parameter along the rays which then are characterized by

$$\frac{d\mathbf{x}}{d\sigma} = l\mathbf{q}, \tag{1.6}$$

l being a function of σ. Then, along the rays,

$$\frac{d\mathbf{q}}{d\sigma} = l\, q\, \nabla q, \qquad \frac{d\tau}{d\sigma} = lq^2.$$

Now we derive the evolution equation for the leading-order amplitude U_0. Consider the transport equation (1.3). Upon multiplication by U_0 we see that $U_0^2 \nabla\tau$ is divergence-free. Hence for any domain D we have

$$0 = \int_{\partial D} U_0^2 \nabla\tau \cdot \mathbf{n}\, da \tag{1.7}$$

where $\mathbf{n}$ is the outward unit normal. Let W_0 be a closed region in a surface of constant τ and consider the tube of rays which pass through the boundary of W_0. Then we let D be the region bounded by the tube, the surface W_0, and the intersection, W say, with another surface of constant, greater, value τ. Observe that $\nabla\tau \cdot \mathbf{n} = 0$ on the lateral surface of the tube, $\nabla\tau \cdot \mathbf{n} = q$ on W, and $\nabla\tau \cdot \mathbf{n} = -q$ on W_0. By (1.7) and the mean value theorem we have

$$(U_0^2 q)(\bar{\mathbf{x}}_0)\, a_0 = (U_0^2 q)(\bar{\mathbf{x}})\, a$$

where $\bar{\mathbf{x}}_0$, $\bar{\mathbf{x}}$ are suitable points of W_0, W and a_0, a are the areas of W_0, W. In this sense we say that $U_0^2 aq$ is constant along ray tubes.

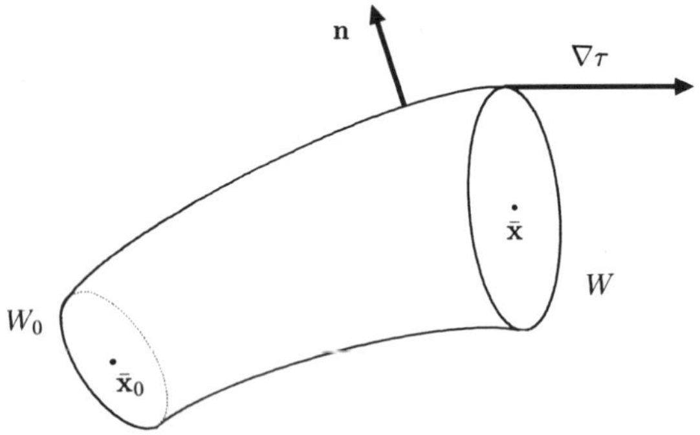

Fig. 9.1 Propagation along a ray tube.

Denote by γ_1, γ_2 a pair of parameters which label the rays through W_0. Then we describe $\mathbf{x}$ as $\mathbf{x} = \mathbf{x}(\sigma, \gamma_1, \gamma_2)$. Letting $\mathbf{i}_1 = \partial\mathbf{x}/\partial\gamma_1$, $\mathbf{i}_2 = \partial\mathbf{x}/\partial\gamma_2$ we can write the surface element as

$$da = |\mathbf{n} \cdot \mathbf{i}_1 \times \mathbf{i}_2|\, d\gamma_1\, d\gamma_2.$$

By definition, the range of the parameters γ_1, γ_2 for W_0 and W is the same. Thus the ratio a/a_0 goes as the ratio of (suitable) mean values of the Jacobian of mapping via rays, $J = |\mathbf{n} \cdot \mathbf{i}_1 \times \mathbf{i}_2|$, whence

$$(U_0^2 Jq)(\tilde{\mathbf{x}}_0) = (U_0^2 Jq)(\tilde{\mathbf{x}})$$

$\tilde{\mathbf{x}}_0$ and $\tilde{\mathbf{x}}$ being suitable points of W_0 and W. At the limit of arbitrarily small surfaces W_0, W we can view $\mathbf{x}$ as a function of σ and write

$$U_0^2(\sigma) = U_0^2(\sigma_0)\frac{(Jq)(\sigma_0)}{(Jq)(\sigma)}$$

where $U_0^2(\sigma_0)$, $J(\sigma_0)$, $q(\sigma_0)$ denote the values at the intersection of the pertinent ray with W_0 and $U_0^2(\sigma)$, $J(\sigma)$, $q(\sigma)$ the values along the ray. Incidentally, differentiation with respect to σ yields

$$\frac{dU_0^2}{d\sigma} = -U_0^2\frac{d}{d\sigma}(\ln\, Jq). \tag{1.8}$$

Consider now the differential equation (1.4) for $U_j, j = 1, 2, ...,$ with U_{j-1} as a known function. By (1.5) and (1.6), along the ray we have

$$\frac{2}{l}\frac{d\mathbf{x}}{d\sigma}\cdot\nabla U_j + U_j\Delta\tau = -\Delta U_{j-1}$$

whence

$$\frac{dU_j}{d\sigma} + \frac{l}{2}\Delta\tau\; U_j = -\frac{l}{2}\Delta U_{j-1}. \tag{1.9}$$

The function $(l\Delta\tau)(\sigma)$ may be written in terms of the function $(Jq)(\sigma)$. By (1.3), (1.5) and (1.6) we can write

$$\frac{2}{l}\frac{d\mathbf{x}}{d\sigma}\cdot\nabla U_0 + U_0\Delta\tau = 0.$$

Hence, upon multiplication by lU_0 we have

$$\frac{dU_0^2}{d\sigma} = -U_0^2 l\Delta\tau.$$

Comparison with (1.8) gives

$$l\Delta\tau = \frac{d}{d\sigma}(\ln\, Jq).$$

Then (1.9) becomes

$$\frac{dU_j}{d\sigma} + \frac{1}{2}U_j\frac{d}{d\sigma}(\ln\, Jq) = -\frac{l}{2}\Delta U_{j-1}.$$

The trivial integration yields

$$U_j(\sigma) = \frac{U_j(\sigma_0)\sqrt{(Jq)(\sigma_0)}}{\sqrt{(Jq)(\sigma)}} - \frac{1}{2\sqrt{(Jq)(\sigma)}}\int_{\sigma_0}^{\sigma} l(s)\sqrt{(Jq)(s)}\;\Delta U_{j-1}(s)\,ds.$$

For later convenience we derive a result concerning rays and wavefronts, that is surfaces of constant phase. Let $\mathbf{t}$ be the tangent vector to a ray. We know from differential geometry that the divergence of $\mathbf{t}$ is related to the principal radii of curvature R_1, R_2, of the wavefront by

$$\nabla\cdot\mathbf{t} = \frac{1}{R_1} + \frac{1}{R_2}. \tag{1.10}$$

Accordingly, the divergence of $\mathbf{t}$ is twice the mean curvature H of the wavefront. A connection of $\nabla \cdot \mathbf{t}$ with the Jacobian J is now established.

Let D be any closed region with volume V and consider the invariant definition of $\nabla \cdot \mathbf{t}$, namely

$$\nabla \cdot \mathbf{t} = \lim_{V \to 0} \frac{1}{V} \int_D \mathbf{t} \cdot \mathbf{n} \, da \tag{1.11}$$

where $\mathbf{n}$ is the unit outward normal. As before, identify D with the region between two wavefronts W and W_0, within a tube of rays. Let the parameter σ correspond to W and σ_0 to W_0. For our considerations the parameter σ is required to be the arclength along the ray. Observe that $\mathbf{t} \cdot \mathbf{n} = 0$ on the lateral surface and $\mathbf{t} \cdot \mathbf{n} = 1, -1$ on W, W_0. Now recall that the surface element da, on W, is related to the corresponding element da_0, on W_0, by

$$da = \frac{J(\sigma)}{J(\sigma_0)} da_0.$$

Then we have

$$\lim_{V \to 0} \frac{1}{V} \int_D \mathbf{t} \cdot \mathbf{n} \, da = \frac{1}{J(\sigma_0)} \lim_{\sigma \to \sigma_0} \frac{J(\sigma) - J(\sigma_0)}{\sigma - \sigma_0} = \frac{1}{J} \frac{dJ}{d\sigma}. \tag{1.12}$$

By (1.10)-(1.12) it follows the desired result

$$\frac{1}{J} \frac{dJ}{d\sigma} = 2H \tag{1.13}$$

which relates the Jacobian J to the mean curvature H of the wavefront.

Sometimes (cf. [105], §5.2) the quantity K is considered such that

$$K(\sigma) da = K(\sigma_0) da_0.$$

We may set $K = 1/J$. Then we have, e.g.,

$$\nabla \cdot \mathbf{t} = -\frac{1}{K} \frac{dK}{d\sigma}. \tag{1.14}$$

9.2 Rays in viscoelastic solids

In homogeneous bodies the ray theory can ultimately be reduced to the analysis of one or more scalar Helmholtz equations. In such cases the theory is well-established and, in this connection, the interested reader is referred to general treatment such as [3]. In the description of continuous media, it usually happens that the behaviour of the body is naturally represented by tensor fields and suitable relations among them. Unless particular symmetries are involved, the tensor relations among the pertinent fields are expressed by more scalar functions and the model equations cannot be reduced to decoupled equations. Accordingly, in heterogeneous bodies the behaviour of the pertinent field cannot reduce to one or more Helmholtz equations.

As we show in a moment, the ray method allows the application of the procedure of §9.1 to more general schemes. In fact, we address our attention to rays in prestressed viscoelastic solids. Differently from what happens in the scalar Helmholtz equation, the recurrence relation involves amplitudes of all lower orders. Moreover, the recurrence relation involves initial derivatives of the relaxation functions of increasing orders as the subscript of the amplitudes involved increases. So in practice the investigation is confined to the lowest order term. It is a favourable feature that such term per se provides a good description of wave propagation as the wavelength is smaller than the characteristic dimensions of the problem.

In this section we derive ray trajectories and amplitude evolutions for wave propagation in prestressed viscoelastic solids. Consider time-harmonic solutions in the usual form $\mathbf{u}(\mathbf{x},t) = \mathbf{U}(\mathbf{x};\omega)\exp(-i\omega t)$. Following the standard ray approach to wave propagation we write the vector $\mathbf{U}(\mathbf{x};\omega)$ through the asymptotic expansion

$$\mathbf{U}(\mathbf{x};\omega) \simeq \exp[i\omega\tau(\mathbf{x})]\sum_{j=0}^{\infty}\mathbf{U}_j(\mathbf{x})(i\omega)^{-j},$$

to within any factor ω^β, here inessential. The phase function τ is taken to be real-valued while the $\mathbf{U}_j$'s are allowed to be complex-valued. As shown in a moment, this view is consistent with the general structure of inhomogeneous waves. Now, on omitting the common factor $\exp[i\omega\tau(\mathbf{x})]$, we have

$$\nabla\mathbf{U} = \sum_{j=0}^{\infty}\{(i\omega)^{1-j}\,\nabla\tau\otimes\mathbf{U}_j + (i\omega)^{-j}\,\nabla\mathbf{U}_j\},$$

$$\nabla\cdot\mathbf{U} = \sum_{j=0}^{\infty}\{(i\omega)^{1-j}\,\nabla\tau\cdot\mathbf{U}_j + (i\omega)^{-j}\,\nabla\cdot\mathbf{U}_j\},$$

$$\begin{aligned}\nabla\nabla\mathbf{U} = \sum_{j=0}^{\infty}\{&(i\omega)^{2-j}\nabla\tau\otimes\nabla\tau\otimes\mathbf{U}_j\\ &+ (i\omega)^{1-j}\nabla\nabla\tau\otimes\mathbf{U}_j + (i\omega)^{1-j}2(\mathrm{sym}\nabla\tau\otimes\nabla)\mathbf{U}_j + (i\omega)^{-j}\nabla\nabla\mathbf{U}_j\},\end{aligned}$$

$$\begin{aligned}\nabla(\nabla\cdot\mathbf{U}) = \sum_{j=0}^{\infty}\{&(i\omega)^{2-j}(\nabla\tau\cdot\mathbf{U}_j)\nabla\tau\\ &+ (i\omega)^{1-j}[(\nabla\nabla\tau)\mathbf{U}_j + (\nabla\mathbf{U}_j)\nabla\tau + \nabla\tau(\nabla\cdot\mathbf{U}_j)] + (i\omega)^{-j}\nabla(\nabla\cdot\mathbf{U}_j)\},\end{aligned}$$

$$\Delta \mathbf{U} = \sum_{j=0}^{\infty} \{(i\omega)^{2-j}(\nabla\tau)^2 \mathbf{U}_j + (i\omega)^{1-j}[\Delta\tau\, \mathbf{U}_j + 2(\nabla\tau\cdot\nabla)\mathbf{U}_j] + (i\omega)^{-j}\Delta\mathbf{U}_j\}.$$

For later convenience we observe that, given the product of two series expansions, we can write the result as the effect of successive sums. Specifically,

$$\sum_{h\geq 0}(-1)^h a_h (i\omega)^{-h} \sum_{k\geq 0} b_k^m (i\omega)^{-k+m} = \sum_{j\geq -m}\sum_{k=0}^{j+m}(-1)^{j-k}(-1)^m a_{j-k+m} b_k^m (i\omega)^{-j}.$$

Moreover we recall the asymptotic behaviour of $\mu(\omega)$ as given by (3.2.17), namely

$$\mu(\omega) = \sum_{h=0}^{n}(-1)^h \mu_0^{(h)}(i\omega)^{-h} + o(\omega^{-n}),$$

and analogously for $\lambda(\omega)$. Substitution in the equation of motion and some rearrangement lead to the vanishing of an asymptotic series with powers in ω^2, ω, ω^{-j}, $j = 0, 1, 2, \ldots$. The vanishing of each coefficient results in

$$-\rho\mathbf{U}_0 + (\nabla\tau\cdot\mathbf{T}^0\nabla\tau)\mathbf{U}_0 + (\mu_0+\lambda_0)(\mathbf{U}_0\cdot\nabla\tau)\nabla\tau + \mu_0\mathbf{U}_0(\nabla\tau)^2 = 0, \tag{2.1}$$

$$\begin{aligned}
&-\rho\mathbf{U}_1 + [\mathbf{T}^0\cdot(\nabla\nabla\tau)]\mathbf{U}_0 + (\nabla\tau\cdot\mathbf{T}^0\nabla\tau)\mathbf{U}_1 + 2(\nabla\tau\cdot\mathbf{T}^0\nabla)\mathbf{U}_0 + [(\nabla\cdot\mathbf{T}^0)\cdot\nabla\tau]\mathbf{U}_0\\
&\quad + (\nabla\mu_0\cdot\nabla\tau)\mathbf{U}_0 + (\mathbf{U}_0\cdot\nabla\mu_0)\nabla\tau + (\nabla\lambda_0)(\nabla\tau\cdot\mathbf{U}_0)\\
&\quad + (\mu_0+\lambda_0)[(\mathbf{U}_0\cdot\nabla)\nabla\tau + (\mathbf{U}_1\cdot\nabla\tau)\nabla\tau + (\nabla\mathbf{U}_0)\nabla\tau + (\nabla\cdot\mathbf{U}_0)\nabla\tau]\\
&\quad + \mu_0[\Delta\tau\,\mathbf{U}_0 + (\nabla\tau)^2\mathbf{U}_1 + 2(\nabla\tau\cdot\nabla)\mathbf{U}_0]\\
&\quad - (\mu_0'+\lambda_0')(\mathbf{U}_0\cdot\nabla\tau)\nabla\tau - \mu_0'(\nabla\tau)^2\mathbf{U}_0 = 0,
\end{aligned} \tag{2.2}$$

and

$$\begin{aligned}
&-\rho\mathbf{U}_{j+2} + (\nabla\tau\cdot\mathbf{T}^0\nabla\tau)\mathbf{U}_{j+2}\\
&\quad + (\mathbf{T}^0\cdot\nabla\nabla\tau)\mathbf{U}_{j+1} + 2(\nabla\tau\cdot\mathbf{T}^0\nabla)\mathbf{U}_{j+1} + [(\nabla\cdot\mathbf{T}^0)\cdot\nabla\tau]\mathbf{U}_{j+1}\\
&\quad + (\mathbf{T}^0\cdot\nabla\nabla)\mathbf{U}_j + [(\nabla\cdot\mathbf{T}^0)\cdot\nabla]\mathbf{U}_j + \sum_{m=0}^{2}\sum_{k=0}^{j+m}(-1)^{j-k}\mathcal{U}_{jk}^m = 0
\end{aligned} \tag{2.3}$$

where

$$\mathcal{U}_{jk}^2 = [\mu_0^{(j-k+2)} + \lambda_0^{(j-k+2)}](\nabla\tau\cdot\mathbf{U}_k)\nabla\tau + \mu_0^{(j-k+2)}(\nabla\tau)^2\mathbf{U}_k,$$

$$\begin{aligned}
\mathcal{U}_{jk}^1 = &-[\nabla\mu_0^{(j-k+1)}\cdot\nabla\tau]\mathbf{U}_k - [\nabla\mu_0^{(j-k+1)}\cdot\mathbf{U}_k]\nabla\tau - \nabla\lambda_0^{(j-k+1)}\nabla\tau\cdot\mathbf{U}_k\\
&-[\mu_0^{(j-k+1)} + \lambda_0^{(j-k+1)}][(\nabla\nabla\tau)\mathbf{U}_k + (\nabla\mathbf{U}_k)\nabla\tau + (\nabla\cdot\mathbf{U}_k)\nabla\tau]\\
&-\mu_0^{(j-k+1)}[\Delta\tau\mathbf{U}_k + 2(\nabla\tau\cdot\nabla)\mathbf{U}_k],
\end{aligned}$$

$$\mathcal{U}^0_{jk} = [\nabla\mu_0^{(j-k)}\cdot\nabla]\mathbf{U}_k + (\nabla\mathbf{U}_k)\nabla\mu_0^{(j-k)} + (\nabla\cdot\mathbf{U}_k)\nabla\lambda_0^{(j-k)}$$
$$+ [\mu_0^{(j-k)} + \lambda_0^{(j-k)}]\nabla(\nabla\cdot\mathbf{U}_k) + \mu_0^{(j-k)}\Delta\mathbf{U}_k.$$

Roughly, the recurrence relation (2.3) provides $\mathbf{U}_{j+2}$ once $\mathbf{U}_{j+1}, \mathbf{U}_j$ and τ are given. So we are left with the problem of determining $\mathbf{U}_0, \mathbf{U}_1$ and τ. In fact the two vectors $\mathbf{U}_0, \mathbf{U}_1$ and the scalar τ cannot be determined by (2.1) and (2.2) only; a component of (2.3) is necessary to complete the procedure.

Inner and vector multiplication of (2.1) by $\nabla\tau$ yields

$$[-\rho + (\nabla\tau\cdot\mathbf{T}^0\nabla\tau) + (2\mu_0+\lambda_0)(\nabla\tau)^2]\mathbf{U}_0\cdot\nabla\tau = 0,$$

$$[-\rho + (\nabla\tau\cdot\mathbf{T}^0\nabla\tau) + \mu_0(\nabla\tau)^2]\mathbf{U}_0\times\nabla\tau = 0.$$

Then two possibilities occur, namely

$$\mathbf{U}_0\cdot\nabla\tau = 0, \qquad -\rho + \nabla\tau\cdot\mathbf{T}^0\nabla\tau + \mu_0(\nabla\tau)^2 = 0 \tag{2.4}$$

and

$$\mathbf{U}_0\times\nabla\tau = 0, \qquad -\rho + \nabla\tau\cdot\mathbf{T}^0\nabla\tau + (2\mu_0+\lambda_0)(\nabla\tau)^2 = 0. \tag{2.5}$$

Quite naturally we call transverse waves and longitudinal waves those characterized by (2.4) and (2.5) respectively. Yet it is to be observed that the transversality or the longitudinality of $\mathbf{U}_0$, relative to $\nabla\tau$, does not imply that the same property holds for $\mathbf{U}_1$ and the next terms in the expansion. This aspect is made clear in a moment. Then, in this context, the adjectives transverse and longitudinal are appropriate inasmuch as the next terms are negligible.

Since $\rho, \mathbf{T}^0, \mu_0$, and λ_0 are taken to be known functions, $(2.4)_2$ and $(2.5)_2$ are first-order differential equations for the unknown function τ in the two allowed circumstances $\mathbf{U}_0\cdot\nabla\tau = 0$ and $\mathbf{U}_0\times\nabla\tau = 0$. In these circumstances they define the surfaces of constant phase, viz the wavefronts. By analogy with the geometrical theory of optics, we can say that $(2.4)_2$ and $(2.5)_2$ are the eikonal equations for transverse and longitudinal waves. The next task is the determination of the transport equations namely the equations which govern the evolution of $\mathbf{U}_0$. Formally, (2.2) is the transport equation for the two cases. Our purpose is to elaborate, on the basis of (2.4) and (2.5), an evolution equation for any wave amplitude U_0 along the pertinent ray.

First consider transverse waves, namely $\mathbf{U}_0\cdot\nabla\tau = 0$. Inner multiplication of (2.2) by $\nabla\tau$ and account of (2.4) yield

$$2[(\nabla\cdot\mathbf{T}^0\nabla)\mathbf{U}_0]\cdot\nabla\tau + (\mathbf{U}_0\cdot\nabla\mu_0)(\nabla\tau)^2 + (\mu_0+\lambda_0)(\mathbf{U}_1\cdot\nabla\tau + \nabla\cdot\mathbf{U}_0)(\nabla\tau)^2$$
$$+ (\mu_0+\lambda_0)[\nabla\tau\cdot(\mathbf{U}_0\cdot\nabla\tau)\nabla\tau + \nabla\tau\cdot(\nabla\tau\cdot\nabla)\mathbf{U}_0] + 2\mu_0\nabla\tau\cdot(\nabla\tau\cdot\nabla)\mathbf{U}_0 = 0. \tag{2.6}$$

Application of the gradient operator to $\mathbf{U}_0 \cdot \nabla\tau = 0$ provides

$$(\nabla \mathbf{U}_0)\nabla\tau + (\mathbf{U}_0 \cdot \nabla)\nabla\tau = 0.$$

Hence inner multiplication by $\nabla\tau$ yields

$$\nabla\tau \cdot (\mathbf{U}_0 \cdot \nabla)\nabla\tau + \nabla\tau \cdot (\nabla\tau \cdot \nabla)\mathbf{U}_0 = 0.$$

Accordingly (2.6) reduces to

$$\begin{aligned}(\mu_0 + \lambda_0)(\mathbf{U}_1 \cdot \nabla\tau + \nabla \cdot \mathbf{U}_0)(\nabla\tau)^2 = -[&2\nabla\tau \cdot (\nabla\tau \cdot \mathbf{T}^0\nabla)\mathbf{U}_0 \\ &+ (\mathbf{U}_0 \cdot \nabla\mu_0)(\nabla\tau)^2 + 2\mu_0\nabla\tau \cdot (\nabla\tau \cdot \nabla)\mathbf{U}_0].\end{aligned} \tag{2.7}$$

Back to (2.2), substitution for $(\mu_0 + \lambda_0)(\mathbf{U}_1 \cdot \nabla\tau + \nabla \cdot \mathbf{U}_0)(\nabla\tau)^2$ from (2.7), account of (2.4), and some rearrangement yield

$$\begin{aligned}(\mathbf{T}^0 \cdot \nabla\nabla\tau)\mathbf{U}_0 &+ 2(\nabla\tau \cdot \mathbf{T}^0\nabla)\mathbf{U}_0 + [(\nabla \cdot \mathbf{T}^0) \cdot \nabla\tau]\mathbf{U}_0 \\ &- \frac{1}{(\nabla\tau)^2}[2\nabla\tau \cdot (\nabla\tau \cdot \mathbf{T}^0\nabla)\mathbf{U}_0 + 2\mu_0\nabla\tau \cdot (\nabla\tau \cdot \nabla)\mathbf{U}_0]\nabla\tau \\ &+ (\nabla\mu_0 \cdot \nabla\tau)\mathbf{U}_0 + \mu_0\,\Delta\tau\,\mathbf{U}_0 + 2\mu_0(\nabla\tau \cdot \nabla)\mathbf{U}_0 - \mu_0'(\nabla\tau)^2\mathbf{U}_0 = 0.\end{aligned} \tag{2.8}$$

Inner multiplication of (2.8) by $\mathbf{U}_0$ and some rearrangement lead to

$$\nabla \cdot [(\mu_0\,\mathbf{1} + \mathbf{T}^0)\nabla\tau\, U_0^2] = \mu_0'(\nabla\tau)^2 U_0^2. \tag{2.9}$$

Equation (2.9) may be viewed as the transport equation for the amplitude U_0 of the transverse waves. Further results on the evolution of U_0 along the rays are derived in §9.3.

Consider the longitudinal waves (2.5). The contribution of $\mathbf{U}_1$ in (2.2) turns out to be

$$-\rho\mathbf{U}_1 + (\nabla\tau \cdot \mathbf{T}^0\nabla\tau)\mathbf{U}_1 + (\mu_0 + \lambda_0)(\mathbf{U}_1 \cdot \nabla\tau)\nabla\tau + \mu_0(\nabla\tau)^2\mathbf{U}_1 = -(\mu_0 + \lambda_0)\nabla\tau \times (\mathbf{U}_1 \times \nabla\tau).$$

This obviously means that the contribution of $\mathbf{U}_1$ is orthogonal to $\nabla\tau$. This feature in turn allows τ and $\mathbf{U}_0$ to be determined by looking at the vector equation (2.1) and the projection of (2.2) along $\nabla\tau$. Now, inner multiplication of (2.2) by $\nabla\tau$ and some rearrangement yield

$$\begin{aligned}(\mathbf{T}^0 \cdot \nabla\nabla\tau)\mathbf{U}_0 \cdot \nabla\tau &+ 2\nabla\tau \cdot (\nabla\tau \cdot \mathbf{T}^0\nabla)\mathbf{U}_0 + [(\nabla \cdot \mathbf{T}^0) \cdot \nabla\tau]\mathbf{U}_0 \cdot \nabla\tau \\ &+ [\nabla(\mu_0 + \lambda_0) \cdot \nabla\tau]\mathbf{U}_0 \cdot \nabla\tau + (\nabla\tau)^2\nabla\mu_0 \cdot \mathbf{U}_0 \\ + \mu_0\{\nabla\tau \cdot (\mathbf{U}_0 \cdot \nabla)\nabla\tau &+ [(\nabla\tau \cdot \nabla)\mathbf{U}_0] \cdot \nabla\tau + (\nabla\tau)^2\nabla \cdot \mathbf{U}_0 + \Delta\tau\,\nabla\tau \cdot \mathbf{U}_0 + 2\nabla\tau \cdot (\nabla\tau \cdot \nabla)\mathbf{U}_0\} \\ &+ \lambda_0\{\nabla\tau \cdot (\mathbf{U}_0 \cdot \nabla)\nabla\tau + [(\nabla\tau \cdot \nabla)\mathbf{U}_0] \cdot \nabla\tau + (\nabla\tau)^2\nabla \cdot \mathbf{U}_0\} \\ &- (2\mu_0' + \lambda_0')(\nabla\tau)^2\mathbf{U}_0 \cdot \nabla\tau = 0\,.\end{aligned} \tag{2.10}$$

Since $\mathbf{U}_0$ and $\nabla\tau$ are parallel we can write

$$\mathbf{U}_0 = \nu\,\nabla\tau$$

where ν is an undetermined scalar function of the position $\mathbf{x}$. Then multiplication of (2.10) by ν provides

$$\begin{aligned}(\mathbf{T}^0\cdot\nabla\nabla\tau)U_0^2 + 2\mathbf{U}_0\cdot(\nabla\tau\cdot\mathbf{T}^0\nabla)\mathbf{U}_0 + [(\nabla\cdot\mathbf{T}^0)\cdot\nabla\tau]U_0^2 + [(2\nabla\mu_0+\nabla\lambda_0)\cdot\nabla\tau]U_0^2 \\ + (\mu_0+\lambda_0)\{\mathbf{U}_0\cdot(\mathbf{U}_0\cdot\nabla)\nabla\tau + [(\nabla\tau\cdot\nabla)\mathbf{U}_0]\cdot\mathbf{U}_0 + \nu(\nabla\tau)^2\nabla\cdot\mathbf{U}_0\} \\ + \mu_0[\Delta\tau U_0^2 + 2\mathbf{U}_0\cdot(\nabla\tau\cdot\nabla)\mathbf{U}_0] - (2\mu_0'+\lambda_0')(\nabla\tau)^2U_0^2 = 0.\end{aligned}$$

Upon some rearrangement we arrive at the equation

$$\nabla\cdot\{[(2\mu_0+\lambda_0)\mathbf{1}+\mathbf{T}^0]\nabla\tau\,U_0^2\} = (2\mu_0'+\lambda_0')(\nabla\tau)^2U_0^2 \tag{2.11}$$

which may be viewed as the transport equation for the amplitude U_0 of the longitudinal waves.

Here we consider transverse waves and longitudinal waves at the same time. Accordingly, let

$$c := \frac{1}{|\nabla\tau|}$$

and regard τ as determined by $(2.4)_2$ or $(2.5)_2$ depending on the type of waves we are dealing with. Letting $c_T(\mathbf{x}), c_L(\mathbf{x})$ denote the two corresponding functions, we also define

$$f := \frac{\mu_0'}{c_T^2},\qquad \frac{2\mu_0'+\lambda_0'}{c_L^2}$$

for transverse and longitudinal waves, respectively; by (2.4.7) - cf. [70] - we have $f \le 0$. Similarly, let

$$\mathbf{w} := (\mu_0\,\mathbf{1}+\mathbf{T}^0)\nabla\tau,\quad [(2\mu_0+\lambda_0)\mathbf{1}+\mathbf{T}^0]\nabla\tau.$$

This allows $(2.4)_2$ and $(2.5)_2$ to be written as

$$-\rho + \mathbf{w}\cdot\nabla\tau = 0 \tag{2.12}$$

and (2.9) and (2.11) as

$$\nabla\cdot(\mathbf{w}U_0^2) = f\,U_0^2. \tag{2.13}$$

We say that rays are the flow of $\mathbf{w}$. Of course $\mathbf{w}$ is known once the eikonal function τ is determined via $(2.4)_2$ or $(2.5)_2$. Observe that, by the anisotropy of $\mathbf{T}^0$, $\mathbf{w}\times\nabla\tau \neq 0$ and then rays and integral curves of $\nabla\tau$ are usually distinct. For any regular region D, integration of (2.13) over D yields

$$\int_{\partial D}\mathbf{w}\cdot\mathbf{n}\,U_0^2\,da = \int_D f\,U_0^2\,dx \tag{2.14}$$

where $\mathbf{n}$ is the unit outward normal. We can regard (2.14) as the energy balance, for the region D, thus ascribing a precise physical meaning to $\mathbf{w}$. Consider unstressed homogeneous materials and identify $\nabla\tau$ with the slowness vector $\mathbf{k}/\omega$, $\mathbf{k}$ being the wave vector of the corresponding plane wave. Then $\mathbf{w} = \rho c\mathbf{k}/k$ and hence $\mathbf{w}U_0^2$ is $2/\omega^2$ times the energy flux vector. We can also say that, according to (2.14), the non-conservation of energy associated with the ray is due to the dissipation $\frac{1}{2}\omega^2 fU_0^2$ per unit time and unit volume.

9.3 Amplitude evolution and energy partition

Interesting consequences of (2.14) follow through appropriate choices of D. Choose any point $\mathbf{x}_0$ and let W_0 be a plane, bounded surface that passes through $\mathbf{x}_0$ and is orthogonal to $\mathbf{w}(\mathbf{x}_0)$. Then consider the tube constituted by the rays that pass through W_0. Follow the ray through $\mathbf{x}_0$ up to a point $\bar{\mathbf{x}}$ and consider the intersection W of the plane, that passes through $\bar{\mathbf{x}}$ and is orthogonal to $\mathbf{w}(\bar{\mathbf{x}})$, with the tube. Let D be the region occupied by the tube between W_0 and W and denote by d_0 the diameter of W_0. We can write

$$\mathbf{w}\cdot\mathbf{n} = \begin{cases} w(\bar{\mathbf{x}}) + O(d_0) & \text{on} \quad W \\ -w(\mathbf{x}_0) + O(d_0) & \text{on} \quad W_0 \end{cases}$$

where $w = |\mathbf{w}|$. Moreover $\mathbf{w}\cdot\mathbf{n} = 0$ on the lateral surface of the tube (between W_0 and W). Accordingly, by (2.13) and the mean value theorem we have

$$(wU_0^2)(\bar{\mathbf{x}})a - (wU_0^2)(\bar{\mathbf{x}}_0)a_0 + a_0 O(d_0) = \int_D fU_0^2\,dx$$

where a_0, a are the areas of W_0, W. The $O(d_0)$ term suggests that we pass to the limit as $d_0 \to 0$, which corresponds to an infinitesimal tube around the pertinent ray. The ratio a/a_0 goes as the ratio of mean values of the Jacobian J of mapping via rays. Moreover at the limit of arbitrarily small surfaces W_0, W we can view $\mathbf{x}$ as a function of the arclength σ only. Then by dividing throughout by a_0 and letting $d_0 \to 0$ we have

$$(wJU_0^2)(\sigma) - (wJU_0^2)(\sigma_0) = \int_{\sigma_0}^{\sigma} f(s)J(s)U_0^2(s)\,ds.$$

Differentiating with respect to σ and dividing by $(wJU_0^2)(\sigma)$ gives

$$\frac{1}{U_0^2}\frac{dU_0^2}{d\sigma} = -\frac{1}{wJ}\frac{d(wJ)}{d\sigma} + \frac{f}{w}.$$

A straightforward integration yields

$$U_0^2(\sigma) = \frac{(wJ)(\sigma_0)}{(wJ)(\sigma)}\,U_0^2(\sigma_0)\exp\Big[\int_{\sigma_0}^{\sigma}\frac{f}{w}(s)\,ds\Big]. \tag{3.1}$$

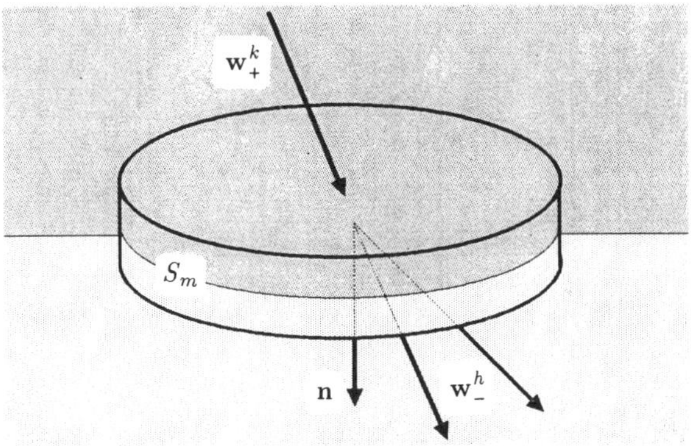

Fig. 9.2 Energy partition at an interface.

By way of comment we can say that, once the family of rays is determined through the eikonal equation, (3.1) yields the amplitude U_0 along the ray tube. Moreover, U_0 is shown to vary along the tube because of two causes. One is the heterogeneity of the material and is reflected in the wJ term and in the denominator w of the integrand. The other one is the dissipation and is expressed by the quantity f in the integrand. Since $f \leq 0$, as expected dissipation results in the decay of the amplitude along the tubes.

It is worth looking at the particular case of isotropically pre-stressed solids where

$$\mathbf{T}^0 = T\,\mathbf{1},$$

T being positive or negative according as $\mathbf{T}^0$ represents a tension or a pressure; we only assume that the possible pressure is not too high, i.e. $T > -\mu_0$. In such a case $\mathbf{w}$ is parallel to $\nabla\tau$ and then the flow of $\mathbf{w}$ coincides with the rays of the usual theory, namely the flow of $\nabla\tau$. Further, $w = \rho c$ and c takes the values

$$c_T = \sqrt{\frac{\mu_0 + T}{\rho}}, \qquad c_L = \sqrt{\frac{2\mu_0 + \lambda_0 + T}{\rho}}$$

according as transverse or longitudinal waves are considered. In particular we have

$$U_0^2(\sigma) = \frac{[\rho c J](\sigma_0)}{[\rho c J](\sigma)} U_0^2(\sigma_0) \exp\Big[\int_{\sigma_0}^{\sigma} \frac{f}{\rho c}(s) ds\Big]. \tag{3.1'}$$

The result (3.1') allows a direct comparison with the analogous result which arises from the analysis of the Helmholtz equation. With regard to §9.1 and, e.g., [15], it looks as if c in this section should be replaced by $1/c$ to get the standard result. Really, this is so because the correct equation of motion comprises terms in $\nabla\mu, \nabla\lambda$ which lead to the occurrence of μ_0 and λ_0 in $\mathbf{w}$. Then, in essence, apart from a factor ρ, the occurrence of μ_0 and λ_0 makes $1/c$ of the standard theory into c of the present approach.

We now apply the result (2.14) to the case when D is a pillbox region. As the thickness of the pillbox approaches zero we have

$$\int_{\partial D} \mathbf{w} \cdot \mathbf{n} U_0^2 da = 0. \tag{3.2}$$

The interpretation that $\mathbf{w}U_0^2$ is ($2/\omega^2$ times) the energy flux vector allows us to view (3.2) as the statement that the net energy entering the slot mean surface S_m equals the net energy leaving S_m. If $\mathbf{w}U_0^2$ comes from different rays, as in reflection and refraction (cf. §9.5), we may write

$$[\sum_k \mathbf{w}_+^k (U_0^k)^2 - \sum_h \mathbf{w}_-^h (U_0^h)^2] \cdot \mathbf{n} = 0 \tag{3.3}$$

where k labels the rays above the surface S_m and h the rays below while $\mathbf{n}$ is the fixed normal to S_m. For any ray, $\mathbf{w}(U_0)^2 \cdot \mathbf{n}$ represents the corresponding energy flowing through S_m per unit time and unit area. More precisely, for each type of ray consider the tube issued from the boundary of S_m, on the appropriate side. According to (3.3), the overall contribution of energy, associated with the pertinent ray tubes, which flows through S_m must vanish. Incidentally, this condition may be regarded as a check of consistency of the results obtained for a reflection-refraction problem at an interface. Also, once we know the amplitudes U_0^k, U_0^h and the vectors $\mathbf{w}_+^k$, $\mathbf{w}_-^h$ we know how energy is partitioned among the various ray tubes due to a reflection-refraction process.

9.4 Rays in viscoelastic fluids

The analysis of this section parallels that for solids of §9.3. Accordingly the presentation of the subject is less detailed here.

Consider the linearized equation of motion for fluids and, for definiteness, identify the body force $\mathbf{b}$ with the acceleration gravity $\mathbf{g}$. Then the equilibrium condition is written as

$$\nabla p(\rho) = \rho \mathbf{g} \tag{4.1}$$

whence we obtain the equilibrium density $\rho = \rho(\mathbf{x})$. As usual, we choose the equilibrium placement as reference, which means that every material particle of the fluid is labelled by the position occupied at equilibrium. Since $\mathbf{b}$ is left unperturbed by a superposed (infinitesimal) motion we write (2.6.9) as

$$\rho \ddot{\mathbf{u}} = \rho(\nabla \mathbf{u})\mathbf{g} + \rho^2 \frac{p_{\rho\rho}}{p_\rho}(\nabla \cdot \mathbf{u})\mathbf{g} + \rho p_\rho \nabla(\nabla \cdot \mathbf{u}) + \nabla \cdot \boldsymbol{\tau} \tag{4.2}$$

where $\boldsymbol{\tau}$ is the viscoelastic stress tensor, relative to the equilibrium placement (cf. (2.5.4) and (2.6.9)).

Take the velocity field to be time-harmonic in that

$$\mathbf{v}(\mathbf{x},t) = \mathbf{V}(\mathbf{x};\omega)\exp(-i\omega t).$$

Of course $\mathbf{U}(\mathbf{x};\omega) = -(i\omega)^{-1}\mathbf{V}(\mathbf{x};\omega)$ is the corresponding displacement vector amplitude. Then, to within the common factor $\exp(-i\omega t)$, the stress $\boldsymbol{\tau}$ is given by

$$\boldsymbol{\tau} = \mu(\nabla\mathbf{V} + \nabla\mathbf{V}^{\dagger}) + \lambda(\nabla\cdot\mathbf{V})\,\mathbf{1} \tag{4.3}$$

where μ, λ depend on ω as

$$\mu = \int_0^{\infty} \tilde{\mu}(s)\exp(i\omega s)\,ds, \qquad \lambda = \int_0^{\infty} \tilde{\lambda}(s)\exp(i\omega s)\,ds.$$

Then the equation of motion (4.2) becomes

$$\rho\omega^2\mathbf{U} + \frac{\rho^2 p_{\rho\rho}}{p_\rho}(\nabla\mathbf{U})\mathbf{g} + \rho p_\rho\nabla(\nabla\cdot\mathbf{U}) + \rho(\nabla\mathbf{U})\mathbf{g}$$
$$+ (\nabla\mu\cdot\nabla)\mathbf{V} + (\nabla\mathbf{V})\nabla\mu + \mu\Delta\mathbf{V} + (\mu+\lambda)\nabla(\nabla\cdot\mathbf{V}) + (\nabla\cdot\mathbf{V})\nabla\lambda. \tag{4.4}$$

For the following calculations we need the asymptotic behaviour of μ, λ. Integrations by parts and application of Riemann-Lebesgue's lemma yield

$$\mu(\omega) = \sum_{j=1}^{n}(-1)^j\tilde{\mu}_0^{(j-1)}(i\omega)^{-j} + o(\omega^{-n})$$

where $\tilde{\mu}_0^{(k)}$ stands for the kth derivative of $\tilde{\mu}(s)$ at $s=0$. An analogous expression holds for λ. In particular,

$$\mu(\omega) = i\frac{\tilde{\mu}_0}{\omega} - \frac{\tilde{\mu}_0'}{\omega^2} + o(\omega^{-2}), \qquad \lambda(\omega) = i\frac{\tilde{\lambda}_0}{\omega} - \frac{\tilde{\lambda}_0'}{\omega^2} + o(\omega^{-2}). \tag{4.5}$$

By thermodynamics, viz (2.5.6), it follows that the real parts of μ and $\frac{2}{3}\mu + \lambda$ are positive, and hence the real part of $2\mu + \lambda$ is positive too. Then by (4.5) we have

$$\tilde{\mu}_0' \leq 0, \qquad 2\tilde{\mu}_0' + \tilde{\lambda}_0' \leq 0. \tag{4.6}$$

No restriction is placed by thermodynamics on the function $p(\rho)$. Since $p_\rho = dp/d\rho$ is the square of the speed of acoustic waves in perfect fluids, it seems reasonable to assume that $p_\rho > 0$.

Represent the vector amplitude $\mathbf{V}(\mathbf{x};\omega)$ as

$$\mathbf{V}(\mathbf{x};\omega) = \exp[i\omega\tau(\mathbf{x})]\sum_{j=0}^{\infty}\mathbf{V}_j(\mathbf{x})(i\omega)^{-j},$$

to within any factor ω^β. Substitution in (4.4) and some rearrangement lead to a power in ω plus a $O(1)$ term plus $O(\omega^{-1})$ as $\omega \to \infty$. The vanishing of the coefficients of ω and of $O(1)$ results in

$$-\rho\mathbf{V}_0 + (\tilde{\mu}_0 + \tilde{\lambda}_0 + \rho p_\rho)(\nabla\tau\cdot\mathbf{V}_0)\nabla\tau + \tilde{\mu}_0(\nabla\tau)^2\mathbf{V}_0 = 0, \tag{4.7}$$

$$\begin{aligned}
&-\rho\mathbf{V}_1 + \rho(\mathbf{V}_0\cdot\mathbf{g})\nabla\tau + \frac{\rho^2 p_{\rho\rho}}{p_\rho}(\mathbf{V}_0\cdot\nabla\tau)\mathbf{g} + \tilde{\mu}_0[\Delta\tau\,\mathbf{V}_0 + 2(\nabla\tau\cdot\nabla)\mathbf{V}_0 + (\nabla\tau)^2\mathbf{V}_1]\\
&\quad + (\rho p_\rho + \tilde{\mu}_0 + \tilde{\lambda}_0)[(\nabla\nabla\tau)\mathbf{V}_0 + (\nabla\mathbf{V}_0)\nabla\tau + (\nabla\cdot\mathbf{V}_0)\nabla\tau + (\nabla\tau\cdot\mathbf{V}_1)\nabla\tau]\\
&+(\nabla\tilde{\mu}_0\cdot\nabla\tau)\mathbf{V}_0 + (\nabla\tilde{\mu}_0\cdot\mathbf{V}_0)\nabla\tau + (\nabla\tau\cdot\mathbf{V}_0)\nabla\tilde{\lambda}_0 - (\tilde{\mu}_0' + \tilde{\lambda}_0')(\nabla\tau\cdot\mathbf{V}_0)\nabla\tau - \tilde{\mu}_0'(\nabla\tau)^2\mathbf{V}_0 = 0.
\end{aligned} \tag{4.8}$$

Inner multiplication and vector multiplication of (4.7) by $\nabla\tau$ yield

$$[\rho - (\rho p_\rho + 2\tilde{\mu}_0 + \tilde{\lambda}_0)(\nabla\tau)^2]\mathbf{V}_0\cdot\nabla\tau = 0,$$

and

$$[\rho - \tilde{\mu}_0(\nabla\tau)^2]\mathbf{V}_0\times\nabla\tau = 0.$$

Then two cases occur.

$$\mathbf{V}_0\cdot\nabla\tau = 0, \qquad (\nabla\tau)^2 = \frac{\rho}{\tilde{\mu}_0}, \tag{4.9}$$

$$\mathbf{V}_0\times\nabla\tau = 0, \qquad (\nabla\tau)^2 = \frac{\rho}{\rho p_\rho + 2\tilde{\mu}_0 + \tilde{\lambda}_0}. \tag{4.10}$$

Quite naturally we call transverse waves and longitudinal waves those characterized by (4.9) and (4.10), respectively. Yet it is to be observed that the transversality or the longitudinality of $\mathbf{V}_0$ does not imply that the same property holds for $\mathbf{V}_1$ and the next terms in the expansion. This aspect is made clear in a moment. Then, in this context, the adjectives transverse and longitudinal are merely conventional.

Consider the transverse waves (4.9). The phase function $\tau(\mathbf{x})$ satisfies the eikonal equation

$$(\nabla\tau)^2 = \frac{\rho}{\tilde{\mu}_0}.$$

Having determined the phase function, we can investigate (4.8) to obtain the transport equation. Observe that the application of ∇ to the transversality condition gives

$$(\nabla\nabla\tau)\mathbf{V}_0 + (\nabla\mathbf{V}_0)\nabla\tau = 0. \tag{4.11}$$

Inner multiplication of (4.8) by $\nabla\tau$ and account of (4.9) and (4.11) yield

$$\nabla\cdot\mathbf{V}_0 + \nabla\tau\cdot\mathbf{V}_1 = -\frac{1}{\rho p_\rho + \tilde{\mu}_0 + \tilde{\lambda}_0}\{\rho\mathbf{V}_0\cdot\mathbf{g} + \nabla\tilde{\mu}_0\cdot\mathbf{V}_0 + \frac{2\tilde{\mu}_0^2}{\rho}[(\nabla\tau\cdot\nabla)\mathbf{V}_0]\cdot\nabla\tau\}.$$

This shows that, in general, $\nabla\tau \cdot \mathbf{V}_1 \neq 0$ albeit $\nabla\tau \cdot \mathbf{V}_0 = 0$. Substitution in (4.8) and account of (4.9) and (4.11) lead to

$$\tilde{\mu}_0 \Delta\tau\, \mathbf{V}_0 + 2\tilde{\mu}_0(\nabla\tau\cdot\nabla)\mathbf{V}_0 - \frac{2\tilde{\mu}_0^2}{\rho}\{[(\nabla\tau\cdot\nabla)\mathbf{V}_0]\cdot\nabla\tau\}\nabla\tau + (\nabla\tilde{\mu}_0\cdot\nabla\tau)\mathbf{V}_0 - \frac{\rho\tilde{\mu}_0'}{\tilde{\mu}_0}\mathbf{V}_0 = 0. \quad (4.12)$$

Inner multiplication by $\mathbf{V}_0$ allows us to write

$$\nabla\cdot[\tilde{\mu}_0 V_0^2 \nabla\tau] = \frac{\rho\tilde{\mu}_0'}{\tilde{\mu}_0} V_0^2. \quad (4.13)$$

Consider the longitudinal waves (4.10). The term $\mathbf{V}_1$ occurs in (4.8) through the expression

$$\rho\mathbf{V}_1 - (\rho p_\rho + \tilde{\mu}_0 + \tilde{\lambda}_0)(\nabla\tau\cdot\mathbf{V}_1)\nabla\tau - \tilde{\mu}_0(\nabla\tau)^2\mathbf{V}_1.$$

Substitution for $(\nabla\tau)^2$ from the eikonal equation yields

$$\rho\mathbf{V}_1 - (\rho p_\rho + \tilde{\mu}_0 + \tilde{\lambda}_0)(\nabla\tau\cdot\mathbf{V}_1)\nabla\tau - \tilde{\mu}_0(\nabla\tau)^2\mathbf{V}_1 = (\rho p_\rho + \tilde{\mu}_0 + \tilde{\lambda}_0)\nabla\tau \times (\mathbf{V}_1 \times \nabla\tau).$$

This means that $\mathbf{V}_1$ contributes to (4.8) through a term orthogonal to $\nabla\tau$. Then we obtain a transport equation for the vector $\mathbf{V}_0$, free from $\mathbf{V}_1$, by confining to the longitudinal part. Inner multiplication by $\nabla\tau$ and some rearrangement provides

$$\rho\Big(1+\frac{\rho p_{\rho\rho}}{p_\rho}\Big)\mathbf{V}_0\cdot\mathbf{g} + (\rho p_\rho + \tilde{\mu}_0 + \tilde{\lambda}_0)\frac{1}{(\nabla\tau)^2}\{[(\nabla\tau\cdot\nabla)\nabla\tau]\cdot\mathbf{V}_0 + [(\nabla\tau\cdot\nabla)\mathbf{V}_0]\cdot\nabla\tau + (\nabla\tau)^2\nabla\cdot\mathbf{V}_0\}$$
$$+\nabla(2\tilde{\mu}_0+\tilde{\lambda}_0)\cdot\mathbf{V}_0 + \frac{\tilde{\mu}_0}{(\nabla\tau)^2}\Delta\tau\,\mathbf{V}_0\cdot\nabla\tau + \frac{2\tilde{\mu}_0}{(\nabla\tau)^2}\nabla\tau\cdot[(\nabla\tau\cdot\nabla)\mathbf{V}_0] - (2\tilde{\mu}_0' + \tilde{\lambda}_0')\mathbf{V}_0\cdot\nabla\tau = 0. \quad (4.14)$$

The parallelism of $\mathbf{V}_0$ and $\nabla\tau$ allows us to write

$$\mathbf{V}_0 = \nu\,\nabla\tau$$

where ν is a function of the position $\mathbf{x}$. Then, multiplication of (4.14) by ν, some rearrangement and eventual substitution of (4.10) yield

$$\nabla\cdot[(\rho p_\rho + 2\tilde{\mu}_0 + \tilde{\lambda}_0)V_0^2\nabla\tau] = \frac{\rho(2\tilde{\mu}_0' + \tilde{\lambda}_0')}{\rho p_\rho + 2\tilde{\mu}_0 + \tilde{\lambda}_0} V_0^2 \quad (4.15)$$

where we have taken into account that, because of (4.1),

$$\nabla(\rho p_\rho) = \rho\Big(1 + \frac{\rho p_{\rho\rho}}{p_\rho}\Big)\mathbf{g}.$$

Now we consider transverse and longitudinal waves at the same time. Let

$$\mathbf{w} := \frac{\tilde{\mu}_0}{\rho}\nabla\tau, \quad \Big(p_\rho + \frac{2\tilde{\mu}_0 + \tilde{\lambda}_0}{\rho}\Big)\nabla\tau,$$

the two values being relative to transverse and longitudinal waves, respectively. Analogously, let

$$f := \frac{\rho\tilde{\mu}_0'}{\tilde{\mu}_0}, \quad \frac{\rho(2\tilde{\mu}_0' + \tilde{\lambda}_0')}{\rho p_\rho + 2\tilde{\mu}_0 + \tilde{\lambda}_0};$$

by (4.6) we have $f \leq 0$. Then (4.13) and (4.15) take the common form

$$\nabla \cdot (\mathbf{w}V_0^2) = fV_0^2 \tag{4.16}$$

while the eikonal equation becomes

$$-\rho + \mathbf{w} \cdot \nabla\tau = 0.$$

For any regular region D, (4.16) and the divergence theorem give

$$\int_{\partial D} \mathbf{w} \cdot \mathbf{n}V_0^2\, da = \int_D fV_0^2\, dx. \tag{4.17}$$

Incidentally, for (viscoelastic) fluids the flow of $\mathbf{w}$ coincides with the flow of $\nabla\tau$ and then here defining rays as the flow of $\mathbf{w}$ is equivalent to defining them as the flow of $\nabla\tau$. In other words, in fluids the rays are orthogonal to the wavefronts.

By paralleling the procedure of the previous section for solids, we conclude that, along an infinitesimal ray tube,

$$V_0^2(\sigma) = \frac{(wJ)(\sigma_0)}{(wJ)(\sigma)}\, V_0^2(\sigma_0) \exp\big[\int_{\sigma_0}^{\sigma} \frac{f}{w}(s)\, ds\big]. \tag{4.18}$$

By the eikonal equation (4.16) it follows that the speed $c := 1/|\nabla\tau|$ takes the values

$$c = \sqrt{\frac{\tilde{\mu}_0}{\rho}}, \quad \sqrt{p_\rho + \frac{2\tilde{\mu}_0 + \tilde{\lambda}_0}{\rho}}.$$

Meanwhile

$$w = \rho c.$$

Then (4.18) can be written as

$$V_0^2(\sigma) = \frac{[\rho cJ](\sigma_0)}{[\rho cJ](\sigma)}\, V_0^2(\sigma_0) \exp\big[\int_{\sigma_0}^{\sigma} \frac{f}{\rho c}(s)\, ds\big]. \tag{4.18'}$$

As with solids, the amplitude V_0 varies along the rays because of two causes. One is the heterogeneity of the material and is reflected in the ρcJ-term and the denominator ρc of the integrand. The other one is the dissipation and is expressed by the quantity f in the integrand. Once the family of rays is determined through the eikonal equation, (4.18') yields the amplitude V_0 along any ray (tube). Again, it looks as though c in our result should be replaced by $1/c$ to obtain standard results [15]. This is so because the correct equation of motion (4.4) comprises terms in $\nabla\tilde{\mu}$, $\nabla\tilde{\lambda}$ which lead to the occurrence of $\tilde{\mu}_0$, $\tilde{\lambda}_0$ in (4.18').

It is of interest to determine the evolution of the vector amplitude $\mathbf{V}_0$ of the transverse wave. Letting

$$n = \sqrt{\rho/\tilde{\mu}_0}$$

we write the eikonal equation as

$$(\nabla\tau)^2 = n^2.$$

We denote by $\mathbf{t} = \nabla\tau/|\nabla\tau|$, $\mathbf{p}$ and $\mathbf{b}$ the unit tangent (vector), the unit normal and the binormal of the ray. Then, if σ is the arclength along the ray, $\nabla\tau\cdot\nabla = nd/d\sigma$ and

$$\begin{aligned}\nabla n^2 &= \nabla(\nabla\tau\cdot\nabla\tau) = 2(\nabla\tau\cdot\nabla)\nabla\tau = 2n\frac{d}{d\sigma}(n\mathbf{t})\\ &= 2n\Big(\frac{dn}{d\sigma}\mathbf{t} + \frac{n}{\varrho}\mathbf{p}\Big)\end{aligned} \tag{4.19}$$

where ϱ is the radius of curvature of the ray and the Frenet formula $d\mathbf{t}/d\sigma = \mathbf{p}/\varrho$ has been used. To obtain the evolution of the vector amplitude $\mathbf{V}_0$ we go back to (4.12) and write

$$\tfrac{1}{2}\Delta\tau\,\mathbf{V}_0 + n\frac{d\mathbf{V}_0}{d\sigma} - n\Big(\frac{d\mathbf{V}_0}{d\sigma}\cdot\mathbf{t}\Big)\mathbf{t} + \frac{1}{2\tilde{\mu}_0}(\nabla\tilde{\mu}_0\cdot\nabla\tau)\mathbf{V}_0 - \frac{\rho\tilde{\mu}_0'}{2\tilde{\mu}_0^2}\mathbf{V}_0 = 0. \tag{4.20}$$

Observe that

$$\Delta\tau + \frac{1}{\tilde{\mu}_0}\nabla\tilde{\mu}_0\cdot\nabla\tau = \frac{1}{\tilde{\mu}_0}\nabla\cdot(\tilde{\mu}_0\nabla\tau)$$

and that the orthogonality of $\mathbf{V}_0$ and $\mathbf{t}$ gives

$$\frac{d\mathbf{V}_0}{d\sigma}\cdot\mathbf{t} = -\frac{1}{\varrho}\mathbf{V}_0\cdot\mathbf{p}.$$

Then (4.20) can be given the form

$$\frac{d\mathbf{V}_0}{d\sigma} + \frac{1}{2\sqrt{\rho\tilde{\mu}_0}}\Big[\nabla\cdot(\tilde{\mu}_0\nabla\tau) - \frac{\rho\tilde{\mu}_0'}{\tilde{\mu}_0}\Big]\mathbf{V}_0 + \frac{1}{\varrho}(\mathbf{V}_0\cdot\mathbf{p})\mathbf{t} = 0. \tag{4.21}$$

We now parallel the procedure of Karal and Keller for elastic solids [101]. Observe that, since $\mathbf{V}_0\cdot\mathbf{t} = 0$, we can write

$$\mathbf{V}_0 = \alpha\mathbf{p} + \beta\mathbf{b}.$$

Then by means of Frenet's formulae we have

$$\frac{d\mathbf{V}_0}{d\sigma} = \Big(\frac{d\alpha}{d\sigma} - \frac{\beta}{\chi}\Big)\mathbf{p} + \Big(\frac{\alpha}{\chi} + \frac{d\beta}{d\sigma}\Big)\mathbf{b} - \frac{\alpha}{\varrho}\mathbf{t} \tag{4.22}$$

where χ is the radius of torsion of the ray. Comparison of (4.21) and (4.22) yields the system of equations

$$\frac{d\alpha}{d\sigma} - \frac{\beta}{\chi} = -\varpi\alpha,$$

$$\frac{d\beta}{d\sigma} + \frac{\alpha}{\chi} = -\varpi\beta,$$

where $\varpi = [\nabla \cdot (\tilde{\mu}_0 \nabla\tau) - \rho\tilde{\mu}_0'/\tilde{\mu}_0]/2\sqrt{\rho\tilde{\mu}_0}$. Letting $\gamma = \alpha + i\beta$ we have

$$\frac{1}{\gamma}\frac{d\gamma}{d\sigma} = -\varpi - i\frac{1}{\chi}.$$

The trivial integration, the definitions

$$\theta(\sigma, \sigma_0) = \int_{\sigma_0}^{\sigma} \frac{1}{\chi(s)} ds, \qquad \phi_0 = \tan^{-1}\frac{\alpha(\sigma_0)}{\beta(\sigma_0)},$$

and some rearrangement yield

$$\mathbf{V}_0 = |\mathbf{V}_0(\sigma_0)| \exp(-\int_{\sigma_0}^{\sigma} \varpi(s)ds)\{\sin[\theta(\sigma,\sigma_0) + \phi_0]\mathbf{p} + \cos[\theta(\sigma,\sigma_0) + \phi_0]\mathbf{b}\}. \qquad (4.23)$$

A more explicit form can be obtained by examining the expression for ϖ. By (1.11) and (1.12) we have $\nabla \cdot \mathbf{t} = (dJ/d\sigma)/J$ and then

$$\nabla \cdot (\tilde{\mu}_0 \nabla\tau) = \nabla \cdot (\tilde{\mu}_0 n\mathbf{t}) = \frac{1}{J}\frac{d}{d\sigma}(\tilde{\mu}_0 nJ).$$

Hence, because $c = \sqrt{\tilde{\mu}_0/\rho}$, we have

$$\frac{1}{2\sqrt{\rho\tilde{\mu}_0}} \nabla \cdot (\tilde{\mu}_0 \nabla\tau) = \frac{d}{d\sigma} \ln\sqrt{J\rho c}.$$

Substitution of

$$-\varpi = \tfrac{1}{2}\frac{f}{\rho c} - \frac{d}{d\sigma}\ln\sqrt{J\rho c}$$

in (4.23) yields

$$\mathbf{V}_0 = |\mathbf{V}_0(\sigma_0)| \sqrt{\frac{(J\rho c)(\sigma_0)}{(J\rho c)(\sigma)}} \exp\left(\tfrac{1}{2}\int_{\sigma_0}^{\sigma} \frac{f}{\rho c}(s)ds\right)\{\sin[\theta(\sigma,\sigma_0) + \phi_0]\mathbf{p} + \cos[\theta(\sigma,\sigma_0) + \phi_0]\mathbf{b}\}. \qquad (4.24)$$

The result (4.24) shows how the direction of $\mathbf{V}_0$ is related to the radius of torsion of the ray. Of course, squaring (4.24) yields (4.18').

Detailed results about the ray trajectory, and then of the amplitude evolution, follow if the fluid is stratified in the sense that the material properties depend on one Cartesian coordinate only. For definiteness, consider the Cartesian coordinates x, y, z, with unit vectors $\mathbf{e}_x, \mathbf{e}_y, \mathbf{e}_z$, and let the material properties depend on z, only. Then by (4.9) and (4.10) we can write

$$(\nabla \tau)^2 = q^2(z) \tag{4.25}$$

where the slowness $q = 1/c$ is a piecewise C^1 function. The solution to (4.25) is determined up to quadratures.

Assume that the far infinity ($z = -\infty$) is homogeneous, which allows incident, plane waves to occur. Consider the plane wavefront passing through a selected point (x_0, y_0, z_0), whose direction numbers are the values of $\partial\tau/\partial x, \partial\tau/\partial y, \partial\tau/\partial z$ at (x_0, y_0, z_0). Let q_0 be the value of q at (x_0, y_0, z_0). To fix ideas, let the value of $\partial\tau/\partial y$ vanish at (x_0, y_0, z_0) and define α such that, at (x_0, y_0, z_0), $\partial\tau/\partial x = q_0 \sin\alpha, \partial\tau/\partial z = q_0 \cos\alpha$. Following [105], §14.4, we can represent the ray that passes through (x_0, y_0, z_0) as

$$\begin{cases} x(\sigma) = (q_0 \sin\alpha)\sigma + x_0, \\ y(\sigma) = y_0, \\ \sigma = \int_{z_0}^{z} [q^2(s) - q_0^2 \sin^2\alpha]^{-1/2} ds, \end{cases} \tag{4.26}$$

whereby the ray is a plane curve. Accordingly, the phase function τ takes the form

$$\tau = q_0(z\cos\alpha + x\sin\alpha) + \int_{z_0}^{z} \left[\sqrt{q^2(s) - q_0^2\sin^2\alpha} - q_0\cos\alpha\right] ds.$$

The first term can be interpreted as the phase function which corresponds to the initial plane wave in the homogeneous region. The second term is the correction due to the heterogeneity of the medium. As expected, $\partial\tau/\partial y = 0$. Observe that here σ is not the arclength. Indeed, $q\,d\sigma$ is the length element.

By means of (4.26) we determine J as follows. Choose two parameters γ_1, γ_2 which describe the initial plane front. It is natural to identify one of them, γ_2 say, with the coordinate y_0. Then we let γ_1 be the coordinate orthogonal to y_0 such that

$$x_0 = \gamma_1 \cos\alpha, \qquad z_0 = -\gamma_1 \sin\alpha.$$

Then we can write (4.26) as

$$\begin{cases} x(\sigma, \gamma_1, \gamma_2) = (q_0 \sin\alpha)\sigma + \gamma_1\cos\alpha, \\ y(\sigma, \gamma_1, \gamma_2) = \gamma_2, \\ \sigma = \int_{-\gamma_1 \sin\alpha}^{z(\sigma,\gamma_1,\gamma_2)} [q^2(s) - q_0^2 \sin^2\alpha]^{-1/2} ds. \end{cases} \tag{4.27}$$

It follows that

$$\frac{\partial \mathbf{x}}{\partial \gamma_1} = \cos\alpha\, \mathbf{e}_x - \frac{\sin\alpha\sqrt{q^2 - q_0^2\sin^2\alpha}}{q_0\cos\alpha}\mathbf{e}_z,$$

$$\frac{\partial \mathbf{x}}{\partial \gamma_2} = \mathbf{e}_y,$$

while the unit vector $\mathbf{n}$ of $\partial\mathbf{x}/\partial\sigma$ is given by

$$\mathbf{n} = \frac{q_0 \sin\alpha}{q}\mathbf{e}_x + \frac{\sqrt{q^2 - q_0^2\sin^2\alpha}}{q}\mathbf{e}_z.$$

Then we have

$$J = \frac{\sqrt{q^2 - q_0^2\sin^2\alpha}}{q\cos\alpha}. \tag{4.28}$$

The mean curvature of the wave front H turns out to be

$$H = \frac{1}{2}\frac{q_0^2\sin^2\alpha}{q^2\sqrt{q^2 - q_0^2\sin^2\alpha}}\frac{dq}{dz}. \tag{4.29}$$

Incidentally, by (4.27) we find that the curvature κ of the ray is given by

$$\kappa = \frac{q_0\sin\alpha}{q^2}\frac{dq}{dz}. \tag{4.30}$$

By (4.27) any ray undergoes a turning point at the plane $z = \bar{z}$ such that $q^2(\bar{z}) = q_0^2\sin^2\alpha$. There, by (4.28) the infinitesimal ray tube shrinks to a vanishing cross-section ($J = 0$). Moreover, by (4.29), H tends to infinity while, by (4.30), the curvature of the ray remains bounded.

Substitution of J, τ and the known functions $q, \rho, p_\rho, p_{\rho\rho}, \tilde{\mu}_0, \tilde{\lambda}_0, \tilde{\mu}_0', \tilde{\lambda}_0'$ in (4.18') provides the evolution of the amplitude V_0 along the ray.

9.5 Reflection and refraction of rays

As with plane waves, when a ray strikes an interface between two media with different material properties, reflected and transmitted rays emanate from the interface. Our purpose is to derive directions and amplitudes of the reflected and transmitted rays when the interface separates two viscoelastic half-spaces. We consider an incident wave of the form

$$\mathbf{U}^i(\mathbf{x};\omega) = \exp[i\omega\tau^i(\mathbf{x})]\sum_{j=0}^{\infty}\mathbf{U}_j^i(\mathbf{x})(i\omega)^{-j}$$

the superscript i denoting quantities pertaining to the incident wave. In general, reflected and transmitted waves consist of both longitudinal and transverse components. They are fully determined through the continuity of displacement and traction at the interface.

Preliminarily we determine the analogue of Snell's law. Letting r, t label quantities pertaining to reflected and transmitted waves, we write $\mathbf{U}^r$ and $\mathbf{U}^t$ as

$$\mathbf{U}^r(\mathbf{x},\omega) = \exp[i\omega\tau_L^r(\mathbf{x})]\sum_{j=0}^{\infty}\mathbf{U}_{Lj}^r(\mathbf{x})(i\omega)^{-j} + \exp[i\omega\tau_T^r(\mathbf{x})]\sum_{j=0}^{\infty}\mathbf{U}_{Tj}^r(\mathbf{x})(i\omega)^{-j},$$

$$\mathbf{U}^t(\mathbf{x},\omega) = \exp[i\omega\tau_L^t(\mathbf{x})]\sum_{j=0}^{\infty}\mathbf{U}_{Lj}^t(\mathbf{x})(i\omega)^{-j} + \exp[i\omega\tau_T^t(\mathbf{x})]\sum_{j=0}^{\infty}\mathbf{U}_{Tj}^t(\mathbf{x})(i\omega)^{-j}.$$

At the point under consideration on the interface S, we have $\mathbf{U} = \mathbf{U}^i + \mathbf{U}^r$ at one side, $\mathbf{U} = \mathbf{U}^t$ at the other. The requirement that the displacement $\mathbf{U}$ be continuous on S leads to the condition

$$\begin{aligned}\exp[i\omega\tau^i(\mathbf{x})]&\sum_{j=0}^{\infty}\mathbf{U}_j^i(\mathbf{x})(i\omega)^{-j}\\ &+ \exp[i\omega\tau_L^r(\mathbf{x})]\sum_{j=0}^{\infty}\mathbf{U}_{Lj}^r(\mathbf{x})(i\omega)^{-j} + \exp[i\omega\tau_T^r(\mathbf{x})]\sum_{j=0}^{\infty}\mathbf{U}_{Tj}^r(\mathbf{x})(i\omega)^{-j}\\ &= \exp[i\omega\tau_L^t(\mathbf{x})]\sum_{j=0}^{\infty}\mathbf{U}_{Lj}^t(\mathbf{x})(i\omega)^{-j} + \exp[i\omega\tau_T^t(\mathbf{x})]\sum_{j=0}^{\infty}\mathbf{U}_{Tj}^t(\mathbf{x})(i\omega)^{-j} \qquad (5.1)\end{aligned}$$

for every $\mathbf{x} \in S$. The arbitrariness of $\mathbf{x} \in S$ implies that (5.1) holds only if

$$\tau^i(\mathbf{x}) = \tau_L^r(\mathbf{x}) = \tau_T^r(\mathbf{x}) = \tau_L^t(\mathbf{x}) = \tau_T^t(\mathbf{x}), \qquad \forall\mathbf{x} \in S. \qquad (5.2)$$

Let $\mathbf{n}$ be the normal to S, directed toward the half-space of the incident and reflected rays. Denote by $\nabla_{\parallel} = \nabla - \mathbf{n}(\mathbf{n}\cdot\nabla)$ the (tangent) gradient operator along the surface. As (5.2) means that the phases $\tau^i, \tau_L^r, \tau_T^r, \tau_L^t, \tau_T^t$ take common values at the surface, then we have

$$\nabla_{\parallel}\tau^i = \nabla_{\parallel}\tau_L^r = \nabla_{\parallel}\tau_T^r = \nabla_{\parallel}\tau_L^t = \nabla_{\parallel}\tau_T^t \qquad \text{on} \qquad S. \qquad (5.3)$$

This means that $\nabla_{\parallel}\tau_L^r, \nabla_{\parallel}\tau_T^r, \nabla_{\parallel}\tau_L^t, \nabla_{\parallel}\tau_T^t$ are determined once the incident wave is given. The vectors $\nabla\tau_L^r, \nabla\tau_T^r, \nabla\tau_L^t, \nabla\tau_T^t$ are then determined by observing that the eikonal equations hold. For both reflected and transmitted waves let τ_L, τ_T denote the phase of longitudinal and transverse components. Of course τ_L, τ_T satisfy the pertinent form of (2.12).

Regard $\mathbf{q} = \nabla\tau$ as the slowness vector of the ray and define $\mathbf{k} = \omega\mathbf{q}$ as the corresponding wave vector. Accordingly, (5.3) means that the tangent part $\mathbf{k}_{\parallel}$ of $\mathbf{k}$ has common values for all waves involved and this can be seen as the strict analogue of Snell's law for plane waves. The normal component $\mathbf{k}_{\perp} = k_{\perp}\mathbf{n}$ is then determined by the eikonal equation (2.12). Observe that

$$k^2 = k_{\parallel}^2 + k_{\perp}^2, \qquad \mathbf{k}\cdot\mathbf{T}^0\mathbf{k} = \mathbf{k}_{\parallel}\cdot\mathbf{T}^0\mathbf{k}_{\parallel} + 2k_{\perp}\mathbf{n}\cdot\mathbf{T}^0\mathbf{k}_{\parallel} + k_{\perp}^2\mathbf{n}\cdot\mathbf{T}^0\mathbf{n}.$$

Let η stand for $2\mu_0 + \lambda_0$ or μ_0 according as longitudinal or transverse rays are considered. By (2.12) we have

$$-\rho\omega^2 + \eta(k_{\perp}^2 + k_{\parallel}^2) + \mathbf{k}_{\parallel}\cdot\mathbf{T}^0\mathbf{k}_{\parallel} + 2k_{\perp}\mathbf{n}\cdot\mathbf{T}^0\mathbf{k}_{\parallel} + k_{\perp}^2\mathbf{n}\cdot\mathbf{T}^0\mathbf{n} = 0$$

whence

$$k_\perp = \frac{-\mathbf{n}\cdot\mathbf{T}^0\mathbf{k}_\parallel \pm \sqrt{(\mathbf{n}\cdot\mathbf{T}^0\mathbf{k}_\parallel)^2 + (\mathbf{n}\cdot\mathbf{T}^0\mathbf{n}+\eta)(\rho\omega^2 - \eta k_\parallel^2 - \mathbf{k}_\parallel\cdot\mathbf{T}^0\mathbf{k}_\parallel)}}{\eta + \mathbf{n}\cdot\mathbf{T}^0\mathbf{n}}, \tag{5.4}$$

the + sign being relative to the reflected rays and the − sign to the incident ray or the transmitted ones. Given the wave vector of the incident ray, the relation (5.4), along with the invariance of $\mathbf{k}_\parallel$, determines the wave vectors $\mathbf{k}_L^r, \mathbf{k}_T^r, \mathbf{k}_L^t, \mathbf{k}_T^t$. By (5.4), in turn, we obtain the appropriate angles θ by simply letting $\tan\theta = |k_\perp|/k_\parallel$. Observe that, though $\mathbf{n}\cdot\mathbf{T}^0\mathbf{n}$ may be negative, we assume $\eta + \mathbf{n}\cdot\mathbf{T}^0\mathbf{n} > 0$. The quantity $\rho\omega^2 - \eta k_\parallel^2 - \mathbf{k}_\parallel\cdot\mathbf{T}^0\mathbf{k}_\parallel$, positive in the half-space of the incident ray, might be negative in the other half-space and render $k_\perp$ in (5.4) complex-valued. Quite naturally this might be viewed as corresponding to complex rays (cf. [91, 71, 172]), namely, in a sense, the analogue of inhomogeneous waves in the context of rays. To our mind this subject deserves further investigation and is beyond the scope of this book. Accordingly, we restrict attention to real values of $k_\perp$.

Remark. We are accustomed to the property that the incidence angle θ^i and the reflection angle θ^r, for the wave of the same type, are equal. By (5.4) it follows that $k_\perp^i$ and $k_\perp^r$ are generally different and then the property is not necessarily true. If, though, $\mathbf{n}$ and $\mathbf{k}_\parallel$ are principal directions of $\mathbf{T}^0$, with eigenvalues $T_\perp^0$ and $T_\parallel^0$, then (5.4) reduces to

$$k_\perp = \pm\sqrt{\frac{\rho\omega^2 - (\eta + T_\parallel^0)k_\parallel^2}{\eta + T_\parallel^0}}$$

and hence incidence angle and reflection angle are equal.

To exploit the boundary conditions for the traction $\mathbf{t}$ at S we preliminarily investigate the expression for the stress tensor $\mathbf{T}$. Following §2.6, the stress perturbation is given by

$$\mathbf{T} - \mathbf{T}^0 = \nabla\mathbf{u}^\dagger\mathbf{T}^0 + 2\mu_0\mathbf{E} + \lambda_0(\mathrm{tr}\,\mathbf{E})\mathbf{1} + \int_0^\infty \{2\mu'(s)\mathbf{E}^t(s) + \lambda'(s)[\mathrm{tr}\,\mathbf{E}^t(s)]\mathbf{1}\}ds.$$

If $\mathbf{u}(\mathbf{x},t) = \mathbf{U}(\mathbf{x};\omega)\exp(-i\omega t)$ we consider $\mathbf{U}(\mathbf{x};\omega)$ in the standard form of asymptotic expansion

$$\mathbf{U}(\mathbf{x};\omega) = \exp[i\omega\tau^i(\mathbf{x})]\sum_{j=0}^{\infty}\mathbf{U}_j(\mathbf{x})(i\omega)^{-j}.$$

Then, to within the common factor $\exp[i\omega(\tau(\mathbf{x}) - t)]$, use of the asymptotic expansions in

§9.2 and some rearrangement yield

$$\begin{aligned}\mathbf{T}-\mathbf{T}^0 =& i\omega[\mathbf{U}_0\otimes(\nabla\tau\mathbf{T}^0)+\mu_0(\nabla\tau\otimes\mathbf{U}_0+\mathbf{U}_0\otimes\nabla\tau)+\lambda_0(\nabla\tau\cdot\mathbf{U}_0)\mathbf{1}]\\ &+\sum_{j\geq 0}(i\omega)^{-j}\Big[\mathbf{U}_{j+1}\otimes(\nabla\tau\mathbf{T}^0)+\nabla\mathbf{U}_j^\dagger\mathbf{T}^0\\ &\quad+(-1)^{j+1}\mu_0^{(j+1)}(\nabla\tau\otimes\mathbf{U}_0+\mathbf{U}_0\otimes\nabla\tau)+(-1)^{j+1}\lambda_0^{(j+1)}(\nabla\tau\cdot\mathbf{U}_0)\mathbf{1}\\ &\quad+\sum_{k=0}^{j}(-1)^{j-k}\mu_0^{(j-k)}(\nabla\tau\otimes\mathbf{U}_{k+1}+\mathbf{U}_{k+1}\otimes\nabla\tau+\nabla\mathbf{U}_k+\nabla\mathbf{U}_k^\dagger)\\ &\quad+\sum_{k=0}^{j}(-1)^{j-k}\lambda_0^{(j-k)}(\nabla\tau\cdot\mathbf{U}_{k+1}+\nabla\cdot\mathbf{U}_k)\,\mathbf{1}\Big].\end{aligned} \tag{5.5}$$

Since $\mathbf{T}^0\mathbf{n}$ is continuous at the interface, the continuity of the traction, viz

$$(\mathbf{T}^i+\mathbf{T}_L^r+\mathbf{T}_T^r)\mathbf{n}=(\mathbf{T}_L^t+\mathbf{T}_T^t)\mathbf{n},$$

results in the continuity of the corresponding perturbation derived by (5.5). Further, by (2.9), $\nabla\tau_T^r\cdot\mathbf{U}_{T0}^r$ and $\nabla\tau_T^t\cdot\mathbf{U}_{T0}^t$ vanish. With this in mind we now determine the amplitudes associated with the reflected and transmitted rays.

By analogy with plane waves, we develop an approach which describes simultaneously the effects of longitudinal and transverse incident rays. Of course, if a single ray is incident then the pertinent formulae are recovered by setting the amplitude of the missing ray equal to zero. A conjugate pair of rays is defined to consist of a longitudinal and transverse ray which hit the surface S, at the point under consideration, with equal parallel components of $\nabla\tau$. Assume that two rays constitute an incident, conjugate pair. Owing to (5.3), reflected and transmitted rays constitute conjugate pairs. Their wave vectors $\mathbf{k}=\omega\mathbf{q}$ are determined by (5.3) and (5.4). The values, at S, of zeroth-order amplitudes of reflected and transmitted rays are then given as functions of the zeroth-order amplitudes of the incident conjugate pair.

We replace $\nabla\tau$ at S with the pertinent wave vector $\mathbf{k}=\omega\nabla\tau$. Then we introduce Cartesian coordinates with origin at the point where the incident ray hits the surface S and z-axis parallel to the normal $\mathbf{n}$ to S, the orientation being chosen so that the incident ray comes from the direction of positive z. The common parallel components of the wave vectors are denoted as k_x and k_y. Consider first the upper half-space; by analogy with inhomogeneous waves we define

$$\beta=\sqrt{k^2-k_\parallel^2}$$

where the usual subscripts L and T are understood. Hence the z-component of $\mathbf{k}$ is equal to β for (upgoing) reflected rays and $-\beta$ for incident rays.

Restrict attention to the zeroth-order amplitude $\mathbf{U}_0$ and observe that the upper value at S is given by

$$\mathbf{U}_0=\mathbf{U}_{L0}^i+\mathbf{U}_{T0}^i+\mathbf{U}_{L0}^r+\mathbf{U}_{T0}^r\ .$$

Since $\mathbf{U}_{L0} \times \mathbf{k}_L = 0$ we can write

$$\mathbf{U}^i_{L0} = \Phi^-(k_x\mathbf{e}_x + k_y\mathbf{e}_y - \beta_L\mathbf{e}_z), \qquad \mathbf{U}^r_{L0} = \Phi^+(k_x\mathbf{e}_x + k_y\mathbf{e}_y + \beta_L\mathbf{e}_z),$$

where Φ^- is given and Φ^+ is unknown. As regards the amplitudes of transverse waves we set

$$\mathbf{U}^i_{T0} = \mathbf{\Psi}^-, \qquad \mathbf{U}^r_{T0} = \mathbf{\Psi}^+,$$

where $\mathbf{\Psi}^-$ is regarded as given while $\mathbf{\Psi}^+$ is unknown. However, owing to the orthogonality between amplitude and wave vector for transverse rays, we have

$$\Psi^+_z = -(k_x\Psi^+_x + k_y\Psi^+_y)/\beta_T, \qquad \Psi^-_z = (k_x\Psi^-_x + k_y\Psi^-_y)/\beta_T.$$

Hence the upper limit of the vector amplitude $\mathbf{U}_0$ reads

$$\begin{aligned}\mathbf{U}_0 = {} & [(\Phi^+ + \Phi^-)k_x + \Psi^+_x + \Psi^-_x]\mathbf{e}_x + [(\Phi^+ + \Phi^-)k_y + \Psi^+_y + \Psi^-_y]\mathbf{e}_y \\ & [(\Phi^+ - \Phi^-)\beta_L - (\Psi^+_x - \Psi^-_x)k_x/\beta_T - (\Psi^+_y - \Psi^-_y)k_y/\beta_T]\mathbf{e}_z. \qquad (5.6)\end{aligned}$$

By continuity, this value is required to be equal to the analogous expression, for the transmitted conjugate pair, which is obtained by merely replacing Φ^- and $\mathbf{\Psi}^-$ with $\breve{\Phi}^-$ and $\breve{\mathbf{\Psi}}^-$ and setting Φ^+ and $\mathbf{\Psi}^+$ equal to zero.

Approximate $\mathbf{T} - \mathbf{T}^0$ in (5.5) by the leading terms given in the first line. Then comparison with (5.6) yields the components of $\mathbf{t}$ as

$$\begin{aligned}-it_x = {} & (T^0_{xz}k_x + T^0_{yz}k_y)[k_x(\Phi^+ + \Phi^-) + \Psi^+_x + \Psi^-_x] + \beta_L k_x(2\mu_0 + T^0_{zz})(\Phi^+ - \Phi^-) \\ & [\beta_T(\mu_0 + T^0_{zz}) - \mu_0 k_x^2/\beta_T](\Psi^+_x - \Psi^-_x) - \mu_0(k_xk_y/\beta_T)(\Psi^+_y - \Psi^-_y),\end{aligned}$$

$$\begin{aligned}-it_y = {} & (T^0_{xz}k_x + T^0_{yz}k_y)[k_y(\Phi^+ + \Phi^-) + \Psi^+_y + \Psi^-_y] + \beta_L k_y(2\mu_0 + T^0_{zz})(\Phi^+ - \Phi^-) \\ & - \mu_0(k_xk_y/\beta_T)(\Psi^+_x - \Psi^-_x) + [\beta_T(\mu_0 + T^0_{zz}) - \mu_0 k_y^2/\beta_T](\Psi^+_y - \Psi^-_y),\end{aligned}$$

$$\begin{aligned}-it_z = {} & (T^0_{xz}k_x + T^0_{yz}k_y)[\beta_L(\Phi^+ - \Phi^-) - k_x(\Psi^+_x - \Psi^-_x)/\beta_T - k_y(\Psi^+_y - \Psi^-_y)/\beta_T] \\ & + \Upsilon(\Phi^+ + \Phi^-) - (2\mu_0 + T^0_{zz})k_x(\Psi^+_x + \Psi^-_x) - (2\mu_0 + T^0_{zz})k_y(\Psi^+_y + \Psi^-_y),\end{aligned}$$

where

$$\Upsilon = (2\mu_0 + \lambda_0 + T^0_{zz})\beta_L^2 + \lambda_0(k_x^2 + k_y^2).$$

Analogous expressions hold for the limit values in the lower half-space. By requiring the continuity of $\mathbf{U}_0$ and $\mathbf{t}$ we can determine the unknown quantities Φ^+, $\breve{\Phi}^-$, Ψ^+_x, $\breve{\Psi}^-_x$, Ψ^+_y, $\breve{\Psi}^-_y$ in terms of Φ^-, Ψ^-_x, Ψ^-_y.

This procedure applies in the general case when the wave vector $\mathbf{k}$ is complex-valued. However if, as we have in mind, $\mathbf{k}$ is real-valued then the Cartesian axes can be so chosen that $k_y = 0$. This in turn makes the original system of equations decouple in two subsystems for the unknowns Φ^+, $\breve{\Phi}^-$, Ψ_x^+, $\breve{\Psi}_x^-$, and Ψ_y^+, $\breve{\Psi}_y^-$. Specifically, the continuity of U_x and t_z yields

$$B\begin{pmatrix}\Phi^+ + \Phi^- \\ \Psi_x^+ + \Psi_x^-\end{pmatrix} = \breve{B}\begin{pmatrix}\breve{\Phi}^- \\ \breve{\Psi}_x^-\end{pmatrix}, \tag{5.7}$$

where

$$B = \begin{pmatrix} k_x & 1 \\ \Upsilon & -(2\mu_0 + T_{zz}^0)\end{pmatrix},$$

while the continuity of U_z and t_x leads to

$$C\begin{pmatrix}\Phi^+ - \Phi^- \\ \Psi_x^+ - \Psi_x^-\end{pmatrix} = -\breve{C}\begin{pmatrix}\breve{\Phi}^- \\ \breve{\Psi}_x^-\end{pmatrix}, \tag{5.8}$$

where

$$C = \begin{pmatrix} \beta_L & -k_x/\beta_T \\ \beta_L k_x(2\mu_0 + T_{zz}^0) & \beta_T(\mu_0 + T_{zz}^0) - \mu_0 k_x^2/\beta_T \end{pmatrix}.$$

Meanwhile the continuity of U_y and t_y results in the system

$$\Psi_y^+ + \Psi_y^- = \breve{\Psi}_y^-,$$

$$\Psi_y^+ - \Psi_y^- = -\frac{\breve{\beta}_T(\breve{\mu}_0 + T_{zz}^0)}{\beta_T(\mu_0 + T_{zz}^0)}\breve{\Psi}_y^-,$$

which may be solved at once to obtain

$$\breve{\Psi}_y^- = \frac{2\beta_T(\mu_0 + T_{zz}^0)}{\beta_T(\mu_0 + T_{zz}^0) + \breve{\beta}_T(\breve{\mu}_0 + T_{zz}^0)}\Psi_y^-,$$

$$\Psi_y^+ = \frac{\beta_T(\mu_0 + T_{zz}^0) - \breve{\beta}_T(\breve{\mu}_0 + T_{zz}^0)}{\beta_T(\mu_0 + T_{zz}^0) + \breve{\beta}_T(\breve{\mu}_0 + T_{zz}^0)}\Psi_y^-.$$

The solution to (5.7) and (5.8) can also be determined. Setting aside inessential details we write the result in the form

$$\begin{pmatrix}\breve{\Phi}^- \\ \breve{\Psi}_x^-\end{pmatrix} = 2(B^{-1}\breve{B} + C^{-1}\breve{C})^{-1}\begin{pmatrix}\Phi^- \\ \Psi_x^-\end{pmatrix},$$

$$\begin{pmatrix}\Phi^+ \\ \Psi_x^+\end{pmatrix} = \left[2B^{-1}\breve{B}(B^{-1}\breve{B} + C^{-1}\breve{C})^{-1} - \mathbb{1}\right]\begin{pmatrix}\Phi^- \\ \Psi_x^-\end{pmatrix}$$

where $\mathbb{1}$ is the 2×2 identity matrix.

An alternative procedure, for the determination of the result of a reflection-refraction process at an interface, may be performed which emphasizes the more customary description in terms of angles. As before, we recall that $\nabla\tau_T^t \cdot \mathbf{U}_{T0}^t$ and $\nabla\tau_T^t \cdot \mathbf{U}_{T0}^t$ vanish and let

$\mathbf{k} = \omega\nabla\tau$. By keeping only the leading terms in the asymptotic expansions we can write the continuity of displacement and traction, at S, as

$$\mathbf{U}_0^i + \mathbf{U}_{L0}^r + \mathbf{U}_{T0}^r = \mathbf{U}_{L0}^t + \mathbf{U}_{T0}^t, \tag{5.9}$$

$$\begin{aligned}
&(\mathbf{k}^i \cdot \mathbf{T}^0\mathbf{n})\mathbf{U}_0^i + \mu_0[(\mathbf{U}_0^i \cdot \mathbf{n})\mathbf{k}^i + (\mathbf{k}^i \cdot \mathbf{n})\mathbf{U}_0^i] + \lambda_0(\mathbf{k}^i \cdot \mathbf{U}_0^i)\mathbf{n} \\
&+ (\mathbf{k}_L^r \cdot \mathbf{T}^0\mathbf{n})\mathbf{U}_{L0}^r + (\mathbf{k}_T^r \cdot \mathbf{T}^0\mathbf{n})\mathbf{U}_{T0}^r \\
&+ \mu_0[2(\mathbf{k}_L^r \cdot \mathbf{n})\mathbf{U}_{L0}^r + (\mathbf{U}_{T0}^r \cdot \mathbf{n})\mathbf{k}_T^r + (\mathbf{k}_T^r \cdot \mathbf{n})\mathbf{U}_{T0}^r] + \lambda_0(\mathbf{k}_L^r \cdot \mathbf{U}_{L0}^r)\mathbf{n} \\
&\quad = (\mathbf{k}_L^t \cdot \mathbf{T}^0\mathbf{n})\mathbf{U}_{L0}^t + (\mathbf{k}_T^t \cdot \mathbf{T}^0\mathbf{n})\mathbf{U}_{T0}^t \\
&\quad + \breve{\mu}_0[2(\mathbf{k}_L^t \cdot \mathbf{n})\mathbf{U}_{L0}^t + (\mathbf{U}_{T0}^t \cdot \mathbf{n})\mathbf{k}_T^t + (\mathbf{k}_T^t \cdot \mathbf{n})\mathbf{U}_{T0}^t] + \breve{\lambda}_0(\mathbf{k}_L^t \cdot \mathbf{U}_{L0}^t)\mathbf{n}.
\end{aligned} \tag{5.10}$$

By Snell's law, the longitudinal vectors $\mathbf{U}_{L0}^r, \mathbf{U}_{L0}^t$, and possibly $\mathbf{U}_0^i$, belong to the "vertical" plane $(\mathbf{n}^i, \mathbf{n})$. Moreover, it follows at once from the system (5.9)-(5.10) that the "horizontal" components of the transverse vectors $\mathbf{U}_{T0}$ are decoupled. Then we can examine separately the case when $\mathbf{U}_0^i$ is transverse, horizontal.

Let $\mathbf{n}_L$ be the pertinent unit vector of $\mathbf{k}$ and $\mathbf{m}_T$ the unit vector orthogonal to transverse rays, $\mathbf{m}_T \cdot \mathbf{n} \geq 0$. Represent any $\mathbf{U}_{L0}, \mathbf{U}_{T0}$ as

$$\mathbf{U}_{L0} = \Phi\mathbf{n}_L, \qquad \mathbf{U}_{T0} = \Psi\mathbf{m}_T.$$

We consider at the same time the case when the incident ray is longitudinal, $\mathbf{U}_0^i = \Phi^i\mathbf{n}^i$, and transverse, vertical, $\mathbf{U}_0^i = \Psi^i\mathbf{m}^i$. We denote by $\theta(< \pi/2)$ the angle between the wave vector and the normal $\mathbf{n}$ (or $-\mathbf{n}$) to S. For instance, $\theta^i, \theta_L^r, \theta_T^t$ are the angle of incidence, the angle of reflection of the longitudinal ray, the angle of transmission of the transverse ray. We have

$$\mathbf{n}^i \cdot \mathbf{n} = \cos\theta^i, \qquad \mathbf{n}_L^r \cdot \mathbf{n} = -\cos\theta_L^r, \qquad \mathbf{n}_T^t \cdot \mathbf{n} = \cos\theta_T^t.$$

Substitution in (5.9)-(5.10), inner multiplication by the unit vector $\mathbf{e}$ of $\mathbf{k}_\parallel$ and the normal $\mathbf{n}$ yield

$$-\sin\theta_L^r\Phi^r + \cos\theta_T^r\Psi^r + \sin\theta_L^t\Phi^t + \cos\theta_T^t\Psi^t = \begin{cases} \sin\theta^i\Phi^i \\ \cos\theta^i\Psi^i, \end{cases} \tag{5.11}$$

$$\cos\theta_L^r\Phi^r + \sin\theta_T^r\Psi^r + \cos\theta_L^t\Phi^t - \sin\theta_T^t\Psi^t = \begin{cases} \cos\theta^i\Phi^i \\ -\sin\theta^i\Psi^i, \end{cases} \tag{5.12}$$

$$\begin{aligned}
&k_L^r[(\mathbf{n}_L^r \cdot \mathbf{T}^0\mathbf{n})\sin\theta_L^r + \mu_0\sin 2\theta_L^r]\Phi^r - k_T^r[(\mathbf{n}_T^r \cdot \mathbf{T}^0\mathbf{n})\cos\theta_T^r + \mu_0\cos 2\theta_T^r]\Psi^r \\
&+ k_L^t[-(\mathbf{n}_L^t \cdot \mathbf{T}^0\mathbf{n})\sin\theta_L^t + \breve{\mu}_0\sin 2\theta_L^t]\Phi^t + k_T^t[-(\mathbf{n}_T^t \cdot \mathbf{T}^0\mathbf{n})\cos\theta_T^t + \breve{\mu}_0\cos 2\theta_T^t]\Psi^t \\
&\quad = \begin{cases} k^i[-(\mathbf{n}^i \cdot \mathbf{T}^0\mathbf{n})\sin\theta^i + \mu_0\sin 2\theta^i]\Phi^i \\ k^i[-(\mathbf{n}^i \cdot \mathbf{T}^0\mathbf{n})\cos\theta^i + \mu_0\cos 2\theta^i]\Psi^i, \end{cases}
\end{aligned} \tag{5.13}$$

$$
\begin{aligned}
&k_L^r[(\mathbf{n}_L^r \cdot \mathbf{T}^0\mathbf{n})\cos\theta_L^r + 2\mu_0\cos^2\theta_L^r + \lambda_0]\Phi^r + k_T^r[(\mathbf{n}_T^r \cdot \mathbf{T}^0\mathbf{n}\sin\theta_T^r + \mu_0\sin 2\theta_T^r]\Psi^r \\
&- k_L^t[-(\mathbf{n}_L^t \cdot \mathbf{T}^0\mathbf{n})\cos\theta_L^t + 2\breve{\mu}_0\cos^2\theta_L^t + \breve{\lambda}_0]\Phi^t + k_T^t[-(\mathbf{n}_T^t \cdot \mathbf{T}^0\mathbf{n})\sin\theta_T^t + \breve{\mu}_0\sin 2\theta_T^t]\Psi^t \\
&\quad = \begin{cases} -k^i[-(\mathbf{n}^i \cdot \mathbf{T}^0\mathbf{n})\cos\theta^i + 2\mu_0\cos^2\theta^i + \lambda_0]\Phi^i \\ k^i[-(\mathbf{n}^i \cdot \mathbf{T}^0\mathbf{n})\sin\theta^i + \mu_0\sin 2\theta^i]\Psi^i. \end{cases}
\end{aligned} \tag{5.14}
$$

If, instead, the incident ray is transverse, horizontal, then the amplitudes Ψ^i, Ψ^r, Ψ^t satisfy the system

$$\Psi^i + \Psi^r = \Psi^t, \tag{5.15}$$

$$k_T^r[(\mathbf{n}_T^r \cdot \mathbf{T}^0\mathbf{n} + \mu_0\cos\theta_T^r]\Psi^r + k_T^t[-(\mathbf{n}_T^t \cdot \mathbf{T}^0\mathbf{n}) + \breve{\mu}_0\cos\theta_T^t]\Psi^t = k^i[-(\mathbf{n}^i \cdot \mathbf{T}^0\mathbf{n}) + \mu_0\cos\theta^i]\Psi^i. \tag{5.16}$$

The solution to (5.15)-(5.16) is given by

$$\frac{\Psi^r}{\Psi^i} = \frac{k^i[(\mathbf{n}^i \cdot \mathbf{T}^0\mathbf{n}) - \mu_0\cos\theta^i] - k_T^t[(\mathbf{n}_T^t \cdot \mathbf{T}^0\mathbf{n}) - \mu_0\cos\theta_T^t]}{k_T^t[(\mathbf{n}_T^t \cdot \mathbf{T}^0\mathbf{n}) - \mu_0\cos\theta_T^t] - k_T^r[(\mathbf{n}_T^r \cdot \mathbf{T}^0\mathbf{n}) + \mu_0\cos\theta_T^r]},$$

$$\frac{\Psi^t}{\Psi^i} = \frac{k^i[(\mathbf{n}^i \cdot \mathbf{T}^0\mathbf{n}) - \mu_0\cos\theta^i] - k_T^r[(\mathbf{n}_T^r \cdot \mathbf{T}^0\mathbf{n}) - \mu_0\cos\theta_T^r]}{k_T^t[(\mathbf{n}_T^t \cdot \mathbf{T}^0\mathbf{n}) - \mu_0\cos\theta_T^t] - k_T^r[(\mathbf{n}_T^r \cdot \mathbf{T}^0\mathbf{n}) + \mu_0\cos\theta_T^r]}.$$

The influence of the prestress $\mathbf{T}^0$ is twofold. One fold is shown explicitly. The other one, presumably less effective, is through the dependence of the wavenumbers k and the angles θ on $\mathbf{T}^0$ as given by (5.4).

It is of interest to consider the special case when reflection of rays occurs at the (traction-free) surface S of a half-space. The argument leading to (5.3) can be repeated step by step in connection with the condition that the traction $(\mathbf{T}^i + \mathbf{T}_L^r + \mathbf{T}_T^r)\mathbf{n}$ vanish at S. We then obtain that $\mathbf{k}_\parallel$ still takes a common value for all rays, which is the content of Snell's law.

Incidentally, the vanishing of $\mathbf{T}^0\mathbf{n}$ makes (5.4) into

$$k_\perp = \pm\sqrt{\frac{\rho\omega^2 - \eta k_\parallel^2 - \mathbf{k}_\parallel \cdot \mathbf{T}^0\mathbf{k}_\parallel}{\eta}} \tag{5.17}$$

whereby $\theta^r = \theta^i$ for rays of the same type. Moreover, for any ray

$$k = \sqrt{\frac{\rho\omega^2 - \mathbf{k}_\parallel \cdot \mathbf{T}^0\mathbf{k}_\parallel}{\eta}}. \tag{5.18}$$

Accordingly, though any k depends on $\mathbf{T}^0$, the ratio

$$\kappa := \frac{k_T}{k_L} = \sqrt{\frac{2\mu_0 + \lambda_0}{\mu_0}}$$

does not.

The traction-free condition $(\mathbf{T}^i + \mathbf{T}^r_L + \mathbf{T}^r_T)\mathbf{n} = 0$ leads to

$$k^r_L \sin 2\theta^r_L \, \Phi^r - k^r_T \cos 2\theta^r_T \, \Psi^r = \begin{cases} k^i \sin 2\theta^i \, \Phi^i \\ k^i \cos 2\theta^i \, \Psi^i, \end{cases}$$

$$k^r_L(2\cos^2\theta^r_L + \lambda_0/\mu_0)\Phi^r + k^r_T \sin 2\theta^r_T \, \Psi^r_T = \begin{cases} -k^i(2\cos^2\theta^i + \lambda_0/\mu_0)\Phi^i, \\ k^i \sin 2\theta^i \, \Psi^i. \end{cases}$$

To fix ideas, let the incident ray be longitudinal, so that $k^i = k^r_L$. By starting from $k^r_L \sin\theta^r_L = k^r_T \sin\theta^r_T$ we prove the identity

$$2\cos\theta^r_L + \frac{\lambda_0}{\mu_0} = \kappa^2 \cos 2\theta^r_T.$$

Then we can write the reflection coefficients as

$$\frac{\Phi^r}{\Phi^i} = \frac{\sin 2\theta^r_L \, \sin 2\theta^r_T - \kappa^2 \cos^2 2\theta^r_T}{\sin 2\theta^r_L \, \sin 2\theta^r_T + \kappa^2 \cos^2 2\theta^r_T}, \tag{5.19}$$

$$\frac{\Psi^r}{\Phi^i} = -\frac{2\kappa \sin 2\theta^r_L \, \cos 2\theta^r_T}{\sin 2\theta^r_L \, \sin 2\theta^r_T + \kappa^2 \cos^2 2\theta^r_T}. \tag{5.20}$$

An analogous result holds for the reflection coefficients when the incident wave is transverse. So at first sight, for reflection at the surface of a half-space, the reflection coefficients seem to be left unaffected by the possible prestress $\mathbf{T}^0$, $\mathbf{T}^0\mathbf{n} = 0$, in that (5.19) and (5.20) are formally the same as those for reflection in unstressed bodies [3]. Really, the angles θ^r_L, θ^r_T are affected by the prestress $\mathbf{T}^0$ even when $\mathbf{T}^0\mathbf{n} = 0$. This is easily seen by observing that, by (5.17) and (5.18),

$$\cos\theta = \sqrt{\frac{\rho\omega^2 - \eta k_\parallel^2 - \mathbf{k}_\parallel \cdot \mathbf{T}^0\mathbf{k}_\parallel}{\rho\omega^2 - \mathbf{k}_\parallel \cdot \mathbf{T}^0\mathbf{k}_\parallel}}, \qquad \sin\theta = \sqrt{\frac{\eta k_\parallel^2}{\rho\omega^2 - \mathbf{k}_\parallel \cdot \mathbf{T}^0\mathbf{k}_\parallel}}. \tag{5.21}$$

As a check of correctness of the results (5.19)-(5.20) we apply the requirement (3.3) about the balance of energy at the interface. We can write

$$\mathbf{w}^i \cdot \mathbf{n}(\Phi^i)^2 + \mathbf{w}^r_L \cdot \mathbf{n}(\Phi^r_L)^2 + \mathbf{w}^r_T \cdot \mathbf{n}(\Psi^r_T)^2 = 0. \tag{5.22}$$

The traction-free condition (at equilibrium) $\mathbf{T}^0\mathbf{n} = 0$ gives

$$-\mathbf{w}^i \cdot \mathbf{n} = \mathbf{w}^r_L \cdot \mathbf{n} = (2\mu_0 + \lambda_0)\mathbf{k}^r_L \cdot \mathbf{n}, \qquad \mathbf{w}^r_T \cdot \mathbf{n} = \mu_0 \mathbf{k}^r_T \cdot \mathbf{n}.$$

Substitution in (5.22), use of (5.17), and some rearrangement yields

$$\begin{aligned}(2\mu_0 + \lambda_0)\,[\rho\omega^2 - (2\mu_0 + \lambda_0)k_\parallel^2 - \mathbf{k}_\parallel \cdot \mathbf{T}^0\mathbf{k}_\parallel] \sin^2\theta^r_T \cos^2\theta^r_T \\ = \mu_0[\rho\omega^2 - \mu_0 k_\parallel^2 - \mathbf{k}_\parallel \cdot \mathbf{T}^0\mathbf{k}_\parallel] \sin^2\theta^r_L \cos^2\theta^r_L.\end{aligned} \tag{5.23}$$

Substitution from (5.21) shows that, as it must be, the relation (5.23) is identically true.

9.6 Remarks on rays in solids

We conclude by appending some remarks on ray description of wave propagation and reflection. In particular our purpose is to establish whether and how connections hold with related topics developed in the literature. A first topic in this sense concerns scalar theories. Much attention has been devoted to ray theory for the scalar Helmholtz equation. A useful reference on this subject is Bleistein's book [15]. Here we summarize the description of reflection and refraction of rays.

Within the standard notation, represent the (scalar) field of the incident wave as

$$U^i(\mathbf{x},\omega) = \exp[i\omega\tau^i(\mathbf{x})]\sum_{j=0}^{\infty} U_j^i(\mathbf{x})(i\omega)^{-j}$$

and similarly for the fields U^r, U^t of the reflected and transmitted waves. The continuity of the solution, namely

$$\begin{aligned}\exp[i\omega\tau^i(\mathbf{x})]\sum_{j=0}^{\infty} U_j^i(\mathbf{x})(i\omega)^{-j} + \exp[i\omega\tau^r(\mathbf{x})]\sum_{j=0}^{\infty} U_j^r(\mathbf{x})(i\omega)^{-j}& \\ = \exp[i\omega\tau^t(\mathbf{x})]\sum_{j=0}^{\infty} U_j^t(\mathbf{x})(i\omega)^{-j},&\end{aligned} \tag{6.1}$$

for every $\mathbf{x} \in S$ leads to the condition that τ^i, τ^r, τ^t take on the same values at S. This in turn means that the tangent part of the slowness $\mathbf{q}$, or the wave vector $\mathbf{k}$, takes on the same values for the three rays, which is the content of Snell's law. The normal part $\mathbf{k}_\perp = k_\perp \mathbf{n}$ is then given by

$$k_\perp^r = -k_\perp^i, \qquad k_\perp^t = \operatorname{sgn} k_\perp^i \sqrt{(k^t)^2 - k_\parallel^2}.$$

Denote by $\partial/\partial n$ the normal derivative. The requirement that the normal derivative be continuous gives

$$\begin{aligned}i\omega\frac{\partial\tau^i}{\partial n}\sum_{j=0}^{\infty} U_j^i(\mathbf{x})(i\omega)^{-j} + \sum_{j=0}^{\infty}\frac{\partial U_j^i(\mathbf{x})}{\partial n}(i\omega)^{-j} + i\omega\frac{\partial\tau^r}{\partial n}\sum_{j=0}^{\infty} U_j^r(\mathbf{x})(i\omega)^{-j} + \sum_{j=0}^{\infty}\frac{\partial U_j^r(\mathbf{x})}{\partial n}(i\omega)^{-j}& \\ = i\omega\frac{\partial\tau^t}{\partial n}\sum_{j=0}^{\infty} U_j^t(\mathbf{x})(i\omega)^{-j} + \sum_{j=0}^{\infty}\frac{\partial U_j^t(\mathbf{x})}{\partial n}(i\omega)^{-j}, \qquad \mathbf{x} \text{ on } S.&\end{aligned} \tag{6.2}$$

By (6.1) and (6.2) we find that

$$U_j^i + U_j^r = U_j^t, \qquad j = 0, 1, 2, \ldots$$

$$\frac{\partial \tau^i}{\partial n}U_j^i + \frac{\partial \tau^r}{\partial n}U_j^i = \frac{\partial \tau^t}{\partial n}U_j^i + \frac{\partial}{\partial n}(U_{j-1}^t - U_{j-1}^i - U_{j-1}^r), \qquad j = 0,1,2,...;$$

of course $U_{-1} = 0$. Solving the equations for the leading order,

$$U_0^i + U_0^r = U_0^t, \qquad \frac{\partial \tau^i}{\partial n}U_0^i + \frac{\partial \tau^r}{\partial n}U_0^r = \frac{\partial \tau^t}{\partial n}U_0^t,$$

we have

$$U_0^r = \mathcal{R}U_0^i, \qquad U_0^t = \mathcal{T}U_0^i$$

where the reflection and refraction coefficients $\mathcal{R}, \mathcal{T}$ are given by

$$\mathcal{R} = \frac{k_\perp^i - \operatorname{sgn}k_\perp^i\sqrt{(k^t)^2 - k_\parallel^2}}{k_\perp^i + \operatorname{sgn}k_\perp^i\sqrt{(k^t)^2 - k_\parallel^2}}, \tag{6.3}$$

$$\mathcal{T} = \frac{2k_\perp^i}{k_\perp^i + \operatorname{sgn}k_\perp^i\sqrt{(k^t)^2 - k_\parallel^2}}. \tag{6.4}$$

Quite naturally we ask whether a particular case of our vector theory leads to (6.3), (6.4). Now, consider the system (5.8)-(5.11) and require that only longitudinal waves be allowed. Also, it is reasonable to consider the projections along $\mathbf{n}$ only. Then we can write

$$\cos\theta^r\,\Phi^r + \cos\theta^t\,\Phi^t = \cos\theta^i\,\Phi^i, \tag{6.5}$$

$$\begin{aligned} k^r[(\mathbf{n}^r\cdot\mathbf{T}^0\mathbf{n})\cos\theta^r + 2\mu_0\cos^2\theta^r + \lambda_0]\Phi^r - k^t[-(\mathbf{n}^t\cdot\mathbf{T}^0\mathbf{n})\cos\theta^t + 2\breve{\mu}_0\cos^2\theta^t + \breve{\lambda}_0]\Phi^t \\ = -k^i[-(\mathbf{n}^i\cdot\mathbf{T}^0\mathbf{n})\cos\theta^i + 2\mu_0\cos^2\theta^i + \lambda_0]\Phi^i. \end{aligned} \tag{6.6}$$

If, further, $\mathbf{T}^0$ is taken to be isotropic, $\mathbf{T}^0 = T^0\mathbf{1}$, then the effect of prestress may be incorporated in the instantaneous elasticities $\mu_0, \breve{\mu}_0$ by formally letting $\mu_0 + T^0/2 \rightarrow \mu_0$, $\breve{\mu}_0 + T^0/2 \rightarrow \breve{\mu}_0$. If we let $\lambda_0, \breve{\lambda}_0 = 0$ and

$$U_0^i = -\Phi^i\cos\theta^i, \qquad U_0^r = \Phi^r\cos\theta^r, \qquad U_0^t = -\Phi^t\cos\theta^t,$$

we can write the system (6.5)-(6.6) as

$$U_0^i + U_0^r = U_0^t,$$

$$k_\perp^i U_0^i + k_\perp^r U_0^r = k_\perp^t U_0^t,$$

whose solution for $\mathcal{R} = U_0^r/U_0^i$, $\mathcal{T} = U_0^t/U_0^i$ is just given by (6.3), (6.4). However this is purely formal because Φ^r/Φ^i and Φ^t/Φ^i are the effective unknowns.

Three-dimensional ray theory analyses have been developed extensively for homogeneous, elastic, unstressed solids; as a remarkable reference we mention [3]. Examine briefly a consequence of the present theory.

For unstressed solids we have

$$\mathbf{w} = \mu_0 \nabla\tau, \qquad \mathbf{w} = (2\mu_0 + \lambda_0)\nabla\tau$$

according as transverse or longitudinal rays are considered. For homogeneous, elastic solids, in both cases (2.13) reduces to

$$\nabla \cdot (U_0^2 \nabla\tau) = 0.$$

Hence (3.1) simplifies to

$$J(\sigma)U_0^2(\sigma) = J(\sigma_0)U_0^2(\sigma_0). \tag{6.7}$$

We know that in homogeneous solids rays are straight lines and the principal radii R_1, R_2 of curvature of the wavefront increase linearly with distance along a ray. Then

$$\frac{dR_1}{d\sigma} = 1, \qquad \frac{dR_2}{d\sigma} = 1. \tag{6.8}$$

Now recall (1.13), viz $d(\ln J)/d\sigma = 2H$, or

$$J(\sigma) = J(\sigma_0)\exp\Big[\int_{\sigma_0}^{\sigma}\Big(\frac{1}{R_1} + \frac{1}{R_2}\Big)d\sigma\Big].$$

By (6.8) we have

$$\int_{\sigma_0}^{\sigma}\Big(\frac{1}{R_1} + \frac{1}{R_2}\Big)d\sigma = \int_{\sigma_0}^{\sigma}\Big(\frac{1}{R_1}\frac{dR_1}{d\sigma} + \frac{1}{R_2}\frac{dR_2}{d\sigma}\Big)d\sigma = \ln\frac{R_1(\sigma)R_2(\sigma)}{R_1(\sigma_0)R_2(\sigma_0)}.$$

Then

$$J(\sigma) = J(\sigma_0)\frac{R_1(\sigma)R_2(\sigma)}{R_1(\sigma_0)R_2(\sigma_0)}.$$

Substitution in (6.7) shows that the product $R_1 R_2 U_0^2$ is constant along a ray, which is a known result (cf. [3], §4.4).

REFERENCES

[1] A. Abo-Zena, *Dispersion function computations for unlimited frequency values,* Geophys. J. Roy. Astr. Soc. 58, 91 (1979).

[2] J.D. Achenbach, *Wave Propagation in Elastic Solids.* North Holland, Amsterdam, 1975.

[3] J.D. Achenbach, A. K. Gautesen and H. McMacken, *Ray Methods for Waves in Elastic Solids.* Pitman, Boston, 1982.

[4] J.D. Achenbach and Z.L. Li, *BIE/BEM approach to scattering of ultrasound by cracks.* In: Mathematical and Numerical Aspects of Wave Propagation Phenomena (G. Cohen, L. Halpern and P. Joly eds.). SIAM, Philadelphia, 1991.

[5] K. Aki and P.G. Richards, *Quantitative Seismology.* Freeman, San Francisco, 1980.

[6] L.E. Alsop, A.S. Goodman and S. Gregersen, *Reflection and transmission of inhomogeneous waves with particular applications to Rayleigh waves,* Bull. Seismol. Soc. Am. 64, 1635 (1974).

[7] J.H. Ansell, *The roots of the Stoneley wave equation for solid-liquid interfaces,* Pure Appl. Geophys. 94, 172 (1972).

[8] F. Bampi and C. Zordan, *Solving Snell's law,* Mech. Res. Comm. 18, 87 (1991).

[9] P.J. Barratt and W.D. Collins, *The scattering cross-section of an obstacle in an elastic solid for plane harmonic waves,* Proc. Camb. Phil. Soc. 61, 969 (1965).

[10] P. Bassanini, *Wave reflection from a system of plane waves,* Wave Motion 8, 311 (1986).

[11] A. Ben-Menahem and S.J. Singh, *Seismic Waves and Sources.* Springer, New York, 1981.

[12] G. Beylkin and R. Burridge, *Linearized inverse scattering problems in acoustics and elasticity,* Wave Motion 12, 15 (1990).

[13] M.A. Biot, *Mechanics of Incremental Deformations.* Wiley, New York, 1965.

[14] D.R. Bland, *The Theory of Linear Viscoelasticity.* Pergamon, Oxford, 1960.

[15] N. Bleistein, *Mathematical Methods for Wave Phenomena.* Academic Press, London, 1984.

[16] N. Bleistein and S.H. Gray, *An extension of the Born inversion method to a depth dependent reference profile,* Geophys. Prosp. 33, 999 (1985).

[17] R.D. Borcherdt, *Energy and plane waves in linear viscoelastic media,* J. Geophys. Res. 78, 2442 (1973).

[18] P. Borejko, *Inhomogeneous plane waves in a constrained elastic body,* Quart. J. Mech. Appl. Math. 40, 71 (1987).

[19] M. Born and E. Wolf, *Principles of Optics.* Pergamon, London, 1959.

[20] P. Boulanger and M. Hayes, *Electromagnetic plane waves in anisotropic media: an approach using bivectors,* Phil. Trans. Roy. Soc. London 330, 335 (1990).

[21] P. Boulanger and M. Hayes, *Inhomogeneous plane waves in viscous fluids,* Cont. Mech. Thermodyn. 2, 1 (1990).

[22] L.M. Brekhovskikh, *Waves in Layered Media.* Academic, New York, 1980.

[23] L. Brekhovskikh and V. Goncharov, *Mechanics of Continua and Wave Dynamics.* Springer, Berlin, 1985.

[24] H. Bremmer, *The W.K.B. approximation as the first term of a geometric-optic series,* Comm. Pure Appl. Math. 4, 105 (1951).

[25] A. Briggs, *An Introduction to Scanning Acoustic Microscopy.* Oxford University Press, Oxford, 1985.

[26] P.W. Buchen, *Plane waves in linear viscoelastic media,* Geophys. J. R. astr. Soc. 23, 531 (1971).

[27] P.W. Buchen, *Reflection, transmission and diffraction of SH-waves in linear viscoelastic solids,* Geophys. J. R. astr. Soc. 25, 97 (1971).

[28] L. Cagniard, *Reflection and Refraction of Progressive Seismic Waves.* McGraw-Hill, New York, 1962.

[29] J.M. Carcione, *Wave propagation in anisotropic linear viscoelastic media: theory and simulated wavefields,* Geophys J. Int. 101, 739 (1990).

[30] J.M. Carcione, D. Kosloff and R. Kosloff, *Wave propagation simulation in a linear viscoelastic medium,* Geophys. J. 95, 597 (1988).

[31] F. Carlini, *Ricerche sulla convergenza della serie che serve alla soluzione del problema di Keplero.* Milan, 1817.

[32] C. Caviglia, P. Cermelli and A. Morro, *Time-harmonic waves in continuously layered media,* Atti Sem. Mat. Fis Univ. Modena (1992).

[33] G. Caviglia and A. Morro, *Inhomogeneous waves and sound absorption in viscous fluids,* Phys. Lett. 134, 127 (1988).

[34] G. Caviglia and A. Morro, *Body force effects on time-harmonic inhomogeneous waves,* Arch. Mech. 42, 337 (1990).

[35] G. Caviglia and A. Morro, *On the modelling of dissipative solids,* Meccanica 25, 124 (1990).

[36] G. Caviglia and A. Morro, *A uniqueness theorem for the scattering of harmonic waves in a fluid-solid medium,* Boll. Un. Mat. Ital. A 4, 113 (1990).

[37] G. Caviglia and A. Morro, *Wave propagation in inhomogeneous viscoelastic solids,* Q. Jl Mech. Appl. Math. 44, 45 (1991).

[38] G. Caviglia and A. Morro, *General behaviour of inhomogeneous waves at interfaces,* Mech. Res. Comm. 18, 319 (1991).

[39] G. Caviglia and A. Morro, *Energy flux in dissipative media,* Acta Mech. 94, 29 (1992).

[40] G. Caviglia and A. Morro, *Ray analysis and wave propagation in nonhomogeneous viscoelastic fluids,* SIAM J. Appl. Math. 52, 315 (1992).

[41] G. Caviglia and A. Morro, *Reflection and refraction of rays in prestressed solids,* Meccanica 27, 47 (1992).

[42] G. Caviglia, A. Morro and E. Pagani, *Time-harmonic waves in viscoelastic media,* Mech. Res. Comm. 16, 53 (1988).

[43] G. Caviglia, A. Morro and E. Pagani, *Surface waves on a solid half-space*, J. Acoust. Soc. Am. 86, 2456 (1989).

[44] G. Caviglia, A. Morro and E. Pagani, *Inhomogeneous waves in viscoelastic media,* Wave Motion 12, 143 (1990).

[45] P. Cervenka and P. Challande, *A new efficient algorithm to compute the exact reflection and transmission factors for plane waves in layered absorbing media (liquids and solids),* J. Acoust. Soc. Am. 89, 1579 (1991).

[46] P. Chadwick and D.A. Jarvis, *Surface waves in a pre-stressed elastic body,* Proc. R. Soc. London A 366, 517 (1979).

[47] P. Chadwick, A.M. Whitworth and P. Borejko, *Basic theory of small-amplitude waves in a constrained elastic body,* Arch. Rational Mech. Anal. 87, 339 (1984).

[48] P.B. Chapman and J.J. Mahony, *Reflection of waves in a slowly varying medium,* SIAM J. Appl. Math. 34, 303 (1978).

[49] J.P. Charlier and F. Crowet, *Wave equations in linear viscoelastic materials,* J. Acoust. Soc. Am. 79, 895 (1986).

[50] P.J. Chen and M.E. Gurtin, *On wave propagation in inextensible elastic bodies,* Int. J. Solid Struct. 10, 275 (1974).

[51] D. Colton, *The inverse scattering problem for time-harmonic acoustic waves,* SIAM Rev. 26, 323 (1984).

[52] D. Colton and R. Kress, *Integral Equation Methods in Scattering Theory.* Wiley, New York, 1984.

[53] H.F. Cooper, *Reflection and transmission of oblique plane waves at a plane interface between viscoelastic media,* J. Acoust, Soc. Am. 42, 1064 (1967).

[54] P.K. Currie, M.A. Hayes, P.M. O'Leary, *Viscoelastic Rayleigh waves,* Quart. Appl. Math. 35, 35 (1977).

[55] G. Dassios and K. Kiriaki, *The low frequency theory of elastic wave scattering,* Quart. Appl. Math. 42, 225 (1984).

[56] G. Dassios and K. Kiriaki, *On the scattering amplitudes for elastic waves,* ZAMP 38, 856 (1987).

[57] G. Dassios and V. Kostopoulos, *The scattering amplitudes and cross sections in the theory of thermoelasticity,* SIAM J. Appl. Math 48, 79 (1988).

[58] A.J. Devaney and G.C. Sherman, *Plane-wave representations for scalar wave fields,* SIAM Rev. 15, 765 (1973).

[59] S. Dey and S.K. Addy, *Reflection and refraction of plane waves under initial stresses at an interface,* Int. J. Non-Linear Mechanics 14, 101 (1979).

[60] M. Dietrich and F. Kormendi, *Perturbation of the plane-wave reflectivity of a wave-dependent elastic medium by weak inhomogeneities,* Geophys. J. Int. 100, 203 (1990).

[61] E.H. Dill, *Simple materials with fading memory.* In: Continuum Physics II (A.C. Eringen ed.). Academic, New York, 1975.

[62] M.A. Dowaikh and R.W. Ogden, *On surface waves and deformations in a pre-stressed incompressible elastic solid,* IMA J. Appl. Math. 44, 261 (1990).

[63] M.A. Dowaikh and R.W. Ogden, *Interfacial waves and deformations in pre-stressed elastic media,* Proc. R. Soc. London A 433, 313 (1991).

[64] M.A. Dowaikh and R.W. Ogden, *On surface waves and deformations in a compressible elastic half-space,* Stab. Appl. Anal. Cont. Media 1, 27 (1991).

[65] J.W. Dunkin, *Computations of modal solutions in layered elastic media at high frequencies,* Bull. Seism. Soc. Amer. 55, 335 (1965).

[66] W.S. Edelstein and M.E. Gurtin, *A generalization of the Lamé and Somigliana stress functions for the dynamic linear theory of viscoelastic solids,* Int. J. Eng. Sci. 3, 109 (1965).

[67] A.C. Eringen and E.S. Suhubi, *Elastodynamics.* Academic, New York, 1974.

[68] R.B. Evans, *The decoupling of seismic waves,* Wave Motion 8, 321 (1986).

[69] M. Fabrizio, *Proprietà e restrizioni costitutive per fluidi viscosi con memoria,* Atti Sem. Mat. Fis. Univ. Modena 37, 429 (1989).

[70] M. Fabrizio and A. Morro, *Viscoelastic relaxation functions compatible with thermodynamics,* J. Elasticity 19, 63 (1988).

[71] L.B. Felsen, *Novel ways for tracking rays,* J. Opt. Soc. Am. A 2, 954 (1985).

[72] S. Feng and D.L. Johnson, *High-frequency acoustic properties of a fluid/porous solid interface. II. The 2D reflection Green's function,* J. Acoust. Soc. Am. 74, 915 (1983).

[73] G.C. Gaunaurd and M.F. McCarthy, *Resonances of elastic scatterers in fluid half-spaces,* IEEE J. Oceanic Engng. 12, 395 (1987).

[74] J. W. Gibbs, *Elements of vector analysis.* In: Scientific Papers. Dover, New York 1961.

[75] F. Gilbert and G.E. Backus, *Propagator matrices in elastic waves and vibration problems,* Geophys. 31, 326 (1966).

[76] J.W. Goodman, *Introduction to Fourier Optics.* McGraw Hill, New York, 1968.

[77] J.E. Gubernatis, E. Domany and J.A. Krumhansl, *Formal aspects of the theory of scattering from ultrasound by flaws in elastic materials,* J. Appl. Phys. 48, 2804 (1977).

[78] M.E. Gurtin, *An Introduction to Continuum Mechanics.* Academic, New York, 1981.

[79] M. Gurtin and E. Sternberg, *On the linear theory of viscoelasticity,* Arch. Rational Mech. Anal. 11, 291 (1962).

[80] W.R. Hamilton, *Lectures on Quaternions.* Hodges and Smith, Dublin, 1853.

[81] W.W. Hager and R. Rostamian, *Reflection and refraction of elastic waves for stratified materials,* Wave Motion 10, 333 (1988).

[82] N.A. Haskell, *The dispersion of surface waves on multilayered media,* Bull. Seism. Soc. Amer. 43, 17 (1953).

[83] M. Hayes, *Inhomogeneous plane waves,* Arch. Rational Mech. Anal. 85, 41 (1984).

[84] M. Hayes, *Energy flux for trains of inhomogeneous plane waves,* Proc. Roy. Soc. London A 370, 417 (1980).

[85] M. Hayes and M.J.P. Musgrave, *On energy flux and group velocity,* Wave Motion 1, 75 (1979).

[86] M.A. Hayes and R.S. Rivlin, *Surface waves in deformed elastic materials,* Arch. Rational Mech. Anal. 8, 358 (1961).

[87] M. Hayes and R.S. Rivlin, *A note on the secular equation for Rayleigh waves,* ZAMP 13, 80 (1962).

[88] M.A. Hayes and R.S. Rivlin, *Propagation of sinusoidal small-amplitude waves in a deformed viscoelastic solid, I,* J. Acoust. Soc. Am. 46, 610 (1969).

[89] M.A. Hayes and R.S. Rivlin, *Plane waves in linear viscoelastic materials*, Quart. Appl. Math. 32, 113 (1974).

[90] J. Heading, *An Introduction to Phase-Integral Methods.* Methuen, London, 1962.

[91] E. Heyman and L.B. Felsen, *Evanescent waves and complex rays for modal propagation in curved open waveguides,* SIAM J. Appl. Math. 43, 855 (1983).

[92] A.T. de Hoop and J.H.M.T van der Hijden, *Generation of acoustic waves by an impulsive line source in a fluid/solid configuration with a plane boundary,* J. Acoust. Soc. Am. 74, 333 (1983).

[93] J.A. Hudson and J.R. Heritage, *The use of the Born approximation in seismic scattering problems,* Geophys. J. R. astr. Soc. 66, 221 (1981).

[94] S.C. Hunter, *Viscoelastic waves.* In: Progress in Solid Mechanics, I. North-Holland, Amsterdam, 1960.

[95] D. Iesan, *Prestressed Bodies.* Longman, Harlow, 1989.

[96] W. Jaunzemis, *Continuum Mechanics.* Macmillan, New York, 1967.

[97] D.S. Jones, *Low-frequency scattering by a body in lubricated contact,* Q. Jl Mech. Appl. Math. 36, 111 (1983).

[98] D.S. Jones, *A uniqueness theorem in elastodynamics,* Q. Jl Mech. Appl. Math. 37, 121 (1984).

[99] D.S. Jones, *An exterior problem in elastodynamics,* Math. Proc. Camb. Phil Soc. 96, 173 (1984).

[100] D.S. Jones, *Acoustic and Electromagnetic Waves.* Clarendon, Oxford, 1986.

[101] F.C. Karal and J.B. Keller, *Elastic wave propagation in homogeneous and inhomogeneous media,* J. Acoust. Soc. Am. 31, 694 (1959).

[102] B.L.N. Kennett, *Seismic Wave Propagation in Stratified Media.* Cambridge University Press, Cambridge, 1985.

[103] J.C. Kirkwood and H.A. Bethe, *Progress report on "The pressure wave produced by an underwater explosion I".* In: Shock and Detonation Waves (W.W. Wood ed.). Gordon and Breach, New York, 1967.

[104] R.E. Kleinman and G.F. Roach, *Boundary integral equations for the three-dimensional Helmholtz equation,* SIAM Rev. 16, 241 (1974).

[105] M. Kline and I.W. Kay, *Electromagnetic Theory and Geometrical Optics,* Interscience, New York, 1965.

[106] E.S. Krebes, *Discrepancies in energy calculations for inhomogeneous waves,* Geophys. J. R. astr. Soc. 75, 839 (1983).

[107] M. Kuipers and A.A.F. van de Ven, *Rayleigh-gravity waves in a heavy elastic medium,* Acta Mech. 81, 181 (1990).

[108] T. Kundu, *Acoustic microscopy at low frequency,* J. Appl. Mech. 55, 545 (1988).

[109] T. Kundu and A.K. Mal, *Acoustic material signature of a layered plate*, Int. J. Eng. Sci. 24, 1819 (1986).

[110] V.D. Kupradze, *Dynamical problems in elasticity.* In: Progress in Solid Mechanics, III. North Holland, Amsterdam 1963.

[111] V.D. Kupradze, *Three-Dimensional Problems of the Mathematical Theory of Elasticity and Thermoelasticity.* North Holland, Amsterdam, 1979.

[112] L. Landau and E. Lifchitz, *Fluid Mechanics.* Pergamon, Oxford, 1959.

[113] L. Landau and E. Lifchitz, *Electrodynamics of Continuous Media.*Pergamon, Oxford, 1960.

[114] W. Lauriks, J.F. Allard, C. Depollier and A. Cops, *Inhomogeneous plane waves in layered materials including fluid, solid and porous layers,* Wave Motion 13, 329 (1991).

[115] M.J. Leitman and G.M.C. Fisher, *The linear theory of viscoelasticity.* In: Encyclopedia of Physics (C. Truesdell ed.), Vol. VIa/3. Springer, Berlin, 1973.

[116] O. Leroy, G. Quentin and J.M. Claeys, *Energy conservation for inhomogeneous plane waves,* J. Acoust. Soc. Am. 84, 374 (1988).

[117] H. Levine, *Unidirectional Wave Motions.* North-Holland, Amsterdam, 1978.

[118] M.J. Lighthill, *Group velocity,* J. Inst. Maths. Applics. 1, 1 (1965).

[119] F.J. Lockett, *The reflection and refraction of waves at an interface between viscoelastic materials,* J. Mech. Phys. Solids 10, 58 (1962).

[120] J.D. Lubahn, *Deformation phenomena.* In: Mechanical Behavior of Materials at Elevated Temperatures (J.E. Dorn, ed.). McGraw-Hill, New York, 1968.

[121] K.J. Marfurt and E. Bielanski McCarron, *Seismic modeling in oil and gas exploration.* In: Mathematical and Numerical Aspects of Wave Propagation Phenomena (G. Cohen, L. Halpern and P. Joly eds.). SIAM, Philadelphia, 1991.

[122] J.E. Marsden and T.J.R. Hughes, *Mathematical Foundations of Elasticity.* Prentice-Hall, Englewood Cliffs, 1983.

[123] G.A. Maugin, *Nonlinear Electromechanical Effects and Applications.* World Scientific, Singapore, 1985.

[124] M.J. Mayes, P.B. Nagy, L. Adler, B.P. Bonner and R. Streit, *Excitation of surface waves of different modes at a fluid-porous solid interface*, J. Acoust. Soc. Am. 79, 249 (1986).

[125] R.E. Meyer, *Gradual reflection of short waves,* SIAM J. Appl. Math. 29, 481 (1975).

[126] S.G. Mikhlin, *Mathematical Physics, an Advanced Course.* North-Holland, Amsterdam, 1970.

[127] J. Milnor, *Morse Theory.* Princeton University Press, Princeton, 1963.

[128] J. Molinero and M. de Billy, *Excitation of leaky guided waves in infinite cylinders by obliquely incident acoustic waves,* J. Appl. Phys. 64, 2894 (1988).

[129] A. Morro, *Thermodynamics and extremum principles in viscoelasticity.* In: Advances in Thermodynamics (S. Sieniutycz and P. Salamon eds.), Vol. 6. Taylor and Francis, New York, 1990.

[130] A. Morro, *Thermodynamics of linear viscoelasticity.* In: Thermodynamics and Kinetic Theory (W. Kosinski, W. Larecki, A. Morro and H. Zorski eds.). World Scientific, Singapore, 1992.

[131] A. Morro and M. Vianello, *Minimal and maximal free energy for materials with memory,* Boll. Un. Mat. Ital. 4A, 113 (1990).

[132] P.M. Morse and K.U. Ingard, *Theoretical Acoustics.* McGraw-Hill, New York, 1968.

[133] G. Mott, *Reflection and refraction coefficients at a fluid-solid interface,* J. Acoust. Soc. Am. 50, 819 (1971).

[134] P.B. Nagy, K. Cho, L. Adler and D.E. Chimenti, *Focal shift of convergent ultrasonic beams reflected from a liquid-solid interface,* J. Acoust. Soc. Am. 81, 835 (1987).

[135] A. Narain and D.D. Joseph, *Classification of linear viscoelastic solids based on a failure criterion,* J. Elasticity 14, 19 (1984).

[136] A.H. Nayfeh and D.E. Chimenti, *Elastic wave propagation in fluid-loaded multiaxial anisotropic media,* J. Acoust. Soc Am. 89, 542 (1991).

[137] A.H. Nayfeh and S. Nemat-Nasser, *Elastic waves in inhomogeneous elastic media,* J. Appl. Mech. 33, 696 (1972).

[138] T.D.K. Ngoc and W.G. Mayer, *Numerical integration method for reflected beam profiles near Rayleigh angles,* J. Acoust. Soc. Am. 67, 1149 (1980).

[139] O'Neill, *Elementary Differential Geometry.* Academic, New York, 1969.

[140] Y. Pao and V. Vasundara, *Huygens principle, radiation conditions, and integral formulas for the scattering of elastic waves,* J. Acoust. Soc. Am. 59, 1361 (1976).

[141] R. Penrose and W. Rindler, *Spinors and space-time, I.* Cambridge University, Cambridge 1984.

[142] L.E. Pitts, T.J. Plona and W.G. Mayer, *Theoretical similarities of Rayleigh and Lamb modes of vibration,* J. Acoust. Soc. Am. 60, 374 (1976).

[143] B. Poirée, *Complex harmonic plane waves.* In: Physical acoustics (O. Leroy and M.A. Breazeale eds.). Plenum Press, New York, 1991.

[144] G. Quentin, A. Derem and B. Poirée, *The formalism of evanescent plane waves and its importance in the study of the generalized Rayleigh wave,* J. Acoustique 3, 321 (1990).

[145] A.G. Ramm, *Scattering by obstacles.* Reidel, Dordrecht, 1986.

[146] P.N.J. Rasolofosaon, *Plane acoustic waves in linear viscoelastic porous media: energy, particle displacement, and physical interpretation,* J. Acoust. Soc. Am. 89, 1532 (1991).

[147] Lord Rayleigh, *Theory of Sound, I.* Dover, New York, 1945.

[148] M. Roseau, *Asymptotic Wave Theory.* North-Holland, Amsterdam, 1976.

[149] M. Rousseau, *Floquet wave properties in a periodically layered medium,* J. Acoust. Soc. Am. 86, 2369 (1989).

[150] A. Schoch, *Der Schalldurchgang durch Platten,* Acoustica 2, 18 (1952).

[151] H. Schmidt and F.B. Jensen, *A full wave solution for propagation in multilayered viscoelastic media with application to Gaussian beam reflection at fluid-solid interfaces,* J. Acoust. Soc. Am. 77, 813 (1984).

[152] M. Schoenberg, *Wave propagation in alternating solid and fluid layers,* Wave Motion 6, 303 (1984).

[153] N. Scott, *Acceleration waves in constrained elastic materials,* Arch. Rational Mech. Anal. 58, 57 (1975).

[154] R.S. Sidhu and S.J. Singh, *Reflection of P and SV waves at the free surface of a pre-stressed elastic half-space,* J. Acoust. Soc. Am. 76, 594 (1984).

[155] V. Smirnov, *Cours de Mathematiques Superieurs IV.* Mir, Moscow, 1984.

[156] R. Snieder, *3-D linearized scattering of surface waves and a formalism for surface wave holography,* Geophys. J. R. astr. Soc. 84, 581 (1986).

[157] C. Somigliana, *Sulle espressioni analitiche generali dei movimenti oscillatori,* Atti Reale Accad. Lincei Roma, Ser. 5, 1, 111 (1892).

[158] A.J.M. Spencer, *Continuum Mechanics.* Longman, London, 1980.

[159] I. Stakgold, *Boundary Value Problems of Mathematical Physics II.* MacMillan, New York, 1968.

[160] E. Sternberg, *On the integration of the equations of motion in the classical theory of elasticity,* Arch. Rational Mech. Anal. 6, 34 (1960).

[161] J.A. Stratton, *Electromagnetic Theory.* McGraw-Hill, New York, 1941.

[162] W.T. Thomson, *Transmission of elastic waves through a stratified solid medium,* J. Appl. Phys. 21, 89 (1950).

[163] J. Tromp and R. Snieder, *The reflection and transmission of plane P- and S-waves by a continuously stratified band: a new approach using invariant imbedding,* Geophys. J. 96, 447 (1989).

[164] C. Truesdell and W. Noll, *The non-linear field theories of mechanics.* In: Encyclopedia of Physics (S. Flügge ed.), vol. III/3. Springer, Berlin, 1965.

[165] E. Van Groesen and F. Mainardi, *Energy propagation in dissipative systems. I. Centrovelocity for linear systems*, Wave Motion 11, 201 (1989).

[166] E. Van Groesen and F. Mainardi, *Balance laws and centro velocity in dissipative systems,* J. Math. Phys. 31, 2136 (1990).

[167] I.A. Viktorov, *Rayleigh and Lamb Waves.* Plenum, New York, 1967.

[168] J.R. Wait, *Electromagnetic Waves in Stratified Media.* Pergamon, Oxford, 1962.

[169] D.J.N. Wall, *Uniqueness theorems for the inverse problem of elastodynamic boundary scattering,* IMA J. Appl. Math. 44, 221 (1990).

[170] H. Weinberg, *Application of ray theory to acoustic propagation in horizontally stratified oceans,* J. Acoust. Soc. Am. 58, 97 (1975).

[171] H. Weinberg and R. Burridge, *Horizontal ray theory for ocean acoustics,* J. Acoust. Soc. Am. 55, 63 (1974).

[172] E.K. Westwood, *Complex ray methods for acoustic interaction at a fluid-fluid interface,* J. Acoust. Soc. Am. 85, 1872 (1989).

[173] A.M. Whitworth and P. Chadwick, *The effect of inextensibility on elastic surface waves,* Wave Motion 6, 289 (1984).

[174] G.B. Witham, *Linear and Non-Linear Waves.* Wiley, New York, 1974.

[175] R. Wu and K. Aki, *Scattering characteristics of elastic waves by an elastic heterogeneity,* Geophys. 50, 582 (1985).

[176] V. Červený, *Synthetic body wave seismograms for laterally varying media containing thin transition layers,* Geophys. J. Int. 99, 331 (1989).

[177] D.G. Crighton and I.P.Lee-Bapty, *Spherical nonlinear wave propagation in a dissipative stratified atmosphere,* Wave Motion 15, 315 (1992).

[178] M. Deschamps and P. Chevée, *Reflection and refraction of a heterogeneous plane wave by a solid layer,* Wave Motion 15, 61 (1992).

[179] M. Fabrizio and A. Morro, *Mathematical Problems in Linear Viscoelasticity.* SIAM, Philadelphia, 1992.

[180] A. Linkov and N. Filippov, *Difference equations approach to the analysis of layered systems,* Meccanica 26, 195 (1991).

[181] D.M. Pai, *Wave propagation in inhomogeneous media: a planewave layer interaction method,* Wave Motion 13, 205 (1991).

INDEX